Apitherapy for Neurodegenerative Diseases

This book explores the use of bee venom therapy, or apitherapy, in managing neurodegenerative diseases within naturopathic and traditional medicine. Once mainly used for pain-related disorders, apitherapy is now recognized for its therapeutic compounds and molecular mechanisms. With a long history in cultures such as Egyptian, Chinese, and Indian medicine, it is gaining renewed interest. However, the absence of standardized procedures has limited its wider adoption in modern healthcare. This book offers an up-to-date overview of apitherapy, covering its history, methods, safety, bioactive components, and clinical applications. It aims to equip medical professionals with the knowledge and skills to integrate apitherapy into holistic patient care and support progress toward global health goals.

Key Features:

- Provides detailed information on the uses of bee venom as therapies for neurodegenerative disorders
- Describes physiological and pharmacological effects of bee products
- Includes modern therapeutic modalities derived from ancient cultures

Neuroscience of Neurodegenerative Disorders
Series Editor: Wael Mohamed

Wael Mohamed Yousef Mohamed
BBCH (Egypt), MMSc (Egypt), MD/PhD (USA)
Dr. Mohamed earned his PhD from Penn State University (USA) and currently is an Associate Professor at International Islamic University Malaysia. He serves as an editor for *Frontiers in Neurology* and *PLOS ONE* journals. His research is well supported by extramural funding. He is the founder and chair of the AfrAbia-PD-Genomic Consortium.

About This Series
This book series aims to publish volumes focusing on both basic and clinical research in the field of Neurodegenerative Disorders with a focus on delineating the recent advances in various disorders including Parkinson's Disease, Alzheimer's Disease, ALS, Huntington's Disease, among other common Neurodegenerative disorders. The books will examine different disorders aspects including pathophysiology, diagnostic, treatment, and precision medicine. Volumes in the series will also highlight the recent advances in the management of Neurodegenerative Disorders especially in OMICS and Genomic era and precision medicine. Volumes will be useful for neuroscientists, neurologists, psychiatrists, psychologists, neurosurgeons, and those interested in preventive medicine.

Integrating Neuroimaging, Computational Neuroscience, and Artificial Intelligence
Edited by Indranath Chatterjee and Nasrallah Moradikor

Non-Pharmacological Interventions for Dementia: A Comprehensive Approach
Edited by Abraham Olufemi Asuku, Ebele Joan Ajagun, and Gbonjubola Oyinlola Ogungbangbe

Apitherapy for Neurodegenerative Diseases
Edited by Olugbemi Tope Olaniyan, Abraham Olufemi Asuku, and Chioma A. Ohanenye

https://www.routledge.com/Neuroscience-of-Neurodegenerative-Disorders/book-series/NND

Apitherapy for Neurodegenerative Diseases

Edited by
Olugbemi Tope Olaniyan
Abraham Olufemi Asuku
Chioma A. Ohanenye

CRC Press
Taylor & Francis Group
Boca Raton London New York

CRC Press is an imprint of the
Taylor & Francis Group, an **informa** business

Designed cover image: ChatGPT

First edition published 2026
by CRC Press
2385 NW Executive Center Drive, Suite 320, Boca Raton FL 33431

and by CRC Press
4 Park Square, Milton Park, Abingdon, Oxon, OX14 4RN

CRC Press is an imprint of Taylor & Francis Group, LLC

ISBN: 978-1-041-16652-8 (hbk)
ISBN: 978-1-041-16653-5 (pbk)
ISBN: 978-1-003-68565-4 (ebk)

DOI: 10.1201/9781003685654

Typeset in Times
by SPi Technologies India Pvt Ltd (Straive)

Contents

Foreword

Neurodegenerative disorders such as Alzheimer's disease, amyotrophic lateral sclerosis, Parkinson's disease, and Huntington's disease are rapidly becoming one of the major medical and societal challenges of our time. Despite impressive advances in neuroscience and pharmacology, truly disease-modifying and effective therapies for these conditions remain limited, and many patients continue to face a progressive loss of function and independence.

Against this background, the present volume, *Apitherapy for Neurodegenerative Diseases*, offers a timely and original perspective. It explores how products derived from honeybees may contribute to neuroprotection and modulation of neurodegenerative processes, opening an intriguing interface between traditional knowledge and modern biomedical science.

Apitherapy—the therapeutic use of bee-derived products such as honey, royal jelly, propolis, pollen, bee venom, and beeswax—has a long history in traditional medicine and is now being revisited through the lens of contemporary research. These products are rich in peptides, enzymes, flavonoids, and diverse bioactive molecules. Their complex biochemistry provides multiple potential mechanisms to influence oxidative stress, neuroinflammation, and neuronal loss, which are central hallmarks of many neurodegenerative diseases.

This book provides a structured and comprehensive overview of apitherapeutic approaches at the molecular, cellular and clinical levels. Across 15 chapters, the contributors integrate mechanistic insights with translational and clinical data, carefully examining how bee-derived compounds may modulate neuroinflammatory pathways, support neuronal survival and function, and potentially improve motor and cognitive outcomes. By bringing together preclinical and clinical findings, the editors succeed in bridging the gap between experimental models and real-world therapeutic perspectives.

An important strength of this work is its genuinely interdisciplinary scope. It draws upon pharmacology, neuroscience, and complementary medicine, and presents apitherapy not as a replacement for established treatments, but as a possible adjunctive and complementary strategy within evidence-based care. The authors also give due attention to dose optimization, safety considerations, interactions with conventional therapies, and the pressing need for rigorous clinical validation.

This volume is more than a collection of isolated studies. It is a call to thoughtfully reconsider the role of natural compounds in addressing the growing burden of neurodegenerative diseases. It invites students, clinicians, and researchers to explore how bee products, when investigated with appropriate scientific rigor, may contribute to future neurotherapeutic strategies that combine the wisdom of traditional practices with the standards of modern evidence-based medicine.

I warmly commend the editors and contributors for their commitment in bringing this book to fruition. They have assembled a resource that will be of interest to pharmacologists, physiologists, neuroscientists, medical practitioners, and all those seeking innovative and integrative approaches to alleviating the devastating impact of neurodegenerative disorders.

Ildar Fakhradiyev, MD, PhD
Vice Rector, S.D. Asfendiyarov Kazakh National Medical University
Almaty, Kazakhstan

Preface

This book provides detailed information about the application of bee venom, known as apitherapy, in the treatment of several neurodegenerative diseases in naturopathic medicine, which is an aspect of alternative and traditional medicine practice. Apitherapy is utilized in the treatment of several disorders linked with pain, but in recent years, several advances have been made in the identification of therapeutic compounds and molecular mechanism of bee venom for healing processes. Apitherapy, using bee toxins which have remarkable physiological and pharmacological effects on organs and tissues, has been utilized as an ancient medical practice in many civilized societies such as Egyptian, Unani, Chinese, Korean, Latin American, and Tibetan cultures. Recently, the Indian medical system has established modern therapeutic modalities of apitherapy. Due to the lack of standard operative procedure (SOPs) for apitherapy in relation to procedures, safety concerns, application points, surgical operative standards, historical perspective, and physiological mechanism of action, many healthcare professionals have doubted the activity and prevented in-depth exploration of apitherapy around the globe in modern-day medical practice. To achieve the United Nations Sustainable Development Goals on good health and well-being, there must be increased access to essential health services, provision of safe and affordable medical services, address of long-standing healthcare shortcomings, and improved investment in health systems. This book thus provides an up-to-date overview of alternative medical practice with apitherapy, exploring the historical background, advantages of apitherapy, methods of injection, safety, mechanisms of actions, bioactive compounds, physiological responses, and therapeutic applications encompassing its utilization in the management of neurodegenerative diseases. The book also equips medical professionals with the necessary knowledge and skills in the use of apitherapy to improve the competence required to make informed decisions for patient care and patient outcomes and to promote a holistic approach to cure and healing process.

The chapters were contributed by 63 academicians and researchers from five different countries (Nigeria, Rwanda, South Africa, the United Kingdom, and South Korea) around the world. This book is made up of 15 chapters: Chapters 1–3 deal with the historical perspectives of apitherapy, basic concepts of apitherapy, and physiological roles of apitherapy on the body systems; Chapters 4–13 discuss the applications of different bee products in neurodegenerative diseases; while Chapter 14 gives a detailed account of apitherapy entrepreneurship with emphasis on innovations and biotechnological advancement, and Chapter 15 discusses the safety and adverse effects of apitherapy.

Our target audiences, who will benefit by reading this book, are clinicians, biomedical scientists, medical professionals, researchers, and academicians. This book will undoubtedly provide knowledge and understanding needed to eliminate bias regarding the use of apitherapy as a cure and adjunct in the treatment of neurodegenerative diseases.

Olugbemi Tope Olaniyan, PhD
Department of Physiology, Faculty of Basic Medical Sciences, Kwara State University, Malete, Nigeria / Department of Physiology, College of Medicine, Korea University, Seoul, South Korea

Abraham Olufemi Asuku, PhD
Bioresources Development Centre, Department of Medical Biotechnology, National Biotechnology Research and Development Agency, Ogbomoso, Nigeria

Chioma A. Ohanenye, PhD
Department of Anatomy, Faculty of Basic Medical Sciences, Rhema University Aba, Nigeria

Acknowledgments

We are deeply grateful for the support of all the contributors, whose prompt response and valuable input have been instrumental in the success of this book. We sincerely appreciate each contributor's enthusiasm and great support in ensuring that this project was completed on time. Their partnership and collaboration are commendable. We also show our immense gratitude to the anonymous reviewers whose constructive feedback helped improve the quality of this book. We deeply appreciate the support of Professor Wael Mohamed, who is the series editor for the volume of this book. We would like to acknowledge the great support of the CRC Press and production team, whose guidance was vital in ensuring the timely completion of this project. In conclusion, we would like to express our immense gratitude to the staff, management, and colleagues at the Department of Physiology, Faculty of Basic Medical Sciences, Kwara State University, Malete, Nigeria; the Bioresources Development Centre, Department of Medical Biotechnology, National Biotechnology Research and Development Agency, Ogbomoso, Nigeria; the Department of Anatomy, Faculty of Basic Medical Sciences, Rhema University Aba, Nigeria; and the Department of Physiology, College of Medicine, Korea University, Seoul, South Korea for their support throughout this book project.

Olugbemi Tope Olaniyan, PhD
Department of Physiology, Faculty of Basic Medical Sciences, Kwara State University, Malete, Nigeria / Department of Physiology, College of Medicine, Korea University, Seoul, South Korea

Abraham Olufemi Asuku, PhD
Bioresources Development Centre, Department of Medical Biotechnology, National Biotechnology Research and Development Agency, Ogbomoso, Nigeria

Chioma A. Ohanenye, PhD
Department of Anatomy, Faculty of Basic Medical Sciences, Rhema University Aba, Nigeria

About the Editors

Olugbemi Tope Olaniyan is Associate Professor, Applied Physiology and Biotechnology, Faculty of Basic Medical Sciences, College of Health Sciences, Kwara State University, Malete, Nigeria, and Visiting Professor, Department of Physiology, College of Medicine, Korea University, Seoul, South Korea. He holds a BSc (University of Ilorin, Nigeria), MSc, PhD (University of Ibadan, Nigeria) in Physiology, and Postdoctoral Fellowship in Biotechnology, Institute de Biotecnologico, Centro de Investigacion Biomedica, University of Granada, Spain. Consistently over the years, he has been ranked among the top 500 scientists in Nigeria with scholarly output from Scopus SCI-VAL/Elsevier from 2020 to 2025, number 7 position in Africa in Medical Sciences and Stanford University World's Top 2% Scientists.

Dr. Olaniyan has experience in cell culture, medical imaging, histology, immunohistochemistry, molecular and cellular biology techniques, ELISA, fluorescence microscopy, Western blot, mitochondria physiology (sea horse), nanotechnology, biomedical statistics (STATA, GRAPH-Pad prism and SPSS), animal handling techniques such as mice, rats, C-elegans, zebra fish, and drosophilae. He also has vast experience in the utilization of modern technology for teaching and research, such as Anatomage table, Powerlab, High and Low fidelity 3G SimMan Mannequins, and learning management system (LMS). He has over 15 years of cognate experience in teaching and research and has trained over 60 graduate students, co-supervising five master's and two PhD postgraduate theses. In addition, he has published over 90 journal articles, including 250 book chapters in reproductive biology, applied physiology, biotechnology, bioinformatics, and biophysics in international reputable journals. He is a member of several professional bodies, including the Physiological Society of Nigeria, Nigerian Bioinformatics and Genomics Networks, Drosophila Africa, AntiVec Network, Genetics Society of America, and West Africa Genetics Society. His future goals are to focus on applied physiology, molecular biology, and biotechnology techniques through the adoption and utilization of high-resolution microarray, next-generation sequencing and current techniques in molecular biology and imaging that could help in the analysis of different factors that affect human health and the environment in order to achieve sustainable development goals.

Abraham Olufemi Asuku earned his Ph.D in Physiology at the Department of Physiology, Faculty of Basic Medical Sciences, University of Ilorin, Ilorin, Nigeria. He also earned his first degree (BSc) and master's degree (MSc) from the same department. Dr. Asuku is a passionate neuroscientist and medical biotechnologist. He currently works with the National Biotechnology Research and Development Agency, Nigeria, as an Assistant Chief Research Officer. His areas of interest are mitigation of neurotoxicity, neurodegenerative diseases, and neurodevelopmental disorders by phytochemicals, as well as the applications of nanotechnology for neurodegenerative disease drug delivery.

He is a member of the Neuroscience Society of Nigeria, the International Parkinson and Movement Disorder Society, USA, and the Physiological Society, United Kingdom. He is an award-winning researcher and has over 60 publications (more than 30 are Scopus-based), including book chapters, research, and review articles of good international repute. He has served as a reviewer for many reputable journals. Dr. Asuku is currently ranked by Scopus SCI-VAL/Elsevier between 2022 and 2025 among the Top 500 Authors in Neuroscience as number 21 and 93 in Nigeria and Africa, respectively.

He has also presented seminars at many conferences, both locally and internationally. He was one of the invited guest speakers at the VII International Conference of Universidad Tecnica De Manabi, Ecuador, in 2023. Dr. Asuku was an editor of the book titled *Non-pharmacological Intervention for Dementia: A Comprehensive Approach*, and he is currently editing two other books, *Bioresources for Managing and Treating Dementia*, and *Evolution and Innovation in Parkinson's Disease Therapies in Under-Resourced Settings*, to be published by CRC Press/Taylor & Francis Group and IGI Global Scientific Publishing, respectively. His long-term goal is to discover therapeutic targets and help create focused treatments for neurodegenerative diseases. By bridging the gap between basic and translational neuroscience, he seeks to further knowledge of neuronal health and provide insights that may result in novel treatments.

Chioma A. Ohanenye earned her B. Med Sc. in Anatomy in 1985 and MBBS in 1989 from University of Portharcourt, Rivers State, Nigeria. Remarkably, during her trips to the United States in 2003 and 2005, where she continued her medical education, she was attached to hospitals for medical workshops for six months. In one of those trips, she observed a renal transplant and followed consultants on ward rounds and reviews in Ocean Springs Hospital–Singing River Health System, Mississippi, USA. She also took medical courses at Baptist Hospital, Pensacola, Florida, USA, during July and August 2005. She returned to Nigeria with a wealth of medical experiences and started a paediatrics residency program in Abia State University Teaching Hospital (ABSUTH) after working at the Emergency Unit for about a year. She continued with the Paediatrics Department as a resident for a few years then she began study for a master's degree in environmental health sciences at Abia State University, where she had a Distinction. She majored in paediatric environmental health from 2017 to 2019. She earned a PhD in anatomy from Abia State University, Uturu in 2024. Dr. Ohanenye is a Fellow of the Society of Environmental and Public Health of Nigeria (FSEPHON). She was a liaison officer in the Society of Environmental and Public Health Professionals of West Africa (SEPHPWA) till date. While at Abia State University Teaching Hospital, she set up the Paediatrics Department at Living Word Mission Hospital, now Rhema University Teaching Hospital, Aba, Nigeria. She found it most rewarding to be the Head of Department of the Paediatric Unit in the Mission Hospital till she retired from both Abia State University Teaching Hospital (ABSUTH) and Rhema University Teaching Hospital in 2021, to set up her own private practice. Dr. Ohanenye's love of teaching spurred her to return to Rhema University where she has been lecturing in anatomy since March 2021. Her special interests are to recognize and treat childhood diseases (including neonatology), prevent reoccurrence, and work with mothers to uplift and improve the health of their children. In the field of anatomy, her special interest is in neuroanatomy. She has authored or coauthored 43 publications in reputable journals. Also, she has written three books of international standards. She has attended several workshops and seminars, locally and internationally.

Contributors

Sulaiman Sime Abdullahi
Department of Anatomy
Kwara State University
Malete, Nigeria

Mudathir Tunde Abdulraheem
Department of Anatomy
Faculty of Basic Medical Sciences
University of Ilorin
Ilorin, Kwara State, Nigeria

Olusegun G. Adebayo
Department of Physiology
Faculty of Basic Medical Sciences
College of Health Sciences
Osun State University
Osogbo, Osun State, Nigeria

Uthman Okikiola Adebayo
Department of Medical Laboratory Science
Neuropsychiatric Hospital, Abeokuta
Ogun State, Nigeria

Eniola Oluwatomisin Adesanya
Department of Human Anatomy
College of Medicine
Lead City University, Ibadan
Oyo State, Nigeria

Bisade Esther Adeyeluwa
Department of Anatomy
University of Ilorin
Ilorin, Kwara State, Nigeria

Babatunde Samuel Adeyemi
Anatomy Unit/Department
McPherson University Seriki-Sotayo
Ajebo, Ogun State, Nigeria

Toluwanimi Omolara Afolabi
Department of Anatomy
Olabisi Onabanjo University
Ago Iwoye, Ogun State, Nigeria

Abdulmajeed Opeyemi Agboola
Department of Public Health Sciences
Faculty of Allied Health Sciences
Kwara State University, Malete, Nigeria

Faith Ewaoluwa Ajayi
FdSc Assistant Practitioner Associate Member
Faculty of Public Health
United Kingdom

Ayodeji Johnson Ajibare
Neuro-Reproductive and Metabolism Unit
Department of Physiology
Faculty of Basic Medical and Health Sciences
College of Medicine
Lead City University
Ibadan, Oyo State, Nigeria

Oluwadamilola Eunice Ajibola
Department of Anatomy
University of Ilorin
Ilorin, Kwara State, Nigeria

Duyilemi Chris Ajonijebu
Department of Human Anatomy & Physiology
Faculty of Health Sciences
Doornfontein Campus, University of Johannesburg
South Africa

Edward Ekundayo Akerele
Department of Tourism and Hospitality Management
Kwara State University
Malete, Nigeria

Akinremi Samuel Akinkunmi
Department of Internal Medicine
Obafemi Awolowo University Teaching Hospital Complex
Ile-Ife, Nigeria

Favour Ayomide Akinpelu
Department of Anatomy
Faculty of Basic Medical Sciences
University of Lagos
Lagos, Nigeria

Ajirioghene E. Akpotu
Department of Pharmacology
Faculty of Basic Medical Sciences
PAMO University of Medical Sciences
Port-Harcourt, Nigeria

Tolulope Oluwatimilehin Alabi
Department of Anatomy
Faculty of Basic Medical Sciences
University of Ilorin
Ilorin, Kwara State, Nigeria

Michael Anozie Amadi
Department of Anatomy
Faculty of Basic Medical Sciences
Pamo University of Medical Sciences
Port Harcourt, Rivers State, Nigeria

Danwahab Olarewaju Ambali
Department of Anatomy
Faculty of Basic Medical Sciences
College of Health Sciences
Kwara State University
Malete, Nigeria

Oluwapelumi Lydia Apanisile
Department of Anatomy
Faculty of Basic Medical Sciences
University of Ilorin
Ilorin, Kwara State, Nigeria

Leviticus O. Arietarhire
Department of Anatomy
School of Basic Medical Science
Babcock University
Ilishan Remo, Ogun State, Nigeria

Abraham Olufemi Asuku
Department of Medical Biotechnology
Bioresources Development Centre
National Biotechnology Research and Development Agency
Ogbomoso, Nigeria

Aminat Temitayo Atoyebi
Department of Human Anatomy
Faculty Basic Medical of Sciences
Al-Hikmah University
Ilorin, Kwara State, Nigeria

Olabisi Elizabeth Ayoola
Department of Physiology
Faculty of Basic Medical Sciences
College of Health Sciences
Kwara State University
Malete, Nigeria

Adebimpe Victoria Babalola
Department of Anatomy
Faculty of Basic Medical Sciences
University of Lagos
Lagos, Nigeria

Abdulwarith Ayomide Bamidele
Department of Public Health
Kwara State University
Malete, Nigeria

Babatunde Joseph Dare
Department of Human Anatomy
Faculty of Basic Medical Sciences
Osun State University
Osogbo, Osun State, Nigeria

Egunjobi Oluwagbemisola Dorcas
Medical Outpatient Department
Cardiac Laboratory
General Hospital Odan Lagos, Nigeria

Sunday Okon Elijah
Department of Anatomy
Faculty of Basic Medical Sciences
College of Medical and Allied Health Sciences
Bingham University
Karu, Nasarawa State, Nigeria

Joseph Igbo Enya
Department of Anatomy
School of Basic Medical Science
Babcock University
Ilishan Remo, Ogun State, Nigeria

Suleiman Muhammed Eze
Department of Human Anatomy
Faculty of Basic Medical Sciences
Al-Hikmah University
Ilorin, Kwara State, Nigeria

Oyesanmi A. Fabunmi
Section Sports Medicine
Faculty of Health Science
University of Pretoria
Pretoria, South Africa

Modupeola Olamide Falabi
Department of Tourism and Hospitality Management
Kwara State University
Malete, Nigeria

Olusola Samson Folorunso
Department of Tourism and Hospitality Management
Kwara State University
Malete, Nigeria

Fatimat Omowumi Hamzat
Department of Human Anatomy
Faculty of Basic Medical Sciences
Al-Hikmah University
Ilorin, Kwara State, Nigeria

Ruqayyah Yetunde Ibiyeye
Department of Anatomy
Faculty of Basic Medical Sciences
College of Health Sciences
Kwara State University
Malete, Nigeria

Solihu Olashile Ibrahim
Department of Human Anatomy
Faculty of Basic Medical Sciences
College of Health Sciences
Al-Hikmah University
Ilorin, Kwara State, Nigeria

Abdulkadir Yaradua Imam-Fulani
Department of Community Medicine and Public Health, Department of Public Health Science
Faculty of Basic Medical Sciences
College of Health Sciences
Al-Hikmah University
Ilorin, Kwara State, Nigeria

Adedokun Kabiru Isola
Department of Physiology
Faculty of Basic Medical Sciences
College of Health Sciences
Osun State University
Osun, Nigeria

Adebayo-Gege Grace Iyabo
Department of Human Physiology
Faculty of Basic Medical Sciences
Baze University
Abuja, Nigeria

Jamiu Sulaiman Jamiu
Department of Anatomy
Kwara State University
Malete, Nigeria

Naomi Joshua
Department of Anatomy
School of Basic Medical Science
Babcock University
Ilishan Remo, Ogun State, Nigeria

Ryu Ki-Jin
Department of Obstetrics and Gynecology
College of Medicine
Korea University
Seoul, South Korea

Ismail Adetayo Lawal
Department of Human Anatomy
Faculty of Basic Medical Sciences
Al-Hikmah University
Ilorin, Kwara State, Nigeria

Imam Abubakar Lekan
Department of Human Anatomy
Faculty of Basic Medical Sciences
College of Health Sciences
University of Ilorin
Ilorin, Kwara State, Nigeria

Abdullahi Mohammed
Department of Human Anatomy
School of Medicine and Pharmacy
College of Medicine and Health Sciences
University of Rwanda
Kigali, Rwanda

Olawunmi Anthonia Odugbesan
Department of Physiology
Faculty of Basic Medical Sciences
Afe Babalola University
Ado-Ekiti, Nigeria

Oluwagbeminiyi Francis Ogunkoya
College of Medicine
Lead City University
Ibadan, Oyo State, Nigeria

Oladimeji Lateefah Olabisi
Department of Community Medicine and Public Health
Faculty of Basic Medical Sciences
College of Health Sciences
Al-Hikmah University
Ilorin, Kwara State, Nigeria

Buhari Abdulhafeez Oladimeji
Department of Community Medicine and Public Health
Faculty of Basic Medical Sciences
College of Health Sciences
Al-Hikmah University
Ilorin, Kwara State, Nigeria

Olugbemi Tope Olaniyan
Department of Physiology
Faculty of Basic Medical Sciences
Kwara State University, Malete, Nigeria
and
Department of Physiology
College of Medicine
Korea University
Seoul, South Korea

Tomilola Morenikeji Olaseinde
Department of Anatomy
University of Ilorin
Ilorin, Kwara State, Nigeria

Samson Ehindero Olorunnado
Department of Human Anatomy
University of Rwanda
Kigali, Rwanda

Akintayo Christopher Oloruntoba
Department of Medicine and Surgery
Obafemi Awolowo University
Teaching Hospital Complex
Ile-Ife, Nigeria

Ijeoma Onyeleonu
Department of Anatomy
Faculty of Basic Medical Sciences
Pamo University of Medical Sciences
Port Harcourt, Rivers State, Nigeria

Kelechi Roselyn Onyema
Department of Anatomy
School of Basic Medical Science
Babcock University
Ilishan Remo, Ogun State, Nigeria

Ayisimaka Adokiye Richard
Department of Human Physiology
Faculty of Basic Medical Sciences
University of Port-Harcourt
Rivers State, Nigeria

Swabirah Enimire Sulaiman
Department of Physiology
Faculty of Basic Medical Sciences
University of Ilorin
Ilorin, Kwara State, Nigeria

Fatimo Ajoke Sulaimon
Department of Anatomy
Faculty of Basic Medical Sciences
College of Health Sciences
University of Ilorin
Ilorin, Kwara State, Nigeria

Salaudeen Faidat Sumisola
Department of Community Medicine and Public Health
Faculty of Basic Medical Sciences
College of Health Sciences
Al-Hikmah University
Ilorin, Kwara State, Nigeria

Kogbara Dinebari Tudor
Department of Surgery
University of Port-Harcourt Teaching Hospital
Rivers State, Nigeria

Nyejirime Young Wike
Department of Human Physiology
State University of Medical and Applied Sciences
Igbo-eno, Enugu State, Nigeria

1 Historical Perspective of Apitherapy

Ruqayyah Yetunde Ibiyeye, Abdullahi Mohammed, Fatimo Ajoke Sulaimon, Danwahab Olarewaju Ambali, Olabisi Elizabeth Ayoola, and Abdulmajeed Opeyemi Agboola

1.1 INTRODUCTION

Apitherapy is a treatment plan that involves apiary yields such as beeswax, bee pollen, honey, bee propolis, royal jelly, and bee venom to promote healing and the management of various illnesses holistically. These methods of treatment are well recognized in traditional and modern medicine. Bee venom therapy is the utmost practiced type of apitherapy. It involves the use of bee venom for treating diseases and promotes healing (Gokulakrishnaa & Thirunavukkarasu, 2020). The word *apitherapy* is a Greek word from the word "apis," which means bee, and "therapeia," which means treatment (Trumbeckaite et al., 2015; Šuligoj, 2021). According to the Apimondia Standing Commission, apitherapy was defined as

> a medical perception that is usually built on scientific grounds in collaboration with traditional knowledge. This concept includes: methods used for bee production that could enhance medical development; change in methods of hive production alone or in conjunction with therapeutic vegetation and their by-products (apipharmacopoeia); and the medical procedures, which incorporates the use of apipharmacopoeia and/or the bees (apimedicine). (Šuligoj, 2021)

It is worthy to note that bee venom, beeswax, and royal jelly are apiary products that are synthesized chemically while honey, bee propolis, pollen, and breads are plant derivatives modified by bees (Weis et al., 2022).

The acts of utilizing bee stings therapy date back to 460 BC, and it is still in use. Various improvements and scientific studies have been utilized to further evaluate and ratify the use of bee by-products in various fields of medicine. Between 460 and 370 BC, Hippocrates used bee stings therapy for various diseases in patients. Ivan IV of Russia was cured of gout using bee stings between 1530 and 1584. In 1859, a French physician named Dr. Desjardins was the first to publish a scientific report on the abilities of apitoxin to cure rheumatic diseases. Dr. Phillip Terc, the forbearer of apitherapy, celebrated World Apitherapy Day on his birthday 30 March 1888. While Dr. Franz Kretsky was the first person to invent injectable bee venom in 1928, it didn't get franchised until 2003, when Dr. Christopher Kim patented it (Gokulakrishnaa & Thirunavukkarasu, 2020). The ancient Egyptians were so keen on the concept of apitherapy that they put bees in the coat of arms of their leaders and also used bees to decorate the walls of Pharaoh's tomb. They were already harvesting honey at this time (Šuligoj, 2021).

Despite the fact that apitherapy is used in the treatment of various illnesses, it takes patience and perseverance to see its positive resultant effect. It must be modified to each person, considering the fact that there are variations in our biological make-up and how each person will react to the treatment. Furthermore, Dr. Stefan Stangaciu, an apitherapist protagonist, prescribed that individual

DOI: 10.1201/9781003685654-1

allergy tests should be carried out before swinging into full action on the use of bee products. Hence, he stated that bee products should be introduced in small doses to observe a person's reaction to it, and that dosage should be increased gradually (Weis et al., 2022). This helps to know that as good as apitherapy is, it has to be administered with caution.

1.2 ORIGIN AND EARLY PRACTICES

1.2.1 Prehistoric Uses

Bee interactions must have existed before the advent of *Homo sapiens*; hence the interaction of bees and humans has been since our ancestors' time. Ethnographic records and rock arts confirmed that apes and other primates in their category used tools to access beehives. Antiquity of apitherapy originates in rupestri canvases, hieroglyphics, archaeological sites, and temples (Weis et al., 2022; Karmakar et al., 2025).

The by-products of apiary have been in the day-to-day lives of humans since ancient times. The graphic evidence of beekeeping records and its use for nutrition, prevention, and treatment of some diseases dates back 15,000 years ago in cave drawings in various places in the world. The Quran, Bible, and Veda also mentioned the use of honey and other apiary products for the management of diseases as well as their serving as nutritious food. Furthermore, it was reported that honey, bee pollen, and its larvae form a part of the food type taken by the *Homo habilis* (Weis et al., 2022; ApiMed Serbia Association, 2025).

According to the King James version of the Bible, "Pleasant words are like honeycombs, Sweetness to the soul and health to the bones" (Proverbs 16:24). This could indicate that honey is a sweet, nourishing food that is healing. It was revealed in the Quran that

> And your Lord revealed to the bee saying: Make hives in the mountains and in the trees and in what they build: Then eat of all the fruits and walk in the ways of your Lord submissively. There comes forth from within it a beverage of many colors, in which there is healing for men; most surely there is a sign in this for a people who reflect. (Quran 16:68–69)

The history of beekeeping records became substantial with the advent of generating synthetic cavities where bees could be vested in and make their products. Our ancestors took honey, bee larvae, and pollen. In the ancient Near East, honey was known as sweetener, food, alcoholic beverage (mead), healing substance, and preserver of a dead body. It also served as a symbol of goodness, abundance, and love. The oldest archaeological beekeeping records can be traced back to the biblical records of King David and King Solomon. The Sumerian bee goddess inscription was discovered in Tel Rehova, Israel, and estimated to be more than 3,000 years old (Weis et al., 2022; ApiMed Sebia Association, 2025; Karmakar et al., 2025).

1.2.2 Ancient Civilization

The experimental study of the use of bees and their by by-products in healing originates in ancient Egypt, China, Greece, and Rome.

The bee used to be the symbol of Lower Egypt. The Egyptians domesticated bees 4,500 years ago. They use honey, propolis, wax, and pollen for food, cosmetics, healing, and preserving mummies. Some authors reported that there are vessels of edible honey that are over 3,000 years old in the tomb of Pharaoh Tutankhamun. Egyptians used propolis for the prevention of dead body decay. Between 2100 and 2000 BC, the first recipe for honey use was documented on a clay tablet in the Euphrates Valley (ApiMed Serbia Association, 2025). Egyptians were able to cure ailments with bee ointments (Karmakar et al., 2025).

In China, bees signify ambition and the success that results from hard work. Bee colonies are a symbol of imperial courts. The Chinese referred to bees as *feng* (fēng) and to honey as *mifeng* (mi fēng). Bees are cited in the oldest written official papers. The Chinese ancient prescription book was documented on a silk scroll, and it was found in the 3rd century BC in the Han tomb residing in Changsha, Hunan province (Zhu & Wongsiri, 2008). Beekeeping started in 221 BC. Beekeeping is said to be a privilege given to dynasties till 25 AD. Ordinary citizens were allowed to keep bees from 25 to 220 AD. Honey was used to display the act of filial loyalty and respect for the elderly, and it was taken as food by monarchs and their descendants (Zhu & Wongsiri, 2008). The earliest professional beekeeper was Jiang-qi (158–167 AD). The first books on beekeeping and collection of honeybee products were published between 1616 and 1911. Ages ago, highly perishable fruits were preserved with honey when being prepared for transport; the composition of honey helps with long-term preservation. China was the world's largest exporter of honey in 2016 (ApiMed Serbia Association, 2025). Furthermore, it was reported that wild pigs, black bears, and some other animals hunted the honeycomb in ancient times, while the primitive man scouted for beehives in rock outcrops and tree holes. These colonies store honey and pollens, which helped primitive men who sought these products and got stung. Hence, they are the first people to be exposed to apitherapy (Zhu & Wongsiri, 2008).

A number of medical scientists and philosophers originated from Old Greece. The father of medicine, Hippocrates (460–377 BC), affirmed that honey acts as an antibiotic, antiseptic, and expectorant, while propolis was used to treat stomach ulcers and internal and external wounds. Aristotle (384–322 BC) believed that eating honey prolongs life, and Galen (130–210) used honey as medicine (ApiMed Serbia Association, 2025).

The consumption of honey serves as a source of strength, healing the body and facilitating fast recovery from war wounds in Greece and Rome. It also serves as a compulsory dietary supplement for athletes during training. It helps enhance their endurance level and prevents them from falling sick. It was also reported that the remains of Alexander the Great were preserved with a combination of honey and beeswax (ApiMed Serbia Association, 2025).

It was in ancient Rome that recipes for making sugar syrups used for feeding bees were found in old writings. Comella describes tools for beekeeping, like combs and hives. They look so much like the modern tools. Honey has been said to enhance digestion and memory and to sharpen the mind (ApiMed Serbia Association, 2025). Furthermore, it should be noted that the earliest medical record of honey being prescribed came from Sumer. Honey was used to treat various diseases in ancient Egypt, Greece, and Rome. It was also known in Chinese folk medicine and Ayurveda. During the medieval era, and the advent of synthetic medicine in the 19th century, apithearpy was rarely used in medicine (Kuropatnicki et al., 2018).

1.2.3 Renaissance to Pre-Modern Era

An Austrian physician, Phillip Terc, published a paper titled "Report on the Unique Relationship Between Bee Stings and Rheumatoid Arthritis" in 1888 (Soares, 2021). The report was on the earliest clinical trials of bee stings. Subsequently, its usage was spread across Europe and the United States. This brought about its popularity in traditional remedies (Soares, 2021).

It was reported that the physicians utilized propolis in treating wounds throughout World War II and the Anglo-Boer war, and it was effective. Propolis cake was also placed on the navels of newborn babies. Traditional medicine made use of propolis in Russia. Propolis is usually inhaled for treating respiratory organ infection, and it also works for burns and angina. It is said to have an efficient anti-microbial property (Gupta & Stangaciu, 2014).

Royal jelly was used in traditional Chinese medicine for centuries, and the Chinese monarchs believed that it helps with longevity and sexual power. Shen Nong (200–100 BC) believed that bee products could be used to treat evil influence, supplement insufficient viscera like the heart and liver, relieve pain in the heart, stomach, and muscles, detoxify, and enhance longevity. Japanese office workers used royal jelly as an energy tonic, and Eastern European countries used it as a substance that helps to increase resistance to stress. It increases endurance level (Gupta & Stangaciu, 2014).

1.2.4 Modern Era

A number of countries, including Nigeria, Germany, Switzerland, Romania, Lithuania, Slovenia, and Bulgaria, have since documented the importance of apitherapy. It is worthy to note that many persons still don't have access to modern medicine and technology. Hence, apitherapy is a modest but effective way of taking care of medical issues in humans. Although, there was a need for the development and advancement, using apitherapy, it is said that "A bee colony is similar to a pharmaceutical factory" (Karmakar et al., 2025). By the 21st century, the medical community had started studying remedies for ailment at the pharmacological level. Proper clinical research and trials were kick-started by the year 2000. There was evidence to support use of bee venom treatment (BVT), and the National Multiple Sclerosis Society started funding research in this line at the Georgetown University Medical Center in Washington. It was discovered that BVT had a substantial effect on individual immunity and reversed arthritis in rodents. Although Professor John Mills of Harvard Medical School did not approve the treatment plan, feeling that a drug treatment will also do justice to the treatment without possible allergic risk posed by the venom, he testified to seeing patients improve after a short-term use of bee venom injection therapy (Soares, 2021). Apitherapy was also an evolving field in Brazil, such that it became a segment in the national policy of integrative and complementary practices under the Ministry of Health (Weis et al., 2022).

In contemporary medicine, apitherapy encompasses a variety of bee products, including honey, propolis, royal jelly, beeswax, and bee venom. Each of the products has been studied and uses various types of mechanisms in different health applications. Summarily, the evolutionary timeline of apitherapy, charted in Figure 1.1, shows the schematic summary of the conceptual milestones in the historical development of apitherapy, moving from ancient practice to contemporary integrative research.

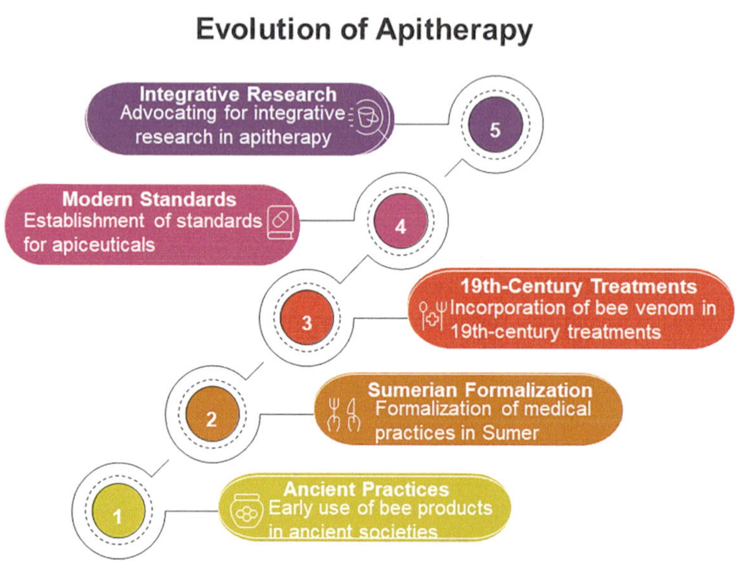

FIGURE 1.1 Evolutionary timeline of apitherapy: a schematic summary of the conceptual milestones in the historical development of apitherapy, moving from ancient practice to contemporary integrative research.

1.3 HISTORICAL PERSPECTIVES OF BEE POLLEN

Bee pollen is the final outcome of the cohesion of flower pollen grains collected by worker bees. They are beehive products from flower pollen grains, nectar, and/or honey and secretions from the salivary gland of the bee. They accumulate at the edges of the hive. Its collection does not destroy the hives where it is collected (El Ghouizi et al., 2023). Usage of bee pollen in indigenous medicine dates back to prehistoric times. The records of Arab and Jewish physicians (i.e. Ibn al-Beithar and Maimonides) define its usage as an aphrodisiac, a sedative, and an effective treatment for the stomach, heart, and intestines since the early 1100s (El Ghouizi et al., 2023). Furthermore, the ancient Egyptians describe pollen as a "life-giving dust," as it is used in supplementary nutrition as a source of vital nutrients such as proteins, carbohydrates, lipids, vitamins, minerals, trace elements, and flavonoids; hence it is referred to as the most complete food in nature (Kacemi & Campos, 2023). Also, the Institute of Apiculture in Russia stated that bee pollen is the richest source of vitamins found in a single food in nature. They could have as much as 2–60% protein, 13–55% of carbohydrates and 1–20% of lipids, which sums up to its being referred to as a complete or functional food (Khalifa et al., 2021; Damto et al., 2024). Bee pollen was referred to as Ambrosia, which means the food of the Gods in ancient Greece; it is said to possess the power of immortality and eternal youth. Also, Hippocrates (the father of modern medicine) and Pythagoras believed in the usage of bee pollen for its healing effects. The Romans referred to bee pollen as panacea; hence their soldiers relied on the use of dried pollen cakes for provision of sustenance and increase in endurance level. Nevertheless, the Māori of New Zealand had an aged custom of taking bee pollen as food. Reverend William Colenso said that "another highly curious article of vegetable food" was pungapunga, which is the yellow pollen of the raupo flower. It is used to make steamed bread called pua.

According to modern researchers, bee pollen contains about 200 biologically active compounds that help it to fulfill its medicinal properties in humans, although their nutritional makeup is influenced by the floral types, geographical source, climate type, and the variant of soil within the area (Damto et al., 2024). Nonetheless, bee pollen helps in the reduction of inflammation and mitigation of menopausal symptoms as well as depression, and aids in digestion, immunity, healing of wounds, and shedding of weight. Also, they possess medicinal values such as antioxidant, antiaging, anticancer, anti-inflammatory and cardioprotective, hepatoprotective, nephroprotective, antinociceptive, and antidiabetic properties (Mohammad et al., 2021; El Ghouizi et al., 2023; Damto et al., 2024). Also, bee pollen prevents obesity, regulates ovarian functions, enhances bone metabolism, enhances cognition, and ameliorates the effect of toxins in the body (Khalifa et al., 2021).

1.4 HISTORICAL PERSPECTIVES AND USES OF BEESWAX IN APITHERAPY

Beeswax is a liquid complex compound secreted from the specialized wax glands (ceriferous glands) in the ventral region of young (12- to 18-day-old) worker bees. It solidifies when it comes in contact with air. The bee's jaw forms the structure of honeycomb that stores both honey and pollen. Once the beeswax comes in contact with honey and pollen, it changes from white to golden yellow, then changes to dark brown after four years, as it contains a cocoon (El-Didamony et al., 2024).

Ebers Papyrus (1550 BC) and the people of ancient Rome stated that beeswax is the major constituent of lots of ointments and creams used for treating burns, wounds, bruises, joint pain, and fractures. Hippocrates prescribed beeswax in treating tonsillitis (El-Didamony et al., 2024). In ancient times, Avicenna stated that different medicinal composition contains beeswax. Archaeological evidence of beeswax ointment has existed since the 16th century (El-Didamony et al., 2024). Beeswax plays a key role in Ayurvedic medicine. It was referred to as Madhuchishtha by those practicing traditional Indian medicine; they use it to nurse wounds from abrasion and burns as a topical application (Fratini et al., 2016b). In the modern era, beeswax is used as an organic lipid component, when preparing nanostructured lipid carriers for various biological activities (El-Didamony et al., 2024).

Beeswax was reported to have antimicrobial activities that enable protection against various bacterial strains such as *Candida albicans*, *Streptococcus aureus*, and so on (Fratini et al., 2016b). It was also reported to be used for wound cleansing and healing. It was used in treating fractures due to its great binding ability; it also treats gouty arthritis and skin diseases like herpes (Mishra et al., 2017).

1.5 HISTORICAL PERSPECTIVES AND USES OF HONEY IN APITHERAPY

The use of honey for its medicinal values dates back to a time when there was no history of medicine. The oldest therapeutic formulation, which has honey as an ingredient, originated from Sumer in 2000 BC. Religious books like the Quran, the Bible, and the Veda mentioned the medicinal and nutritious effects of honey (Šuligoj, 2021). In 25–88 AD, a 92-piece bamboo manuscript on prescriptions for diseases has on record that honey was the major ingredient in a honey pill used to treat asthma-symptomatic cough. Zhang Zhongjing also reported in a book that honey suppository can be used to treat constipation, while another book stated that honey soup with licorice powder can help take care of stomach ache caused by worms (Zhu & Wongsiri, 2008). Then, Aristotle (384–332 BC) believed that taking honey helps humans to keep their health and vitality and is a good ointment for sore eyes (El-Didamony et al., 2024). Homer, Pythagoras, Ovid, and Democritus also believed that honey helps preserve health and stamina. According to Hippocrates (460–370 BC), honey helps to generate heat, could be used for cleaning sore wounds, softens and cleans ulcers, and heals boils and running sores. Around 1550 BC, the Ebers papyrus prescribed honey for external use: "Mix honey, red ochre, powdered alabaster to cure spotted baldness." Then, similar mixtures were used to dress wounds, burns, abscess, sores, and other skin conditions. Honey was used after surgery including circumcision, as a suppository, and to reduce inflammation and loosen stiff joints (Kuropatnicki et al., 2018).

Aside from the ancient Egyptians putting bees into their coat of arms, it was also documented that honey was suggested for treatment of eye, renal, and stomach disorders. Honey was reported to be used against animal, plant, and mineral poisonings in ancient Eastern medicine, while the Buddhists used honey for longevity. The Tang dynasty in ancient China used honey for skin care. The Indians, Greeks, and Romans also utilized honey for its medicinal properties (El-Didamony et al., 2024; Šuligoj, 2021).

Honey is a liquid, viscous, light- or dark-amber-colored, natural sweet substance produced by bees from the flower nectar or secretions from plants. The secretions are gathered and modified by bees. They are collected, desiccated, and kept in the honeycomb to ripen and mature.

The main constituents of honey include sugars (i.e. glucose, fructose, maltose, sucrose, isomaltose, maltotriose, maltulose, melezitose, and threalose), water, and enzymes (invertase, alpha and beta amylase, catalase, and glucose oxidase). Other minor constituents include proteins, vitamins (B and C), minerals (sodium, potassium, calcium, magnesium, iron, copper, manganese, and zinc) and other phytochemicals (El-Didamony et al., 2024).

The specie of bee, floral origin, geographical locations, and season of production all have a huge effect on the quality, color, flavor, and smell of honey. The handling, mode of storage, and climate change also affect it. It is important to note that dark-colored honeys have a high antioxidant capacity, as they usually have a high quantity of phenols (El-Didamony et al., 2024).

Generally, various scientific studies have proven that honey has antioxidant, anti-inflammatory, anticancer, antimicrobial, antiparasitic, cardioprotective, and tissue regeneration properties (El-Didamony et al., 2024). Honey has been reported to have ameliorative effects on oxidative stress in the brain tissue of sleep-deprived experimental rats (Akintoye et al., 2021). Furthermore, honey was also used in the management of diabetic foot sores, dental hypersensitivity, and respiratory infections (Kafle & Joshi, 2023).

1.6 HISTORICAL PERSPECTIVES AND USES OF BEE VENOM

The use of bee venom for its medicinal values is one of the earliest bee products utilized for apitherapy. Around 3000 BC, traditional Eastern medicine used bee venom for treating inflammatory illnesses. During the period of ancient civilizations, Hippocrates used bee venom to treat arthritis and inflammation, and Charlemagne used it to cure gout. In the 19th century (1897–1899), Langer was the first to attempt harvesting bee venom without killing the bee. The venom was used in treating rheumatism (Bava et al., 2023). Dr. Filip Terc (1844–1917) achieved a breakthrough in modern medicine through studying the effects of bee venom on humans. He realized bee stings reduced the symptoms of rheumatism. Breaking records in the study of allergic reactions and immunology, he carried out a number of studies on bee stings; hence his birth date (March 30) was pronounced as World Apitherapy Day by the International Apitherapy Congress, Passau, in 2006 (Šuligoj, 2021). The venom first became available for sale in 1930 by the Mack Company in southern Germany. The method of venom extraction was perfected in 1960, in Czechoslovakia (Bava et al., 2023).

Dr. Bodog Beck was the pioneer of bee venom therapy, having published a powerful book named *Bee Venom Therapy* in 1930. Dr. Joseph Broadman started using bee venom therapy to tackle arthritis in the early 1950s and authored the book *Bee Venom: The Natural Curative for Arthritis and Rheumatism* in 1962, in which he shared his experience. About 12 nations in the European region, 3 Asian countries, and South, Central, and North American nations have been using apitoxin since 1971 (Bava et al., 2023).

Bee venom is also referred to as apitoxin. It is a clear, bitter transparent acidic liquid that is secreted from bee glands. The substance has a higher level of production in its first two weeks of life. About 0.3 mg of bee venom is produced in the venom sac. It gets crystallized and dried when in contact with air. Bee venom contains about 18 pharmacologically active components. It is composed of various active constituents such as peptides (melittin, adolapin, and apamin), enzymes (hyaluronidase and phospholipase A2), and mast cell degranulating peptide; non-peptide molecules (lipids, carbohydrates, free amino acids, histamine, norepinephrine, and dopamine); and micronutrients like formic acid hydrochloride, phosphoric acid, acetate, sulfur, calcium, and magnesium. Melittin makes up about 50% of dry venom, then phospholipase A2, which makes up about 12% (Wehbe et al., 2019; El-Didamony et al., 2024; Jaafar & Albushabaa, 2025). Hence, the bee venom has therapeutic effects such as anti-arthritic, anti-mutagenic, anti-nociceptive, radioprotective, hepatoprotective, neuroprotective, antioxidant, anti-inflammatory, antimicrobial, antiviral, and anticancer properties (Wehbe et al., 2019; El-Didamony et al., 2024). Furthermore, bee venom is usually used in treating neurodegenerative disorders such as Parkinson's disease, Alzheimer's disease, amyotrophic lateral sclerosis, and so on; ovarian and prostate cancer; and HIV (Azam et al., 2019; Wehbe et al., 2019).

Bee sting therapy, also known as apipuncture, is carried out by licensed practitioners of acupuncturists using a live bee. However, bee sting therapy could be applied through direct or indirect ways (El-Didamony et al., 2024).

1.7 HISTORICAL PERSPECTIVES OF BEE PROPOLIS

As with honey, propolis has been used in traditional medicine from time immemorial. The ancient Egyptians used propolis to mitigate various diseases, and it was also used for embalming. It was also confirmed by scientists in the 1960s that propolis limits infection spread and lowers bacterial infection with hives. The ancient Jews referred to propolis (also known as tzori in Hebrew) as a medicine. This is evident in the Old Testament Bible, when it was stated that Queen Sheba gifted King Solomon tzori. Aristotle was said to have used propolis to cure bruises and suppurating sores, while Hippocrates used it for wounds and ulcers. Pedanios Dioscorides (50 AD) stated that it could be used for old coughs. According to Pliny the Elder, propolis can be used to extract stings and foreign substances from the flesh, disperse boils, and mitigate pains and ulceration, as written in natural history. Cornelius Celsus wrote in his book *De Medicina Celsus* that propolis is a drug that promotes suppuration and is used for treating open wounds and abscesses. The Arabs prescribed propolis for treating eczema, myalgia, and rheumatism (Rojczyk et al., 2020).

In the 12th century, propolis was used for treating oral and pharyngeal infections including dental caries. In 17th-century England, propolis was listed as the major constituent of healing ointments. Furthermore, propolis is a Greek word, which means defense of the city. In Greece and Rome, propolis was used to disinfect the mouth and heal wounds as a result of its antiseptic effect. Propolis was also used as an anti-febrile agent by the Incas; Persians prescribed it for treating eczemas, myalgia, and rheumatism. Propolis was used in managing tuberculosis and healing wounds during World War II (Weis et al., 2022). Dr. Karl Lund Aagaard was referred to as "Dr. Propolis," having researched about it for over 20 years in about 50,000 patients between 1967 and 1973. He found out that propolis could be used in the treatment of cancer, urinary tract infection, tonsillitis, inflamed gout, open wounds, blocked sinuses, cold, influenza, bronchititis, gastritis, ear infection, periodontal diseases, ulcers, eczema, Parkinson's disease, arthritis, pneumonia, chronic colitis, pharyngitis, and conjunctivitis (Kuropatnicki et al., 2013; Rojczyk et al., 2020).

Propolis is a natural wax-like resin substance found in beehives. It is a viscous, sticky, dark yellow or brownish substance collected by honeybees from plant buds and shoots. It is used as cement to seal cracks and protect bee colonies (El-Didamony et al., 2024). At high temperatures it is usually soft, pliable, and sticky, and it gets hard and brittle in low temperatures (Kuropatnicki et al., 2013; Rojczyk et al., 2020).

Raw propolis contains about 300 compounds; these include phenolic compounds, flavonoids (flavones, flavanones), aromatic aldehydes, amino acids, stilbenes, chalone, esters, diterpenes, sesquiterpenes, steroids, lignans, and minerals. Hence, it is said to have pharmacological activities such as antimicrobial, antioxidant, anti-inflammatory, anti-tumor, immunomodulatory, and a cytoprotective effect on human breast cancer cells (El-Didamony et al., 2024). It is also said to have antipsoriatic, estrogenic, antihypertensive, and analgesic, hepatoprotective, antidiabetic, hypolipidemic, neuroprotective and antidepressant effects (Weis et al., 2022).

1.8 HISTORICAL PERSPECTIVES AND USES OF ROYAL JELLY

The ancient Greeks believed that royal jelly is part of the constituent of the "ambrosia"; nectar gives immortality to the gods of Olympus. The first person to discover the benefits of royal jelly was Aristotle. He found out that it increases physical strength and improves intellectual capacity. It was the unique beauty secret of Cleopatra; hence it was well used for cosmetics in ancient Egypt and as a symbol of strength and majesty of the Pharaohs. Royal jelly was used in Chinese folk medicine for about 70–80 years. It was of great importance to Chinese monarchs; they believed it increases the life span and enhances sexual prowess. The Japanese use it to curb fatigue. Hence, it is referred to as a secret of life or an elixir (Fratini et al., 2016a; El-Didamony et al., 2024).

Royal jelly is a viscous, milky white or yellow acidic substance. It has a minimal pungent odor and taste, with a pH of 3.6. They are secreted from hypo-pharyngeal and mandibular glands of young worker bees. The queen bees depend on royal jelly as their superfood for their entire life cycle (El-Didamony et al., 2024).

Royal jelly is composed of water, protein (antibacterial protein, apisin, and the major royal jelly proteins, MRJP1), sugars, lipids, vitamins (A, B complex, C, and E), aminoacids (Leucine, Valine, Threonine, Methionine, Trypsin, and Lysine), minerals (Calcium, iron, manganese, potassium, sodium, zinc, magnesium and copper), peptides (specific peptides such as apisimin, royalisin, royalactin, apidaecin, defensin-1, and jelleins), hormones, polyphenols, and enzymes. Because of the special protein type in it, it improves memory and learning, serves as an antioxidant, neuroprotective, cardiprotective, antiaging, antiapoptotic, and immunomodulatory agent. It also helps to modulate hormonal balance and gut microbiota composition (El-Didamony et al., 2024; Oršolić & Jazvinšćak Jembrek, 2024; Varol et al., 2024). Table 1.1 shows the summary of major bee products.

TABLE 1.1
Comparative Summary of Major Bee Products

Bee Product	Historical Applications	Primary Bioactive Components	Contemporary Evidence	Clinical Uses/ Indications	Safety Considerations and Contraindications	Knowledge Gaps and Recommended Future Research
Honey	Employed as a therapeutic agent for digestive and respiratory ailments, ancient wound dressings, and embalming practices across diverse cultures such as Egypt, Greece, China, and India.	Sugars, water, vitamins, minerals, phenolic compounds, enzymes, and proteins.	Modern scientific inquiry, encompassing both clinical and laboratory studies, has substantiated its antioxidant, antimicrobial, wound-healing, and anti-inflammatory properties.	Consequently, honey finds contemporary application in topical wound care, cough suppressants, and as an oral supplement.	Nevertheless, its use necessitates caution due to the risk of infant botulism, potential adverse effects in diabetic individuals, rare allergic reactions, and concerns regarding contamination.	Future research should prioritize the standardization of raw materials, conduct dose-response randomized controlled trials for specific indications, and establish robust quality control and regulatory guidelines.
Propolis	Employed in wound care, embalming, traditional respiratory remedies, and as a mouth disinfectant.	It comprises a complex mixture of phenolic compounds, flavonoids, aromatic aldehydes, esters, diterpenes, and sesquiterpenes.	Preclinical and small clinical studies indicate its efficacy in wound healing, with additional antimicrobial, anti-inflammatory, and immunomodulatory properties.	Its applications include topical antiseptics, an adjunct in dental care, mouthwashes, and topical wound adjuncts.	Potential adverse effects include contact dermatitis, and its composition varies by region, necessitating standardization. Allergy is also a possibility.	Future research should focus on large-scale randomized controlled trials, chemical standardization, advanced formulation research (including bioavailability and delivery), and comprehensive safety profiling.
Royal jelly	Historically revered as an "ambrosia" or elixir of longevity in ancient cultures, royal jelly has been utilized as an energy tonic and cosmetic ingredient.	It comprises major royal jelly proteins, peptides (e.g., royalism, apidaecin), 10-HDA (a unique fatty acid), as well as vitamins, minerals, and amino acids.	Preclinical investigations indicate antioxidant, immunomodulatory, and potential endocrine and neuroprotective effects; however, robust clinical evidence remains limited.	It is incorporated into nutraceuticals and cosmetics and is traditionally used as a tonic, with ongoing research exploring its metabolic and neuroprotective benefits.	Reported adverse effects include allergic reactions and potential hormonal impacts, necessitating caution in pregnant individuals or those with hormone-sensitive conditions.	Future research should focus on rigorous randomized controlled trials, standardized extraction and delivery methods, comprehensive safety evaluations, pharmacokinetic studies, and detailed mechanistic investigations.

(Continued)

Table 1.1 *(Continued)*

Bee Product	Historical Applications	Primary Bioactive Components	Contemporary Evidence	Clinical Uses/ Indications	Safety Considerations and Contraindications	Knowledge Gaps and Recommended Future Research
Beeswax	Historically utilized as a foundation for ointments, wound salves, and as a preservative; frequently employed as an excipient in cosmetic and pharmaceutical preparations.	A complex lipid mixture comprising esters, hydrocarbons, long-chain alcohols, and fatty acids; also contains minor phenolic constituents.	Primarily functions as a formulation excipient; some reports indicate antimicrobial barrier effects and wound-protective characteristics when used as a vehicle.	Serves as a base for ointments, protective wound dressings, various topical formulations, and cosmetic items.	Exhibits low intrinsic toxicity; potential for contamination or allergenicity exists; generally well tolerated when applied topically.	Further research is needed to characterize its therapeutic contributions versus its vehicle effect, standardized sourcing and purity, antimicrobial activity confirmation.
Bee pollen	Historically revered as a functional food and tonic, bee pollen was incorporated into traditional remedies and consumed for sustenance.	Composed of proteins, carbohydrates, lipids, vitamins, minerals, flavonoids, and various phenolic compounds, its exact composition varies significantly based on floral origin and geographical region.	Preclinical studies suggest antioxidant, anti-inflammatory, and potential metabolic benefits; however, human clinical data remain scarce and inconsistent.	Currently marketed as a dietary supplement and functional food, it is also explored as an adjunctive treatment for fatigue and inflammatory conditions.	A significant allergen, bee pollen poses a risk of severe allergic reactions, particularly in pollen-sensitive individuals; product contamination is another concern.	Future research should focus on standardization according to floral origin, conducting controlled clinical trials to assess metabolic and immune outcomes, allergenicity profiles, and optimal dosing standards.
Bee venom	Employed for treating arthritis and various rheumatic conditions; has a long lineage of use in bee-sting therapy, spanning from ancient times to the contemporary era.	Its primary active constituents include melittin, phospholipase A2, apamin, adolapin, hyaluronidase, biogenic amines, and various peptides.	Preclinical investigations suggest anti-inflammatory, neuroprotective, and anticancer properties; however, clinical trial data for rheumatological and neurological disorders remain limited and exhibit mixed results.	Current applications involve the experimental management of arthritis, multiple sclerosis, and acupuncture techniques, as well as its exploration as an adjuvant in oncology research.	A significant drawback is the high potential for allergic reactions and anaphylaxis, necessitating allergy testing and medical supervision. Furthermore, dosing and delivery methods lack standardization.	Future research priorities include well-designed randomized controlled trials, the standardization of extracts and delivery systems, rigorous safety and pharmacokinetic studies, and detailed mechanistic investigations.

Note: This table synthesizes historical uses, key bioactive constituents, modern evidence, clinical applications, safety considerations, and research gaps for major bee products.

1.9 CONCLUSION

Apitherapy is a treatment plan that involves apiary yields such as beeswax, bee pollen, honey, bee propolis, royal jelly, and bee venom to promote healing and the management of various illnesses holistically. The advent of apitherapy took place during prehistorical times, and its usage keeps improving in the modern era. Today, apitherapy stands at the connection of traditional knowledge and modern biomedical research. Its renewed global interest highlights the continuing relevance of natural therapies and underscores the importance of integrating historical wisdom with contemporary scientific validation to ensure safe, effective, and evidence-based use of bee products in medicine.

REFERENCES

Akintoye, O. O., Ajibare, A. J., Fabunmi, O. A., Asuku, A. O., & Tajudeen, R. A. (2021). Effect of honey on oxidative stress in brain tissue of sleep deprived rats. *Nigerian Journal of Basic and Applied Science*, *29*(1), 83–89, https://doi.org/10.4314/njbas.v29i1.10

ApiMed Serbia Association. (2025). *Apitherapy through History and Today*. https://apimedsrbija.rs/en/apitherapy/apitherapy-through-history-and-today/

Azam, Md. N. K., Ahmed, Md. N., Biswas, S., Ara, N., Rahman, Md. M., Hirashima, A., & Hasan, Md. N. (2019). A review on bioactivities of honey bee venom. *Annual Research & Review in Biology*, *30*(2), 1–13. https://doi.org/10.9734/ARRB/2018/45028

Bava, R., Castagna, F., Musella, V., Lupia, C., Palma, E., & Britti, D. (2023). Therapeutic use of bee venom and potential applications in veterinary medicine. *Veterinary Sciences*, *10*(2), 119. https://doi.org/10.3390/vetsci10020119

Damto, T., Gemeda, M., & Kebaba, D. (2024). Chemical and mineral composition of the mono-floral pollen of Honeybees (Apis mellifera) in Ethiopia. *Bee Studies- Apiculture Research Institute*, 2. https://doi.org/10.51458/BSTD.2024.42

El Ghouizi, A., Bakour, M., Laaroussi, H., Ousaaid, D., El Menyiy, N., Hano, C., & Lyoussi, B. (2023). Bee pollen as functional food: Insights into its composition and therapeutic Properties. *Antioxidants*, *12*(3), 557. https://doi.org/10.3390/antiox12030557

El-Didamony, S. E., Gouda, H. I. A., Zidan, M. M. M., & Amer, R. I. (2024). Bee products: An overview of sources, biological activities and advanced approaches used in apitherapy application. *Biotechnology Reports*, *44*, e00862. https://doi.org/10.1016/j.btre.2024.e00862

Fratini, F., Cilia, G., Mancini, S., & Felicioli, A. (2016a). Royal Jelly: An ancient remedy with remarkable antibacterial properties. *Microbiological Research*, *192*, 130–141. https://doi.org/10.1016/j.micres.2016.06.007

Fratini, F., Cilia, G., Turchi, B., & Felicioli, A. (2016b). Beeswax: A minireview of its antimicrobial activity and its application in medicine. *Asian Pacific Journal of Tropical Medicine*, *9*(9), 839–843. https://doi.org/10.1016/j.apjtm.2016.07.003

Gokulakrishnaa, R., & Thirunavukkarasu, S. (2020). Apitherapy: A valuable gift from honey bee. *Journal of Entomology and Zoology Studies*, *8*(5), 2317–2323. https://www.entomoljournal.com/archives/2020/vol8issue5/PartAF/8-5-138-346.pdf

Gupta, R. K., & Stangaciu, S. (2014). Apitherapy: Holistic healing through the honeybee and bee products in countries with poor healthcare system. In R. K. Gupta, W. Reybroeck, J. W. Van Veen, & A. Gupta (Eds.), *Beekeeping for Poverty Alleviation and Livelihood Security* (pp. 413–446). Springer Netherlands. https://doi.org/10.1007/978-94-017-9199-1_15

Jaafar, F. S., & Albushabaa, S. H. H. (2025). General study of honey bee venom Apis mellifera: A review. *Journal of Applied Health Sciences and Medicine*, *5*(1), 12–18. https://doi.org/10.58614/jahsm512

Kacemi, R., & Campos, M. G. (2023). Translational research on bee pollen as a source of nutrients: A scoping review from bench to real world. *Nutrients*, *15*(10), 2413. https://doi.org/10.3390/nu15102413

Kafle, L., & Joshi, A. C. (2023). Apitherapy: An overview. *Journal of Applied Zoological Researches*, *34*(2), 1–7.

Karmakar, R., Soni, S., Soni, V., & Patil, U. K. (2025). Potential of apitherapy in the management of cardiovascular diseases. *European Journal of Medicinal Chemistry Reports*, *13*, 100256. https://doi.org/10.1016/j.ejmcr.2025.100256

Khalifa, S. A. M., Elashal, M. H., Yosri, N., Du, M., Musharraf, S. G., Nahar, L., Sarker, S. D., Guo, Z., Cao, W., Zou, X., Abd El-Wahed, A. A., Xiao, J., Omar, H. A., Hegazy, M.-E. F., & El-Seedi, H. R. (2021). Bee pollen: Current status and therapeutic potential. *Nutrients*, *13*(6), 1876. https://doi.org/10.3390/nu13061876

Kuropatnicki, A. K., Szliszka, E., & Krol, W. (2013). Historical aspects of propolis research in modern times. *Evidence-Based Complementary and Alternative Medicine*, *2013*, 1–11. https://doi.org/10.1155/2013/964149

Kuropatnicki, A. K., Kłósek, M., & Kucharzewski, M. (2018). Honey as medicine: Historical perspectives. *Journal of Apicultural Research*, *57*(1), 113–118. https://doi.org/10.1080/00218839.2017.1411182

Mishra, A. K., Krishnapriya, & Ramesh, N. V. (2017). Therapeutic uses of bees wax in ayurveda and its physico chemical analyses: A review. *International Journal of Research in Ayurveda & Pharmacy*, *8*(1), 25–26. https://doi.org/10.7897/2277-4343.08131

Mohammad, S. M., Mahmud-Ab-Rashid, N.-K., & Zawawi, N. (2021). Stingless Bee-collected pollen (Bee Bread): Chemical and microbiology properties and health benefits. *Molecules*, *26*(4), 957. https://doi.org/10.3390/molecules26040957

Oršolić, N., & Jazvinšćak Jembrek, M. (2024). Royal Jelly: Biological action and health benefits. *International Journal of Molecular Sciences*, *25*(11), 6023. https://doi.org/10.3390/ijms25116023

Rojczyk, E., Klama-Baryła, A., Łabuś, W., Wilemska-Kucharzewska, K., & Kucharzewski, M. (2020). Historical and modern research on propolis and its application in wound healing and other fields of medicine and contributions by Polish studies. *Journal of Ethnopharmacology*, *262*, 113159. https://doi.org/10.1016/j.jep.2020.113159

Soares, C. (2021). Origin and development of apitherapy: Focus on bee venom therapy and apitourism. *Journal of Apitherapy*, *8*(7).

Šuligoj, M. (2021). Origins and development of apitherapy and apitourism. *Journal of Apicultural Research*, *60*(3), 369–374. https://doi.org/10.1080/00218839.2021.1874178

Trumbeckaite, S., Dauksiene, J., Bernatoniene, J., & Janulis, V. (2015). Knowledge, attitudes, and usage of apitherapy for disease prevention and treatment among undergraduate pharmacy students in Lithuania. *Evidence-Based Complementary and Alternative Medicine*, *2015*, 1–9. https://doi.org/10.1155/2015/172502

Varol, E., Balkanska, R., & Yücel, B. (2024). Royal Jelly; Biochemical Properties, Activity and Medical Use. *Hayvansal Üretim*, *65*(2), 196–205. https://doi.org/10.29185/hayuretim.1571944

Wehbe, R., Frangieh, J., Rima, M., El Obeid, D., Sabatier, J.-M., & Fajloun, Z. (2019). Bee Venom: Overview of main compounds and bioactivities for therapeutic interests. *Molecules*, *24*(16), 2997. https://doi.org/10.3390/molecules24162997

Weis, W. A., Ripari, N., Conte, F. L., Honorio, M. D. S., Sartori, A. A., Matucci, R. H., & Sforcin, J. M. (2022). An overview about apitherapy and its clinical applications. *Phytomedicine Plus*, *2*(2), 100239. https://doi.org/10.1016/j.phyplu.2022.100239

Zhu, F., & Wongsiri, S. (2008). A brief introduction to apitherapy healthcare. *Journal of Thai Traditional and Alternative Medicine*, *6*(3), 303–312.

2 Basic Concept of Apiary and Apitherapy

Modupeola Olamide Falabi, Olusola Samson Folorunso, and Edward Ekundayo Akerele

2.1 INTRODUCTION TO APIARY AND APITHERAPY

Beekeeping and apitherapy are two related concepts that center on ecological practices, agricultural endeavor, and medicinal importance in the use of honeybees. Lately, it has been found also to have touristic importance. An apiary is the professional name given to the space where honeybee colonies (*Apis mellifera*) are kept for purposes of honey production, crop pollination, and the harvesting of other hive products. Traditionally, apiaries are associated with nutrition and health; however, they are now gaining traction as cultural and educational spaces through apitourism, which integrates beekeeping into tourism activities. Scholars have associated apitourism with rural development, environmental sustainability, heritage conservation, and experiential learning opportunities (Topal et al., 2021).

Beekeeping contributes to food production by providing honey and other nutritional substances; it sustains ecosystem services by enhancing crop pollination, a process critical for biodiversity, and supplies material resources such as beeswax and other related products known for commercial and therapeutic applications. Within the tourism sector, however, these functions extend beyond production to serve as tools for storytelling, education, and sustainable destination development. Apitourism therefore positions beekeeping as both an economic and a cultural resource, offering rural communities alternative pathways to development and resilience (Gascón, 2025).

In addition, apitherapy, from the Latin *apis* meaning "bee" and the Greek therapeia meaning "healing," involves the medicinal use of hive products. Honey has been used as an antimicrobial wound dressing, bee venom has shown potential in the treatment of inflammatory and autoimmune conditions, and propolis has demonstrated immunomodulatory and antioxidant properties. These therapeutic practices not only contribute to human health but also strengthen the experiential dimension of apitourism, as visitors engage with the healing and wellness traditions associated with bee products (Topal et al., 2021). Consequently, apiary and apitherapy demonstrate a long-standing human relationship with honeybees that now extends beyond agriculture and medicine into sustainable tourism. Apiary and apitherapy can now be situated within the realm of regenerative tourism, thereby contributing to ecological awareness, rural livelihoods, and holistic well-being (Topal et al., 2021; Gascón, 2025). In this chapter, these fundamental ideas will be explored in greater detail.

2.2 THE SCIENCE AND HISTORY OF BEEKEEPING (APIARY)

2.2.1 Definition and Importance of Apiaries

An apiary serves the primary purpose of honey production, pollination services, and the harvesting of other hive products that link agriculture, biodiversity, and rural livelihoods. Animal pollination plays an important role in global food production, with honeybees playing a central role (Khalifa et al., 2021). In the United States alone, their pollination services contribute billions of dollars annually to agricultural production, underscoring their critical economic importance (Jordan et al., 2021).

DOI: 10.1201/9781003685654-2

Commercial apiaries serve the purposes of honey production, pollination, and biodiversity conservation by supplementing the ecological role of declining wild pollinator populations. Apiaries also provide pharmaceutical and industrial raw materials (Topal et al., 2021). These functions illustrate the multilayered contributions of apiaries along the value chain that transcend simple food production and extend into economic, ecological, and cultural domains.

2.2.2 Historical Background of Beekeeping

The history of beekeeping dates back to Ancient Egypt around 2600 BCE, where honey was harvested not only as a sweetener but also for medicinal purposes and mummification processes (Kuropatnicki et al., 2018). Honey to the Egyptians was not just for dietary purposes; it was crucial to everyday ritual, signaling its cultural and spiritual value. In Ancient Greece and Rome, bees were similarly revered. The ancient Greek civilization also associated honey with divine origins, while Aristotle's *Historia Animalium* (350 BCE) represents one of the earliest systematic efforts to study bee behavior from a naturalistic perspective (Šuligoj, 2021).

In medieval Europe, the use of honey in religious rituals is very well established as monasteries maintained apiaries, underscoring the integration of apiculture into religious life and material culture. The invention of the Langstroth movable-frame hive in 1851 revolutionized beekeeping by allowing non-destructive inspections, efficient honey harvesting, and systematic study of colony dynamics. This innovation marked the transition from traditional, often destructive methods to scientific apiculture that prioritized both productivity and sustainability (Topal et al., 2021).

Recent scholarship such as that of Topal et al. (2021) shows how apiaries functioned as spaces where ecological knowledge, agricultural practice, and social organization intersected, while Šuligoj (2021) highlights the evolution of apitherapy and apitourism as modern extensions of this ancient tradition. These developments suggest that apiary and apitherapy should be understood not only as a food production system but also as a cultural practice that has historically adapted to shifting human needs and that continues to generate new opportunities for sustainability, health, and tourism.

2.3 COMPONENTS OF AN APIARY

An apiary is also known as a beehive; beehives can be in various forms. Types of beehives are as follows:

1. *Langstroth Hive* – This is most common, with removable frames. Its major advantages are standardized equipment and high honey yield. However, it also becomes heavy supers when full.
2. *Top-Bar Hive* – This is characterized with low-cost, horizontal design and natural comb building. Its disadvantages include lower honey production.
3. *Warre Hive* – This is a natural beekeeping approach, with minimal intervention. However, it comes with difficulty in honey extraction.

2.3.1 Honeybee Species and Their Roles

Worker bees – Females that gather nectar, build combs, and defend the hive.
Drones – Males that mate with the queen.
Queen bee – The sole egg-layer in the colony.

The honeybee caste system is illustrated in Table 2.1 and Figure 2.1.

TABLE 2.1
Honeybee Caste System

Caste	Population	Lifespan	Function	Reference
Queen	1 per hive	2–3 years	Egg laying (2000/day)	Fèvre and Dearden (2024)
Worker	20,000–60,000	6 weeks (Summer)	Foraging, nursing, defending	Prado et al. (2020)
Drone	200–300	90 days	Mating with virging queens	Schaumann et al. (2024)

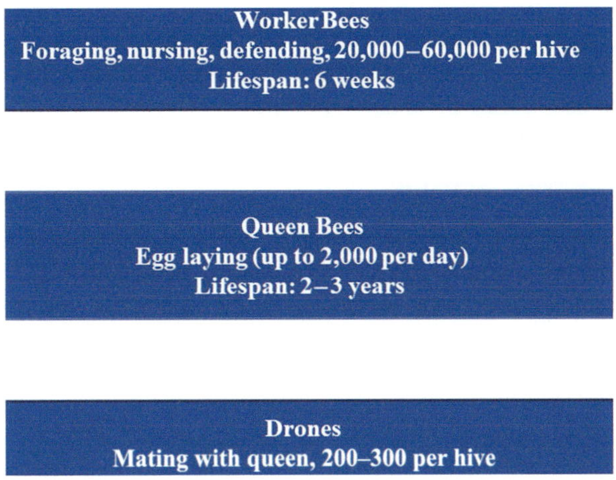

FIGURE 2.1 Honeybee caste system.

2.3.2 Essential Beekeeping Equipment

Beekeeping equipment includes the following:

Protective gear – Ventilated suits with integrated veils, that is, Bee suit, gloves, and veil.
Hive tools – J-hook design for frame prying, and for inspecting and maintaining hives.
Honey extractors – Machines to harvest honey efficiently through radial and tangential honey extraction

2.4 BEEKEEPING MANAGEMENT AND PRACTICES

2.4.1 Setting Up an Apiary

The process of establishing an apiary requires multiple essential steps which lead to successful apiary management through proper planning and execution. The selection of a location with abundant floral resources and pure water supply stands as the essential first step in this process. The second requirement involves proper hive spacing to prevent overcrowding. These include:

Temperature: The ideal hive temperature is between 32°C and 35°C, when bees receive morning sunlight followed by afternoon shade. The importance of sunlight exposure for hives becomes clear through this information.
Wind protection: Natural windbreaks or artificial barriers.

The water source on site includes floating corks, which serve as a safety measure against drowning accidents.

The bees need to access foraging areas within 3–5 kilometers of diverse floral sources, which can be determined through USDA Plant Hardiness Zone calculations.

The following details the requirements for hive placement.

- The hive entrance needs to face in a southeast direction to support early foraging activities.
- The hives need to maintain 1 m distance from each other and 3 m distance between rows.
- The hive base should rest at 30–45 cm height above the ground to protect against dampness and pests.

The colony installation process follows the package bee setup and nucleus colony transfer. The direct release method replaces the queen cage method during package bee transfers, while we use 1:1 sugar syrup for spraying the bees during the transfer process.

2.4.2 Seasonal Beekeeping Activities

Seasonal beekeeping involves a variety of procedures. Firstly, through frame-by-frame inspection for brood pattern quality as an aid for queen introduction protocols, nucleus colony transfer can be accomplished. This can be achieved by scheduling based on the seasonality of beekeeping.

Seasonal beekeeping activities are as follows:

Spring management (March–May): Swarm prevention via artificial swarm techniques and demaree method for population control; hive expansion through the addition of supers when 70% of frames are drawn; and checkerboarding technique for comb building.

Summer operations (June–August): By using a refractometer (optimal moisture: 17–18%) for honey extraction and harvesting, through uncapping methods (cold vs. hot knife); and pest monitoring with alcohol wash for varroa mite counts, while monitoring for small hive beetles (trap designs) are performed.

Fall/winter preparation (September–February): This is the time when winter stores calculation is done with 15–20 kg honey reserves for temperate winters and fondant feeding systems are installed. Then hive wrapping through insulation materials comparison to satisfy ventilation requirements, which is called hive winterizing, is done.

2.4.3 Common Challenges and Solutions

Beekeeping is closely related to a set of common problems. Nevertheless, through time, studies have contributed significantly to finding better solutions for these problems. Such problems are:

- Varroa Mites infestation is a huge problem that goes hand in hand with beekeeping. Treatment of the problem generally involves, but is not limited to, the application of organic acids like oxalic acid and formic acid, or the use of chemical miticides such as apivar and checkmite. Besides, employing biotechnical methods like drone brood removal, PCR assay for pesticide resistance genes is suggested for the purpose of resistance monitoring.
- Colony Collapse Disorder (CCD) might cause loss of habitat, with danger indicators of neonicotinoid exposure to pesticides and monoculture of foraging backgrounds. Mitigation strategies involve various pollen supplement mixes and probiotic supplementation (*Lactobacillus spp.*) But intentional preservation and protection of the apiary is the best prevention. Implementation of environmentally friendly practices will provide sustainability of the apiary.

Ropey test to test *Paenibacillus* larvae and PCR diagnostic kits proved to be an effective containment measure, with biosecurity measures including apiary quarantine protocols and equipment sanitation (160°C for 1 hour).

2.5 PRODUCTS OF THE APIARY

The main apiary product is honey that is made directly by the bees, although they can differ by region and ecosystem. There are various types of honey, such as manuka, acacia, and wildflower. Honey is a demand-economic product that is sold across climates worldwide. Achieving the highest commercial benefit requires smooth processing and quality control across the process flowchart. Adulteration testing of C4 sugars and pollen analysis to prove origin during quality control are also important components. Beeswax forms many by-products with value chain and economic return to the players therein.

Two major modes of preserving apiary products are through the use of lyophilization at 50°C for freeze-drying, and fresh storage at 4°C with 1% lactic acid. Honey has rich and verifiable health benefits. Studies show that honey is used for antibacterial purposes, and wound healing from nectar conversion and invertase enzyme activity (Yupanqui Mieles et al., 2022). Beeswax is yet another useful apiary product; it can be utilized for making candles, cosmetics, and food wraps. Propolis, royal jelly, and bee pollen are also essential products of the apiary. Propolis is an antimicrobial resin that is used in throat lozenges. Royal jelly is a nutrient-rich secretion for queen bees. Bee venom is also collected and used in apitherapy for arthritis and as anti-inflammatory effects.

2.5.1 Apitherapy

Apitherapy has been used in traditional Chinese medicine (TCM) and Ayurveda for centuries. The scientific bases of apitherapy are as follows:

- *Melittin (in bee venom)*: Melittin has anti-inflammatory properties, making it a very useful apiary product for apitherapy. Ongoing cancer research has positioned bee venom peptides as a possible anticancer medicine.
- *Propolis*: Propolis has the capacity to boost immunity due to flavonoids. Propolis has been used in modern applications in medicine for pharmacology purposes. It has wound-healing (medical-grade honey) elements.

2.5.2 Therapeutic Uses of Bee Products

Ancient medical texts documented that Ebers Papyrus in 1550 BCE discovered 147 honey-based remedies, while Charaka Samhita in 300 BCE submitted that bee venom is used for arthritis. Modern scientific validations have been done through the clinical trial landscape. Honey processed 87 RCTs for wound healing as affirmed by Jull et al. (2015) in *Cochrane Review*. On the other hand, Kwon et al. (2022) documented that bee venom phase III has been trialed for osteoarthritis.

Mechanistic studies concluded that melittin, produced from processed honey with NF-κB pathway inhibition, contains anticancer cytotoxicity (IC50 values). Other medicinal properties of honey are as follows:

As an antibacterial (effective against MRSA).
As a cough suppressant (better than some OTC syrups).
In bee venom therapy (BVT) (used for multiple sclerosis, Lyme disease, and pain relief).

In addition:

Propolis is used in immune support to fight colds, flu, and oral infections.
Royal jelly has anti-aging benefits (rich in B vitamins and collagen-boosting proteins).
Beeswax is used in dermatology and cosmetics (including lip balms, lotions, and natural skincare).

2.5.3 Medicinal Honey: Beyond Folk Remedy

Honey has been an age-old traditional treatment, but evidence-based practice more recently confirms its clinical utility. In wound management and chronic illness, for instance, Leptospermum honey (currently commercially marketed as Medihoney®) has been found to exhibit powerful antimicrobial activity. Randomized controlled trials attest to its efficacy in debridement of *Staphylococcus aureus* biofilm (Jenkins & Cooper, 2012). Routine treatment is an application of approximately 20 g/cm^2 of honey covered by a hydrocolloid dressing that is replaced every 24–48 hours.

In child gastroenteritis, honey is administered as an adjuvant to oral rehydration. An oral rehydration solution with 10 mL/kg of buckwheat honey, which has been shown to reduce the duration of diarrhea by approximately 18 hours from that of placebo (Alharbi et al., 2024), is recommended by the World Health Organization.

The pharmacological effects of medicinal honey are attributed to multiple mechanisms. Its osmotic effect, with a water activity of 0.56, inhibits bacterial growth. Enzymatic hydrogen peroxide generation, facilitated by glucose oxidase activity, provides antibacterial action. Synergism of the phytochemicals, particularly the methylglyoxal-leptosin interaction activity of manuka honey, enhances antimicrobial efficacy.

2.5.4 Bee Venom Therapy (BVT): Precision Medicine Approach

Bee venom therapy is increasingly being embraced as a specific therapy modality, particularly for inflammatory and autoimmune conditions. In osteoarthritis, treatment involves acupoint injection with a concentration of 0.1 mg melittin per milliliter, administered twice weekly. For multiple sclerosis, the protocol prescribes subcutaneous injections of 50 micrograms of whole venom, given three times weekly.

Table 2.2 shows how bee venom therapy is used for osteoarthritis and multiple sclerosis, while bee products and their therapeutic applications are shown in Figure 2.2.

Neuropharmacologically, bee venom exerts its effects through active compounds. Apamin, for instance, blocks SK channels, leading to central nervous system stimulation. Additionally, phospholipase A2 demonstrates anti-inflammatory potential by inhibiting COX-2 with an IC50 of 0.8 μM.

Propolis, another bee-derived product, has diverse pharmaceutical applications. In oropharyngeal conditions such as streptococcal pharyngitis, a 30% ethanolic extract spray is commonly prescribed

TABLE 2.2
Bee Venom Therapy

Condition	Administration Route	Dosage	Frequency	Reference
Osteoarthritis	Acupoint injection	0.1mg melittin/mL	Two times a week	Conrad et al. (2019), Shi et al. (2022)
Multiple sclerosis	Subcutaneous	50 μg whole venom	Three times a week	Wesselius et al. (2005), Zhang et al. (2024)

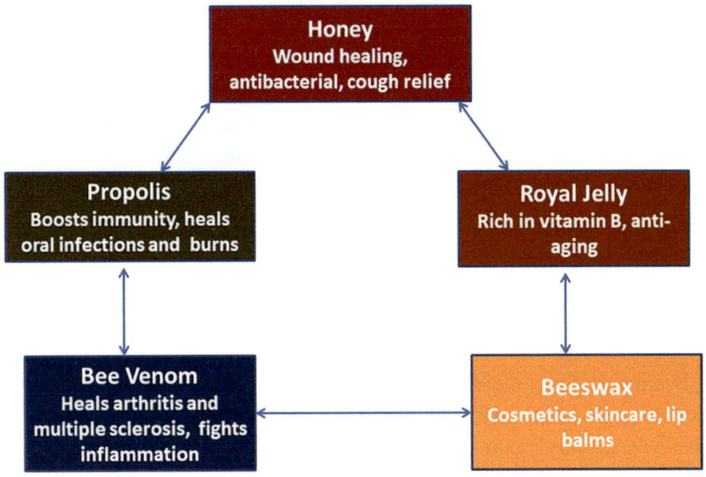

FIGURE 2.2 Bee products and their therapeutic uses.

at four puffs, four times daily. For dermatological applications, propolis is formulated as a topical cream—typically 5% propolis combined with 1% silver sulfadiazine—to enhance burn healing.

The immunomodulatory activity of propolis has been extensively proven. It immunomodulates through immune response rebalancing of TH1/TH2, increasing interleukin-10 (IL-10) secretion and inhibiting tumor necrosis factor-alpha (TNF-α). It also immunomodulates the ARG-1 vs. iNOS pathway to activate macrophages, thereby enhancing host defenses.

2.6 SAFETY AND ETHICAL CONSIDERATIONS: RISK MITIGATION STRATEGIES

2.6.1 Apitherapy Adverse Effect Management

On the one hand, products made out of the honey bee are really good for the health of patients; on the other hand, a high level of safety must be assured due to the risk of allergic reactions and other side effects. The anaphylaxis protocol is the emergency response plan. Always, the immediate intervention procedure begins with epinephrine (0.3 mg IM in the anterolateral thigh) due to its life-saving effects, and then diphenhydramine (50 mg IV) to neutralize the histamine effects is administered. Pre-treatment precautionary measures are equally important in the process. Generally, the patients are usually tested by means of the CAP-RAST method to locate the level of Api m 1 IgE that is a marker for the bee venom allergy. However, a graded venom challenge can be followed by an injection starting at 0.001 μg and then by continuing to stepwise increase the dose up to 1 μg for setting the tolerance threshold before the full treatment is initiated, in some occurrences.

2.6.2 Sustainable Apiculture Practices

Generally, bee colonies and their sustainability are a major concern apart from things like ethical considerations in apiculture. Colony stress can be mitigated through such ways of forage augmentation as *Phacelia tanacetifolia*, which can produce up to 500 kg of pollen per hectare and thus be a major source of nectar. Again, an 18% protein content is the basis of artificial pollen patties, which are relied upon during resource-scarce periods. Harvesting is allowed only with complete protection of the colony. For example, the honey that is consumed is controlled not to exceed 30% of the reserves, while at the same time at least 15 kg of honey is left for the winter stores. On the other hand, the royal jelly harvesting is managed in such a manner that only 5 g per day per hive is estimated so as not to starve the queen and also to keep the colony vigorous.

2.6.3 Future Frontiers: Disruptive Technologies in Apiculture and Tourism

The future of apitourism is increasingly linked to science and technology, which is the main reason for the rise of Apiculture 4.0. This new scenario blends digital innovations and biotechnological and pharmaceutical discoveries into beekeeping and tourism; hence, apitourism is no longer a sector of the traditional rural community, but a highly innovative and global one. These actions not only facilitate the removal of environmental problems, such as the lack of pollinators, but also cater to the shifting discourse of sustainable development and wellness tourism, which locate bees and their derivatives at the core of future solutions (Asma et al., 2024; Kirdat et al., 2025).

2.7 APICULTURE 4.0

Digitalization in beekeeping is a major innovation that has changed the management of apiaries, the way they can be integrated into rural economies, and also into tourism circuits. Using AI, hive monitoring systems nowadays can forecast a swarm by sound frequency analysis with 85% accuracy; therefore, the prevention of a swarm can be more exact, and the necessary interventions are reduced. Besides that, infrared thermography is being employed to find the concealed infestations of the Varroa mite at an early stage to assist the health of the colony and to lower the number of chemical interventions (Asma et al., 2024). Pollination robots are not only doing the job of a mere observer, but they are stepping in to fill gaps in the places of declining bee populations. One of the best examples is the RoboBee of Harvard, a flapping wing microdrone that weighs only 120 milligrams, designed to support the natural pollination systems in stressed ecosystems. While these inventions are making Apiculture 4.0 more efficient and practical, they can also be seen as the basis for Apitourism 4.0, when visitors can watch the implementation of a smart hive and robotic pollination as part of an immersive educational tourism experience.

2.7.1 Pharmaceutical Innovations

Alongside the products of nature, the pharmaceutical industry is being revolutionized to showcase the market opportunity of the future for apitourism—health and wellness. For example, melittin, a peptide of bee venom origin, has already been put to use in drug conjugates as targeted therapeutics. One of the directions in which bee-inspired biomedical technology is going is the use of HER2-targeted nanoparticles in breast cancer as a breakthrough (Kirdat et al., 2025). Besides, propolis is not only a product of ethnomedicine, but is also undergoing state-of-the-art productions in the form of caffeic acid phenethyl ester (CAPE) analogs with ten times higher bioavailability than the natural extract. The biggest challenge behind this is that the wellness industry not only enhances its pharmaceutical sector through such initiatives, but it also plays a major role in health-tourism growth, as apitourism destinations become areas of transition from traditional biomedical knowledge to apitherapy, ancient healing arts, and contemporary science.

2.8 STRATEGIC PROSPECTS FOR TOURISM INTEGRATION

The cross point of phama innovation and Apiculture 4.0 with apitourism is the fertile land for repurposing rural and health tourism. As Antonie (2024) markedly indicate in the context of tourism futures; the step of investing in new types of tourism is most appropriately done by structural foresight and scenario planning. Thus, apitourism destinations can be turned into living laboratories where the visitor gets to know bees as an ecological entity as well as a medical and technological catalyst. Therefore, the apitourism transformation will no longer be confined to the present rural and environmental niche but will become an equally diversified niche, simultaneously, serving sustainability, health, and science education beyond the economy.

New disruptive beekeeping technologies open up new tourist development pathways that can be based on the digital convergence of innovation, ecological sustainability, and biomedical innovation. By merging Apiculture 4.0 and pharmacological innovation into tourist products, destinations can create new experiential value differentiation that attracts global regenerative tourism and well-being travel trends. This allows apitourism to be at the forefront of both tourist theory and practice and highlights its potential to reconceptualize the human, bee, and sustainable futures nexus.

2.9 TOURISTIC VALUE OF APITHERAPY

2.9.1 Apitourism

Apitourism is a new segment of tourism that is a mixture of beekeeping, sustainable travel, and associated health tourism. It is a shift from traditional beekeeping towards a global fascination with experiences, nature, and health products. Practically through research, it has been established that apitourism is not just beekeeping and apiary, but also the use of bee products in a manner that merges several sectors, like education, food, health, and cultural exchange (Šuligoj, 2021; Fusté-Forné et al., 2024). Consequently, it is a multifaceted tourism platform that can promote rural society, conserve biodiversity, and promote well-being. Specifically when the worldwide development of tour systems has apparently been rerouted towards regenerative and sustainable paradigms, this is largely necessary. Apitourism looks like a genuine path through which native people are able to increase environmental protection without giving up economic sustainability in the process, thus making honeybees not only environmental actors but also cultural and tourism assets (Izquierdo-Gascón & Rubio-Gil, 2023a; Justs et al., 2024).

Through Europe rural, apitourism has evolved into a regenerative model. To illustrate the point, the setting up of apiary routes, made in Spain, accommodates both cultural corridors and ecological educational trails that revitalize rural life and mitigate depopulation (Izquierdo-Gascón & Rubio-Gil, 2023b). In such a way, apitourism becomes a comprehensive farm visit product accompanied by honey tastings, bee-watching experiences, and wellness retreats. Villages that keep bees in the wild in Romania interweave traditional beekeeping with hospitality services, cultural heritage interpretation, and community educational efforts. Rural sustainability is described as the main outcome of this, according to a piece of research (Antonie, 2024). Likewise, sympathetically, Ukraine has supported apitourism as a probable driver of rural diversification in the areas where agricultural tourism has traditionally been the dominant sector of the rural economy (Roman et al., 2024). The point of this chapter is that those initiatives in various parts of Europe convey the message that apitourism is a resource for sustainable tourism which can strengthen community identity and rural economy simultaneously.

In Asia and the Middle East, apitourism is becoming a new idea that could help diversify the local economies and reinforce the area culture. Iranian scholars interpret the concept of apitourism as a solution to the escalating socioeconomic risks in Iran (Fathi et al., 2022). In Kazakhstan, scholars have linked the concept of apitourism with rural development opportunities of the area (Mukhambetova et al., 2025). In the Indian context, the apitourism is projected as the perfect development model where the apiculture industry and wellness-centered heritage tourism merge together to provide a new ecosystem for the planet (Singh et al., 2022). Such examples show how apitourism is adaptive in different ecological and cultural zones with the potential not only to improve ecological conservation but also to empower rural economic diversification and wellness tourism. Before gaining the global perspective on apitourism, a thorough analysis of denied keywords and extensive research pose significant issues regarding the connection between tourism and agriculture in the countries of the Global South.

A significant feature of apitourism is nothing but a relationship with wellness tourism through the procedure of apitherapy. The thought of a sensual lifestyle, including gastronomy that accepts honey as the main product, spa therapy, and traditional remedies, is rapidly becoming more prominent in apitourism, following the world trend towards wellness and experiential travel (Justs et al., 2024; Šuligoj, 2021). The things that were once only ideas in the folklore of the mountain side in Bosnia and Herzegovina have now been turned into factual points of a scientific conference. The studies there illustrate how wellness activities integrated in apitourism can be the driving force that makes rural areas more competitive and also economically viable, in terms of agribusiness activity (Jašic & Šubaric, 2022). In the republic of Macedonia, apitourism is at present being supported as a future developmental resource combining rural tourism with organically grown agriculture, functional food markets, and organic cultivation (Jovanović et al., 2023). The merging of apitourism with the apitherapy is what will set its attraction for foreign tourists to a higher level. This is because it caters to the modern trend of health-conscious, simple, and nature-based lifestyle that the consumers demand. The duality makes it one of the health forms and a cultural innovation at the same time, and it can become a part of the larger wellness economy.

Despite its promise, apitourism is still constrained by a number of factors that impede its widespread adoption as well as its mainstreaming into global tourism policy. The major ones are seasonality that limits tourism activities to particular climatic windows, environmental dependency on bee populations that are highly vulnerable to environmental pollution, and institutions that are less capable of supporting most countries (Uldemolins et al., 2025). Moreover, consumer awareness of apitourism remains at a low level, and marketing has not been effectively organized to reach the targeted tourists. For example, in Tanzania where the environmental conditions around Ruaha and Udzungwa National Parks are very good, the statistics indicate that the lack of infrastructure and community training are still challenges for the development of the area (Mmassy et al., 2024). Additionally, promotional consumer behavior studies also suggest that advertising and visitor interaction techniques have a pivotal position in creating interest, which means without more professional marketing, the results would not be achieved and apitourism would be limited to the tourism sector (Uldemolins et al., 2025).

Similarly, apitourism in Nigeria is extremely lagging as far as the country's biological assets, its heritage, and the new tourism economy are concerned. In the past, beekeeping was a tradition in each of the three biomes of the savannah, forest, and montane, but the honey produced was neither for human consumption nor for medicine, and the practice was not yet part of the tourism circuits. This is a major drawback for the country when it is compared to well-documented organizational structures in Spain, Romania, and Tanzania, where beekeeping is under the rural and wellness tourism sector (Izquierdo-Gascón & Rubio-Gil, 2023b; Antonie, 2024; Mmassy et al., 2024).

The biological and cultural capital of Nigeria is an untapped platform that could be used to lift apitourism to a higher level, but due to poor institutional support, bad infrastructure, and poor public awareness, apitourism has not yet been utilized. This is a glaring gap, as apitourism could not only give a boost to the nation's existing ecotourism initiative but also be a means of tourism economy and rural development in the country. A complete pathway is now open for the integration of apitourism in Nigeria ecotourism destinations. Nature- and wellness-based Obudu Mountain Resort in Cross River State can be converted to highland apiculture and made accessible through a combination of tourism activities such as guided tours of apiaries, honey tasting, and therapeutic wellness products. Yankari Game Reserve in Bauchi State, which is ranked among the most visited wildlife parks in West Africa, has local communities that are already engaged in small-scale beekeeping and can thus be developed apitourism products. Osun Sacred Grove, a UNESCO World Heritage site, also stands a chance of bringing together cultural heritage and apitourism through ritual, craft, and food made from beeswax and honey. Just by looking at these examples, one can conclude that Nigeria is endowed with so many environmental, cultural, and tourism assets that can be good enough to enable the country to create apitourism that is competitive at both regional and global levels. The question is, however, how to channel policy assistance, infrastructure investment, and training at the local level to make the potential visible.

Apitourism is a potential game-changer for rural areas in Nigeria, as it can lead to the creation of more jobs and help in the conservation of biodiversity while diversifying the country's tourism product to the global regenerative and sustainable tourism trend (Fusté-Forné et al., 2024). The situation in Kazakhstan and India provides the proof that apitourism can still become a breakthrough niche even if the area has poor infrastructural and policy conditions, by applying a well-thought-out strategy (Mukhambetova et al., 2025; Singh et al., 2022). The components contained in the aforementioned scenarios are important for Nigeria, as apitourism should first and foremost be acknowledged as a major branch of the national tourism policy of the country, the setting up of apri-routes and cultural api-villages and investing in the rejuvenation of the rural part of the country to complete the scene with apitherapy. Consequently, Nigeria can become an African apitourism destination with the green ecological space being the place where cultural tourism and wellness tourism cohabit.

2.10 CONCLUSION: INTEGRATED BEE-BASED SOLUTIONS

The dependency of area under agriculture on bees has never been more transparent than at present, and the involvement of apiaries, apitherapy, and apitourism are among the most ecologically sustainable ways of achieving sustainable development. Contemporary apiaries have been so successful that they are no longer just honey-producing units; they have become micro-food security ecosystems, which serve as living biodiversity and having socioeconomic potential. According to the Food and Agriculture Organization, about 75% of all crops that need pollination are service agreements for pollination priced at hundreds of dollars a hive per growing season that have now become strategically valuable for direct farm productivity and climate resilience. Apiculture, along with its journey, can still have bigger roles ecologically by helping with the decline of wild pollinator levels through the urban and rural biodiversity. To be exact, biodiversity within the city has been gained through urban beekeeping by up to 40%, which means that the bee has the potential to include ecological stewardship in human systems. Therefore beekeeping is no longer just an agricultural business, but the pillar of environmental resilience and a launch of economic diversification and innovation for peri-urban and rural communities (Asma et al., 2024; Kirdat et al., 2025).

Apitherapy's health and wellness potential changes the way bees are described from being only ecologically significant to positioning the hive products as the new "keepers" of human health and well-being. Scientific breakthroughs have greatly supported the history of the archaic usage of honey, propolis, royal jelly, and bee venom, a usage that goes even to the present day for the mentioned products as wound dressings with the property of being antimicrobial, in immunomodulatory therapy and as coadjuvants in cancer therapy. For instance, bee venom is on its way through clinical trials for Parkinson's and some cancers, whereas one of the propolis-derived compounds, caffeic acid phenethyl ester analogs, is now far more bioavailable than those in their natural forms, which consequently opens new and more relevant pharmaceutical applications.

Such progress provides a twofold opportunity: the first is to become a source of evidence-based complementary therapies that would satisfy the already-existing demand for natural and preventive medicine across the globe, and the second is to become a source of wellness tourism products intending apitherapy as a part of health, culture, and recreation packages. The connections of these are already there in some areas of Asia and Europe, where honey gastronomy, bee venom therapy, and apitherapy clinics are not only physically but also programmatically intertwined in the apitourism circuits; thus they are even able to testify that the innovation of health care and tourism can be carried out simultaneously (Šuligoj, 2021; Justs et al., 2024).

Apitherapy continues to be one of the main causes for ethical and environmental issues alongside risks such as anaphylaxis and colony loss caused by overharvesting. These concerns must be resolved through an integrated plan that includes responsible harvesting, scientific management, and public awareness as part of the apitherapy production process to ensure that health care utilized does not interfere with ecological balance (Topal et al., 2021). Furthermore, apitourism represents the connectivity that brings together the agricultural, ecological, and medical aspects into one integrated experience of tourism which is educational, cultural, and regenerative at the same time.

Basically, as a niche market, it is responsible for the revival of the rural economy and provides urban travelers with unique close nature experiences. As a result, it is worth mentioning that Spain and Romania are not the only two European countries which have been forerunners of the numerous api-routes designs. These routes serve as systematic combinations of apiary tours, honey sampling, well-being therapies, and cultural rituals as packaged modes of regenerative tourism. Positive testimonies of this practice can be found in empirical evidence that proves how these measures attract to rural communities the economic aspect of not only the production of goods but also the reduction of the depopulation rate through the introduction of new livelihood opportunities, the enhanced local sense of identity, and the linking of localities to global discourses of sustainability (Izquierdo-Gascón & Rubio-Gil, 2023a; Antonie, 2024). Moreover, there is apitourism development as a distinctive product of ecotourism in Asia and Africa, along with pilot projects in Iran, Kazakhstan, and Tanzania for the establishment of mixed models in apiculture and conservation and management of cultural heritage (Fathi et al., 2022; Mmassy et al., 2024; Mukhambetova et al., 2025).

Nigeria is a country with a high level of ecological intensity and rich cultural heritage. It stands at a very strategic point where apitourism not only could diversify the country's tourist market portfolio, but also could help in promoting rural development and introducing wellness practice in community-based tourism. Places like Obudu Mountain Resort, Yankari Game Reserve, and the Osun Sacred Grove are full of unexploited opportunities waiting for apiculture to be correctly positioned in ecotourism and heritage tourism circuits. What is lacking is the absence of policy instruments that are investment reflective in kind, infrastructure, and market-oriented packaging efforts that would transform beekeeping from a rural activity into a rural tourism and wellness innovation driver.

The survival of apitourism, apitherapy, and apiaries, therefore, depends on the conjunction of science, culture, and society. Such an intersection will require three converging routes to a sustainable future. First of all, a joint effort of veterinary science, pharmacology, tourism research, and material science is obligatory so that not a single potenial of bee-related innovations is left behind. Secondly, the philosophy of the circular economy has to be implemented in the apiculture sector to give value to each part and at the same tme to use a minimum of resources. An example of this could be the upcycling of beeswax, propolis, and other by-products for biomedical, cosmetic, and craft purposes. The third principle is that participatory beekeeping and tourism community science enterprise projects will have a huge role in knowledge disclosure, equality promotion, and development of the local economies that are resilient. These models are no longer in a theoretical phase; rather, they are proven solutions that are being piloted in various locations across Europe and Asia, and they are potentially practically relevant for Nigeria along with other African regions if they are implemented via research-based policy and institutional legitimacy (Izquierdo-Gascón & Rubio-Gil, 2023b; Fusté-Forné et al., 2024). By means of innovation, we can turn such integrated models into apiculture and apitherapy keys that open the door to sustainable rural development and global well-being.

Finally, a bee is an energy and a sign of a close connection between beings. The soundness of honeybees affects directly not only the agricultural sector, but also the production of new medicines, as well as rural people's health. The linking of apitourism, apitherapy, and apiaries depicts a transition to a new model where environmental resilience, human life quality, and economic diversification overlap. Nigeria and the developing economies can be innovation leaders only if they recognize apitourism as a niche market opportunity, apitherapy as health tourism, and beekeeping as a food security and conservation practice.

Through the moral leader, who will act as lighthouse, not only will the innovations centering on the bee save the pollinators, which humanity is dependent on, but also they will be at the forefront of the regenerative tourism, rural regeneration, and health revolution. The future of beekeeping is still far from being the hive, but it is very close to the front of sustainable human development, and bees are still one of the most valuable coauthors of the book on which the resilient and victorious future is written.

REFERENCES

Alharbi, A. A. D., Alalawi, F. M., Dabash, S. H. Y., Alotaibi, M. D. A., Alsadah, A. M. A., Alshammari, M. N., & Alhrbi, S. O. (2024). Electrolyte imbalances and dehydration in children: Biochemical insights and nursing interventions in ICU settings—An updated review. *Journal of Medicinal and Chemical Sciences*, 7(11), 1708–1721.

Antonie, I. (2024). Api-tourism, an important component of the sustainable development of the Romanian village: Case study: The locality of Sibiel (Mărginimea Sibiului), Romania. *Scientific Papers Series Management, Economic Engineering in Agriculture & Rural Development*, 24(4), 53–62.

Asma, S. T., Acaroz, U., Alim, H., & Pandohee, J. (2024). The origin and evolution of beekeeping. In A. A. Waxhkoo, G. A. Nayik, J. Uddin, & M. J. Ansari (Eds.), *Honey bees, beekeeping and bee products* (pp. 1–9). CRC Press.

Conrad, V. J., Hazan, L. L., Latorre, A. J., Jakubowska, A., & Kim, C. M. (2019). Efficacy and safety of honey bee venom (Apis mellifera) dermal injections to treat osteoarthritis knee pain and physical disability: A randomized controlled trial. *The Journal of Alternative and Complementary Medicine*, 25(8), 845–855.

Fathi, M. R., Torabi, M., & Razi Moheb Saraj, S. (2022). The future of apitourism in Iran based on critical uncertainty approach and DEMATEL/COPRAS techniques. *Journal of Tourism Futures*. Advance online publication. https://doi.org/10.1108/JTF-09-2021-0215

Fèvre, D. P., & Dearden, P. K. (2024). Influence of nutrition on honeybee queen egg-laying. *Apidologie*, 55(4), 53.

Fusté-Forné, F., Noguer-Juncà, E., & Crespi-Vallbona, M. (2024). Bee tourism: Apiculture and sustainable development in rural areas. *Journal of Apicultural Research*, 63(1), 1–12.

Gascón, M. I. (2025). Beyond the beehives: Local development through api-tourism. In F. Fusté-Forné & A. Hussain (Eds.), *The Routledge handbook of regenerative tourism* (pp. 273–288). Routledge.

Izquierdo-Gascón, M., & Rubio-Gil, Á. (2023a). Theoretical approach to api-tourism routes as a paradigm of sustainable and regenerative rural development. *Journal of Apicultural Research*, 62(4), 751–766.

Izquierdo-Gascón, M., & Rubio-Gil, Á. (2023b). Api-tourism as a regenerative tourism modality against rural depopulation in Spain. *Journal of Rural Studies*, 98(2), 45–57.

Jenkins, R. E., & Cooper, R. (2012). Synergy between oxacillin and manuka honey sensitizes methicillin-resistant Staphylococcus aureus to oxacillin. *Journal of Antimicrobial Chemotherapy*, 67(6), 1405–1407.

Jašic, M., & Šubaric, D. (Eds.). (2022). *7th Congress of beekeeping and bee products with international participation* (Sarajevo, Bosnia and Herzegovina), 12–13.

Jull, A. B., Cullum, N., Dumville, J. C., Westby, M. J., Deshpande, S., & Walker, N. (2015). Honey as a topical treatment for wounds. *Cochrane Database of Systematic Reviews*, 6(3), Article CD005083. https://doi.org/10.1002/14651858.CD005083.pub4

Justs, I., Poiša, L., Vīksna, T., Pole, V., Ivanova, K., Vīksna, J., & Skromulis, A. (2024). Characteristics of the beekeeping industry in Latvia. In *Environment. Technology. Resources. Proceedings of the International Scientific and Practical Conference* (Vol. 1, pp. 182–188).

Jordan, A., Patch, H. M., Grozinger, C. M., & Khanna, V. (2021). Economic dependence and vulnerability of United States agricultural sector on insect-mediated pollination service. *Environmental Science & Technology*, 55(4), 2243–2253.

Jovanović, S. S., Simicevic, D., & Mihailovic, D. (2023). Apitourism as a development opportunity for health tourism—A case study of Surčin city municipality. In *The first International conference CASB: Health tourism and hospitality* (pp. 81–87).

Khalifa, S. A., Elshafiey, E. H., Shetaia, A. A., El-Wahed, A. A. A., Algethami, A. F., Musharraf, S. G., ... & El-Seedi, H. R. (2021). Overview of bee pollination and its economic value for crop production. *Insects*, 12(8), 688.

Kirdat, D. A., Waykar, B. B., & Gawali, A. R. (2025). Honeybees: An adaptive and resilient solution for sustainable global food security. In *Environmental landscape and sustainable biodiversity for healthy green growth* (pp. 155–161). Springer Nature Switzerland.

Kuropatnicki, A. K., Kłósek, M., & Kucharzewski, M. (2018). Honey as medicine: Historical perspectives. *Journal of Apicultural Research*, 57(1), 113–118.

Kwon, N. Y., Sung, S. H., Sung, H. K., & Park, J. K. (2022). Anticancer activity of bee venom components against breast cancer. *Toxins*, 14(7), 460.

Mmassy, E., Ntiniwa, K., Lesio, N., Keyyu, J., Bukombe, J., Mjingo, E. E., & Marealle, W. (2024). Api-tourism potential of Southern Tanzania: A case study of villages around Ruaha and Udzungwa National Parks. *Journal of Apicultural Research*, 63(2), 211–223.

Mukhambetova, N., Moldagalieva, A. E., Kozak, M., Aktymbayeva, A. S., Assipova, Z. M., Temirbayeva, K. A., & Aktymbayeva, B. I. (2025). Current condition and development prospects of apitourism in Kazakhstan. *Journal of Geography & Environmental Management*, 77(2), 155–170.

Prado, A., Requier, F., Crauser, D., Le Conte, Y., Bretagnolle, V., & Alaux, C. (2020). Honeybee lifespan: The critical role of pre-foraging stage. *Royal Society Open Science*, 7(11), 200998.

Roman, M., Kudinova, I., Samsonova, V., & Kawęcki, N. (2024). Innovative development of rural green tourism in Ukraine. *Tourism and Hospitality*, 5(3), 537–558.

Schaumann, F., Norrström, N., Niklasson, M., & Leidenberger, S. (2024). Ecological comparison of native (Apis mellifera mellifera) and hybrid (Buckfast) honeybee drones in southwestern Sweden indicates local adaptation. *PLOS ONE*, 19(8), e0308831. https://doi.org/10.1371/journal.pone.0308831

Shi, P., Xie, S., Yang, J., Zhang, Y., Han, S., Su, S., & Yao, H. (2022). Pharmacological effects and mechanisms of bee venom and its main components: Recent progress and perspective. *Frontiers in Pharmacology*, 13, 1001553.

Singh, N., Negi, R. S., Singh, S., Sati, A., Singh, E., Tiwari, S., & Singh, A. (2022). The potential of api-tourism in India: A sustainable development opportunity. *BioThink E–Magazine*, 2(3), 290–304.

Šuligoj, M. (2021). Origins and development of apitherapy and apitourism. *Journal of Apicultural Research*, 60(3), 369–374.

Topal, E., Adamchuk, L., Negri, I., Kösoğlu, M., Papa, G., Dârjan, M. S., & Mărgăoan, R. (2021). Traces of honeybees, api-tourism and beekeeping: From past to present. *Sustainability*, 13(21), 11659.

Uldemolins, P., de-Magistris, T., & Maza, M. T. (2025). How promotional formats affect consumer preferences for apitourism experiences in rural areas. *Journal of Rural Studies*, 119, 103784.

Wesselius, T., Heersema, D. J., Mostert, J. P., Heerings, M., Admiraal-Behloul, F., Talebian, A., . . . & De Keyser, J. (2005). A randomized crossover study of bee sting therapy for multiple sclerosis. *Neurology*, 65(11), 1764–1768.

Yupanqui Mieles, J., Vyas, C., Aslan, E., Humphreys, G., Diver, C., & Bartolo, P. (2022). Honey: An advanced antimicrobial and wound healing biomaterial for tissue engineering applications. *Pharmaceutics*, 14(8), 1663.

Zhang, H. Q., Sun, C., Xu, N., & Liu, W. (2024). The current landscape of the antimicrobial peptide melittin and its therapeutic potential. *Frontiers in Immunology*, 15, 1326033.

3 Physiological Roles of Apitherapy on the Body System

Olugbemi Tope Olaniyan, Ryu Ki-Jin, and Duyilemi Chris Ajonijebu

3.1 INTRODUCTION

Apitherapy is described as a natural remedy currently utilized for the therapeutic management of diseases due to the presence of active ingredients (vitamins, enzymes, minerals, amino acids, water, carbohydrates, protein, antioxidants, and fatty acids) in bee products such as honey, wax, royal jelly, pollen, bee venom, and propolis. Weis et al. (2022) reported that bee products have been utilized for the management of several diseases due to the presence of active ingredients.

The nutraceutical and dietary supplement value placed on bee products is very high as integrative and complementary medicine in therapeutic management of diseases such as arthritis, multiple sclerosis, and infections. Furthermore, the active molecules such as melittin, phospholipase A_2, apamine, histamine, norepinephrine, and dopamine are found in bee venom with physiological applications such as anti-inflammatory, anti-diabetes, antiparkinsonian, anticancer, and respiratory disorders. Lekhnath and Ravindra (2023) reported that apitherapy has shown promising results as anticancer agents, particularly in the use of bee venom due to the apitoxin content. Bee venom is able to induce cellular apoptosis and suppress tumor development. El-Gammal et al. (2018) revealed that honey bee products used as apitherapy relieved several human ailments. For instance, the utilization of propolis in the treatment of psoriasis has shown positive results according to the authors. The authors recommended that palmoplantar psoriasis could be treated with propolis. Shramana and Saswata (2015) revealed that bee venom contains the following components: proteins and enzymes (phospholipase B, phospholipase A_2, hyaluronidase, alpha-glucosidase, and phosphatase); peptides (melittin, MCD peptide, apamine, secapine, minimine, pamine, Adolapine, protease inhibitor, procamine A and B, tertiapine, melittin F, and cardiopep); biogenic amines (dopamine, histamine, and noradrenalin); amino acids (aminobutyric acid); sugars (fructose, glucose); volatiles (complex ethers); and minerals (P, Mg, Ca^{2+}). Adrian (2023) reported the role of apitherapy in the treatment of autoimmune diseases. It was demonstrated that the medicinal applications of bee products for therapeutic management of autoimmune diseases is an emerging research area.

The anti-inflammatory and immunomodulatory role of apitherapy makes it very unique for the treatment of rheumatoid multiple sclerosis, arthritis, and lupus. Mahmoud (2006) revealed that propolis has a lot of health benefits in the body. The physiological applications reported by the author include antimicrobial, anti-ulcer, antioxidative, anti-tumor, and anti-inflammatory activities. These physiological activities have been associated with the presence of sequiterpene quinines, polyphenols, coumarins, phenolic aldehydes, steroids, inorganic salt, and amino acids in propolis. Azza (2021) reported that bee venom has beneficial effects in several disorders like neurological, hematological, cardiovascular, dermatological, and musculoskeletal dysfunction. The author demonstrated that bee venom significantly reduces hypercholesterolemia and improve blood parameters. The hypocholesterolemic effects are reported to be linked to the presence of peptides such as adolapin, melittin, mast cell

DOI: 10.1201/9781003685654-3

degranulating polypeptide, and apamin; enzymes like α-glucosidase, phosphatase, hyaluronidase, and phospholipase A_2 and B; and non-peptide portion (norepinephrine, dopamine, and histamine). Bava et al. (2023) reported that the health and medicinal value of bee products include anti-inflammatory, antioxidant, nervous system inhibition, radioprotective, antifungal, antiviral, and antibacterial activity. Zhijie (2023) showed that bee venom's biologically active constituents like phospholipase A_2, melittin, and apamin possess anti-inflammatory, analgesic and neuroprotective properties in the central nervous system neurodegenerative diseases, viral diseases, allergic diseases, organic-specific inflammatory diseases, and immune diseases. Sidra and Tasawar (2023) reported that bee venom possesses the ability to manage inflammatory diseases and autoimmune disorder like rheumatoid arthritis due to the presence of bioactive constituents like adolapin, hyaluronidase and melittin.

Apitherapy is a method utilized to improve metabolism, blood circulation, and tissue trophicity. Mohsin et al. (2023) reported that oxidative stress is the most prevailing risk factor for several diseases. The authors in their study revealed that the phytochemical constituents in honey can mitigate oxidative stress with free-radical scavenging ability. Weis et al. (2022) reported that the benefit of bee products to human health for the management of various ailments is enormous. It was revealed that the practice of apitherapy cuts across several continents. The authors demonstrated that propolis has been utilized for the treatment of cold and flu symptoms, psoriasis, skin injury (acne, burns, wounds), obstetric and gynecologic concerns, otorhinolaryngologic issues, alleviating caries, proctologic diseases, and managing stomatitis or gingivitis. Bee venom is reported to be effective in managing liver injury, airway inflammation, and acne due to the presence of melittin bioactive compound within it. The apamine in bee venom is an important neuropeptide in the SK family of calcium-sensitive potassium channels in Alzheimer's patients.

Melittin is reported to enhance membrane activity; stabilize surface tension reduction capacity and anti-inflammatory activity; promote the inhibition of vascular smooth muscle cell proliferation, apoptosis, and migration; enhance capillary permeability, thus elevating blood flow and reducing blood pressure; act as an immunostimulatory agent; reduce blood coagulation; cause immunosuppressive activity; activate the hypophysis and adrenal glands; and possess radiation prevention, anticancer, modulate central nervous system, antiviral, antibacterial, antifungal, endosomolytic, and antiatherosclerosis properties. Also, phospholipase A_2 is revealed to have immunomodulatory and anti-inflammatory effects against several inflammatory diseases, like lupus nephritis, cisplatin-induced hepatotoxicity, and nephrotoxicity; against neurodegenerative diseases such as Parkinson's, Alzheimer's, allergic asthma, and multiple sclerosis; against lung inflammation-induced radiotherapy, anticancer, antibacterial, and antinociceptive, immunotherapeutic role terminates phospholipids, anti-parasitic, breaks the cell membrane of blood; stops neuronal cell apoptosis caused by prion peptides, and lowers the blood pressure and blood coagulation. Phospholipase B is capable of detoxification activity while hyaluronidase is able to hydrolyze and break down proteins, thereby resulting in the penetration of bee venom across the cell and tissue enhanced permeability, and vasodilation of blood vessels resulting in increased blood circulation.

Apamin is an anti-inflammatory agent which can also stimulate the secretion of cortisone. It also conserves red blood cells, with anti-serotonin action; activates the adrenal glands and hypophysis; increases the defense capability as an immunosuppressor; and is capable of managing Alzheimer's, multiple sclerosis, and Parkinson's diseases. The active components of honey, such as adolapin and mast cell degranulating peptide, are capable of lysing mast cells; releasing serotonin, histamine, and heparin; elevating capillary permeability; enhancing anti-inflammatory, melittin-like effect; and activating the central nervous system. Other ingredients like protease inhibitors, secapin, cardiopep, tertiapin, minimin, histamine, and procamine, are able to inhibit the action of various proteases such as thrombin, plasmin, trypsin, and chymotprypsin, thus reducing inflammation, dilating blood vessels, increasing permeability and blood circulation, and stimulating smooth muscle contraction. This chapter focuses on the physiological roles of apitherapy on the body system.

3.2 PHYSIOLOGICAL ROLES OF APITHERAPY ON THE BODY SYSTEM

3.2.1 Antiplaque, Pulp, Chronic Periodontitis, and Oral Candidiasis Therapies

Ahuja and Ahuja (2011) revealed the application of propolis for the management of pathogenic microorganisms. In their study, they revealed that propolis is a known viscous and sticky compound made by the bee through the mixture of beeswax and plant buds, sap, or leaves. The therapeutic application of propolis involves antiviral, anti-inflammatory, antibacterial, anti-carcinogenesis and antioxidant mechanisms. Propolis improved the body's premature aging, weakness, weight loss, and constipation. Ahuja and Ahuja (2011) showed that over 180 varieties of chemicals exist in propolis with several pharmacological and physiological applications like antimicrobial, anesthetic, anti-inflammatory, cariostatic, and cytostatic properties. In traditional Chinese medicine, propolis is described as a therapeutic material with enormous medical application, including the treatment and management of aphthous ulcers, acute necrotizing ulcerative gingivitis (ANUG), candidiasis, periodontitis, pulpitis, and gingivitis in dental medicine. The application of propolis in dentistry, like cariology, pathology, oral surgery, periodontics, pedodontics and endodontics, is due to the anti-inflammatory and antimicrobial physiological activities. Ahuja and Ahuja (2011) demonstrated that the chemical component of propolis is rich in terpenes, phenolic compounds like aromatic and flavonoid compounds) and essential oil. The propolis cinnamic acid and flavonoids like citrin and vitamin P are the main active ingredients, which are secondary metabolites and ketone-rich compounds. In human diets, flavonoid-rich polyphenolic compounds in propolis exhibit anti-inflammatory activity with the hyaluronidase potential. In addition, acetophenone, lignans, diterpenes, and p-coumaric acid are found in propolis with antifungal and antibacterial properties. Furthermore, the antibacterial role of propolis against *Escherichia coli*, *Staphylococcus aureus*, and *Enterococcus sp*s has been recorded. Propolis's active ingredients can significantly block bacterial mobility, alter the permeability of electrolytes across the inner membrane of bacteria, and promote protein synthesis and cytoplasm disorganization inhibition. Propolis's antioxidant mechanism involves reduction in oxidative stress and tissue damage by decreasing superoxide anion, hyaluronidase activity, and xanthine oxidase. Propolis improves mitochondrial tissue, thus enhancing the respiratory system. The anticancer activity of propolis has been reported as a result of the presence of artepillin C, CAPE, and propolin C and A. These active ingredients have the capacity to inhibit the stimulation of the nuclear factor transcription, block angiogenesis VEGF expression and colon cancer, and reduce prostaglandin E_2 synthesis and induction. Ahuja and Ahuja (2011) showed that propolis contains substances (flavonoid and phenolic compounds) that can be utilized as antiplaque, oral candidiasis, pulp therapy, and periodontal diseases such as gingivitis, calculus, bleeding, gingival recession, fluid accumulation, subgingival irrigation, tooth mobility, bone loss, and pus formation. Bee venom is capable of preventing arthritis (rheumatic arthritis and osteoarthritis),

3.2.2 Antimicrobial Role

Pasupuleti et al. (2016) revealed that honey exerts antimicrobial properties against many intestinal microorganisms like *Shigella spp.*, *Escherichia coli*, *Salmonella spp.*, and *Helicobacter pylori*. In the pulmonary system, the beehive is utilized for the management of microbial pathogens responsible for respiratory diseases. The beehive contains active ingredients like secondary metabolites and apilarnil. Patel and Cichello (2013) revealed that honey is capable of inhibiting microbial activity and pathogenic growth.

3.2.3 Anti-Inflammatory, Antioxidant, and Anti-Arthritis Actions

Pasupuleti et al. (2016) revealed that honey from Manuka possesses the ability to produce various inflammatory cytokines when exposed to human monocytes. The inflammatory cytokines are interleukin-6 and tumor necrosis factor and interleukin-1 via the toll-like receptor 4 dependent pathway (Seema & Simon, 2013). Lee et al. (2005) demonstrated that bee venom was utilized for the treatment of arthritis as apitherapy. The venom was able to reduce lameness score, leukocytosis, hind paw edema volume, inflammatory edema, bone change, and polyarthritis. Lekhnath and Ravindra (2023) reported that bee venom is commonly utilized for the management of arthritis, chronic pain, and rheumatism. Bee venom is important for anti-inflammatory properties through aspirin and glucocorticoid effects on arthritis. The bee honey contains fructose, glucose, proteins and amino acids, trace amounts of enzymes (invertase, glucose oxidase, α- and β-amilase, peroxidase, and catalase), vitamins, phenolic compounds, ferulic acid, caffeic, ellagic, and p-coumaric acid; flavonoids like galangin, hesperetin, chrysin, apigenin, quercetin, pinocembrin, and kaempferol; and antioxidants like superoxide dismutase (SOD), tocopherol, ascorbic acid, reduced glutathione (GSH), and catalase (CAT). Bee venom generates an inflammatory response resulting in increased circulation and stimulation of anti-inflammatory hormones for the reduction of pain sensation, particularly for arthritic joints (Patel & Cichello, 2013). The bioactive compounds found in propolis, like pinocembrin, phenolic acids, and galangin, are effective antioxidant agents. Bee venom is able to regulate TNF-α activity, causing downregulation and reduction of pro-inflammatory proteins like cPLA2, COX-2, IL-1β, and iNOS. This physiological property is facilitated by depression of DNA from binding with NF-κB via alleviating IκBα, inhibiting p50 translocation into the cell nucleus. Amjad et al. (2023) reported the physiological properties of bee venom. In their study, it was revealed that bee venom has biological effects like anticancer, anti-inflammatory, antimicrobial, antinociceptive, and anti-arthritis owing to the presence of apitoxin, melittin, and phospholipase A_2. Sanyal et al. (2023) revealed that bee products like bee pollen have a potential role in the production of biomaterial for healthcare needs. In their study, it was demonstrated that these biomaterials have physiological functions like anti-inflammatory, antimicrobial, and antioxidant action.

3.2.4 Anticancer Effects

Ajay et al. (2019) revealed that apitoxin has been widely utilized for the treatment of cancer globally. Phospholipase A_2 and melittin in bee venom can treat renal, liver, lung, prostate, mammary, and bladder cancer cells. Bindlish and Sawal (2024) revealed that the aggressive nature of cancer makes treatment and management difficult. The authors demonstrated that a new alternative and complementary therapy approach using bee venom has received global attention in recent times. Their study showed that the impact of bee venom on cancer biology as an anticancer agent due to the presence of bioactive ingredient requires more research. It was revealed that bee venom can trigger apoptosis, inhibit metastasis, prevent cell migration and invasion, and suppress the existing breast cancer cells through the bioactive chemical in bee venom like phospholipase A_2, melittin, apamin, hyaluronidase, epinephrine, histamine, and amino acids. The authors revealed the physiological role of melittin on cancer through the induction of cell cycle arrest, apoptosis, reduction of proliferation or angiogenesis, blocking of cancer motility, metastasis, migration, and invasion. Apoptosis by melittin is via mitochondrial mechanisms, permeability induction, pore formation, cell shrinkage, cellular separation, cell shape irregularity, and membrane damage. In complementary and alternative medicine, bee venom is currently being utilized for the treatment of cancer. Several types of cancers, like liver, breast, ovarian, and prostate cancer, have been treated using honey. The biological compound in honey (phenolic compounds like quercetin and kaempferol) has antiproliferative, anti-inflammatory, antioxidant, antitumor, anticancer, and metastatic properties. The anticancer role of honey is linked to targeting the estrogen receptor (ER)-signaling pathway, induction of apoptosis, and reduction of membrane potential of mitochondrial through elevated lactate dehydrogenase leakage from the plasma membrane. The general physiological mechanisms of honey's anticancer properties involve inhibition of the cell cycle, stimulation of mitochondrial pathways, activation of apoptosis, and regulation of oxidative stress as shown in Figure 3.1. Patel and Cichello (2013) reported that honey possesses antiproliferative activity through the stimulation of apoptosis via activation of caspase 3, activation of caspase 7, caspase 9 pathway, DNA fragmentation, inhibition of protein survival, and reduction in glutathione.

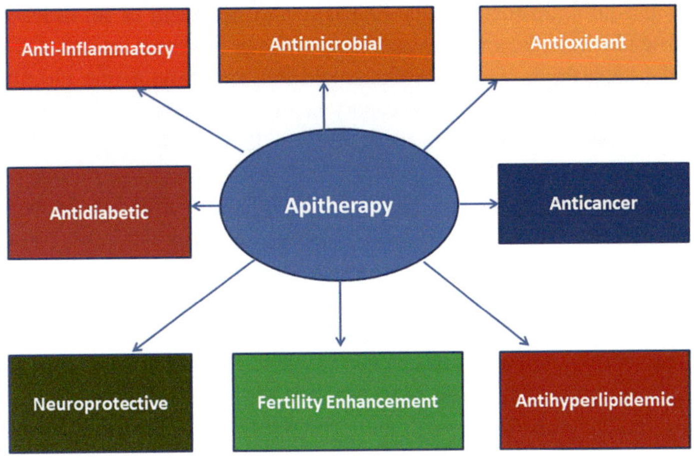

FIGURE 3.1 The therapeutic efficacies of honey.

3.2.5 Antidiabetic and Antihyperlipidemic Effects

Weis et al. (2022) revealed that the hyperglycemic effects from diabetes mellitus as a result of dysfunctional proteins, carbohydrates, and fats metabolism and damaged cells of the pancreas, kidney, and liver can be treated by bee products which are natural products, such as propolis. Propolis is known to have exhibit biological activities such as antioxidant as a result of flavonoids and phenolics' active ingredients. The authors demonstrated that the consumption of propolis in type II diabetes mellitus can suppress long-term microvascular and macrovascular complications. Propolis can reduce fasting hyperglycemia in type II diabetes mellitus. Bee venom containing melittin, phospholipase A_2, apamin, and adolapin help reduce pain, degenerative diseases, and inflammation in the body. Basa et al. (2016) revealed that honeybee products like propolis and pollen are commonly utilized for the management or treatment of diabetes as a result of their anticoagulant, anti-inflammatory, and antioxidant actions.

3.2.6 Cardioprotective Properties

El-Didamony et al. (2024) demonstrated that apitherapy is beginning to gain public interest as a means for enhancing circulation, reducing inflammation, and promoting immunological response. Furthermore, the authors revealed that bee products containing active ingredients like melittin are developed into various nanoparticle nanofibers, scaffolds, and others to enhance their bioavailability, easy-to-use pharmaceutical and assessable drugs for various applications. Chrysin, a green propolis, is reported to be effective against inflammation by reducing keratinocyte reactivity (Weis et al., 2022). Basa et al. (2016) revealed that the use of honey for the management of cardiovascular diseases, such as ischemic heart disease, chronic fatigue, and arteriosclerosis, has gained more attention recently. The efficacy of honey is linked to phenolic bioactive constituents capable of displaying antithrombotic, antioxidant (flavonoids, vitamin C, beta-glycosidase enzymes, polyphenolics, and monophenolics), anti-ischemic, and vasorelaxant properties.

The general actions of honey on the cardiovascular system include enhancing coronary vasodilatation, stopping low density lipoproteins from oxidation and reducing the capacity of platelets for blood clotting. According to Basa et al. (2016), bee pollens can suppress allergies as a result of the actions of phenolic extracts and flavonoid myicetin. Bee propolis possesses very important biological activity incuding as an anti-inflammatory, hypotensive agent, bacteriostatic, bactericidal agent, and immune system stimulant. These physiological functions are possible as a result of the presence of flavonoids, esters, coumarin, cinnamic acid, and aromatic acids in its composition. Bee venom possesses anticoagulant and anti-inflammatory actions linked to the treatment of arthritis, tendinitis, bursitis, dissolving scar tissue like keloids, joint disease, Herpes zoster, rheumatoid arthritis, multiple sclerosis, Lyme disease, and osteoarthritis. Gokulakrishnaa and Selvamuthukumaran (2020) reported that the flavonoid component of honey can reduce coronary heart disease, reduce the capability of platelets to form clots, improve coronary vasodilation, reestablish oxidation of low-density lipoproteins, improve endothelial functions, and enhance high-density lipoproteins. Ajay et al. (2019) showed that honey has the ability to balance cardiovascular and metabolic health thus, offers cardioprotective effects like vasodilation, improvements in lipid profile and maintaining vascular homeostasis. The bioactive compound in honey can reduce blood pressure. El-Hakam et al. (2022) demonstrated that apitherapy can be utilized for the treatment of hypertension, heart failure, and myocardial infarction. The apitherapy products are known to contain phytochemicals like flavonoids, phenolic elements, and aromatic acids. Bee venom is involved in increasing peripheral and coronary blood flow, enhancing the blood microcirculation, slowing down heart rate at reduced doses but showing ability to activate it at increased doses, reducing blood pressure and arrhythmia, and preventing blood clotting through fibrinolytic activation of the budding of red blood cells. Bee venom is revealed to be effective in managing hypertension, endarteritis (severe inflammation of arteries inner layer), arteriosclerosis, angina pectoris, and arrhythmia. Münstedt and Männle (2020) revealed that apitherapy using bee products like bee venom, propolis, and honey as alternative and complementary medicine can be utilized for the management of seasonal allergic rhinitis.

3.2.7 Gastrointestinal Tract Disease Management

Basa et al. (2016) reported that bee products such as honey can be utilized for the inhibition of microbial growth, and thus is potent in the treatment of gastritis, duodenal and gastric ulcers, and gastroenteritis. *Helicobacter pylori* are known to induce duodenal and gastric ulcers, gastritis, and gastroenteritis, which is a Gram-negative microaerophilic bacterium with a lifelong, life-threatening chronic disease. If left untreated, this microorganism may cause the induction of gastric adenocarcinomas and gastric lymphomas. Basa et al. (2016) reported that beeswax can be utilized for the treatment of gastrointestinal conditions through increasing saliva and strengthening the gingival and stomach juices due to its antibacterial properties. Gokulakrishnaa and Selvamuthukumaran (2020) reported that honey can prevent microbial growth and promote oral health. The antioxidant and anti-inflammatory actions of honey are able to protect pancreatic and liver health. In addition, the authors revealed that honey has an apoptotic effect on colon cancer via the active ingredients in honey such as quercetin and flavonoids, causing activation of mitochondrial outer membrane layer permeabilization and malignant cell apoptosis. Ajay et al. (2019) demonstrated that honey can increase the absorption molecules due to the presence of prebiotics. Honey can block liver fibrosis and inhibit bloating, diarrhea, pain, and inflammation.

3.2.8 Management of Neurological Disorders

El-Menabbawy et al. (2014) revealed that bee venom is gaining significant attention in the treatment of neurological conditions such as cerebral palsy. The authors reported that cerebral palsy is a movement and non-progressive syndrome as a result of injury of the central nervous system. Bee venom therapy as an apitherapy was shown to improve vertebral column deformity angles of deviation and nutritional intake among patients. Bee venom contains apamin, monamine, mast-cell degranulating peptide, and melittin; it is capable of activating the immune system and reducing contractures in cerebral palsy. Gokulakrishnaa and Selvamuthukumaran (2020) reported that royal jelly can improve mental health and offer neuroprotection. Bee venom is known to contain at least 18 bioactive compounds including enzymes phospholipases, amino acid compounds like melittin, and peptide. Melittin can inhibit the pathways of toll-like receptors like TLR4, TLR2, CD4, PDGFRβ, and NEMO, thus blocking the function of pro-inflammatory genes. Honey contains proteins (catalase and glucose oxidase), water, organic acids (butyric, aspartic, citric, formic, acetic, fumaric, gluconic, galacturonic, glutamic, glyoxylic acid, glutaric), vitamins (vitamin B6, niacin, thiamine, pantothenic acid, riboflavin), minerals (copper, iron, calcium, manganese, phosphorus, magnesium, sodium, zinc, potassium), flavonoids, phenolic acids, carotenoids, ascorbic acid, and α tocopherol). These bioactive chemicals are effective against many neurological disorders. Bee venom has been reported to be utilized for nervous system dysfunctions like multiple sclerosis, Alzheimer's, and Parkinson's through the inhibition of cellular toxicity of glutamate across microglia and neuronal cells. In addition, bee venom apamin, a peptide molecule, has a neuroprotective effect, capable of increasing blood supply to the neurons and reducing blood coagulation among patients with Parkinson's disease. Bee venom has been reported to be highly effective in the management of dementia, multiple sclerosis, polyneuritis, post-stroke paralysis, cerebellar ataxy, ganglion nerve inflammation, syringomyelia (pain of headache and extremities), myopathy (neuromuscular disease), inflammation of facial nerve, posttraumatic inflammation of the plexus nerve, trigeminal neuralgia, Parkinson's, lower back pain, and inflammation of the arachnoid central nervous system membrane. The combination of pollen and honey produces royal jelly, a milky thick mixture that has been revealed to be effective for so many physiological functions. Weis et al. (2022) revealed that royal jelly is very effective as nutraceutical compound and for improving fertility due to the presence of bioactive constituents like 10 hydroxy-trans-2-decenoic acid with strong immunomodulatory activity and ability to counteract oxidative stress, mitochondrial dysfunction, and neuroinflammation.

Its properties enable royal jelly to improve motor functions in neurodegenerative diseases and to mitigate neuronal damage and inflammation. Stela et al. (2024) reported that bee venom can be utilized in the management or treatment of central nervous disorders like Parkinson's disease. In their study, it was found that bee venom inhibits the degeneration of the dopaminergic nerve in the striatum and substantia nigra, thus increasing the level of dopamine.

Amira and Hiroshi (2020) reported the capability of apitherapy in the treatment of Parkinson's disease. Their study found that royal jelly and propolis as bee products are capable of promoting health and treating age-related diseases like Parkinson's disease. The mechanism of royal jelly and propolis in the management or treatment of Parkinson's disease involves the scavenging effects of the bioactive compounds found in them, such as chrysin, pinocembrin, caffeic acid phenethyl ester, flavonoids, phenols, terpenoids and phenolic aldehydes, vitamins, resin, and essential oils. Hegazi et al. (2015) revealed that apitherapy is utilized for the novel therapeutic management of multiple sclerosis. The authors demonstrated that apitherapy such as bee venom containing active compounds like monamine, melittin, mast-cell degranulating peptide, and apamin is able to reduce the symptoms of multiple sclerosis.

3.2.9 Management of Fertility Disorders

El-Didamony et al. (2024) further revealed that royal jelly is developed into a health care gel capsule and utilized as vaginal lotion for the treatment and management of genitourinary disease. Weis et al. (2022) reported that royal jelly possess the capability to mitigate side effects of dehydroepiandrosterone sulfate sex hormone among individuals, regulate hormonal imbalance, and reduce menopausal symptoms. Furthermore, the authors reported beneficial effects of royal jelly in the treatment of male infertility, thus increasing sperm motility and testosterone levels. Royal jelly can interact with endoplasmic reticulum and stimulate mRNA gene expression of estrogen-responsive elements such as palindromic sequences of DNA. Apilarnil found in bee products is very rich in steroidogenic hormones like progesterone, prolactin, testosterone, E-dec-2-enedioic acid, and estradiol, which are involved in reproductive functions. Weis et al. (2022) reported that apilarnil can stimulate sexual maturation; increase the weight of the penile organ and seminal glands, and the density of germ cells, epididymis, and the ejaculated volume; and exert androgenic effect and sperm mobility, which are indicative of increased male reproductive capacity.

Furthermore, apilarnil has the ability to suppress pro-inflammatory cytokines, reduce neuronal damage, apoptosis, and oxidative stress, and promote neuroprotective effects. In general, apilamil is reported to be utilized for the treatment of male infertility, ovarian dysfunction, malnutrition, and thyroid and immune disorders. Basa et al. (2016) reported that bee pollen can be utilized for the treatment of prostate issues such as swelling, infections, and cancer. In their study, it was revealed that other phytonutrients in bee products like lycopene, phytosterols, flavonoids, and beta sitosterol in bee pollen can inhibit the growth and development, inflammation, and pain in the prostate gland due to cancer, tumor, swelling, and enlargement. Al-Shaeli et al. (2022) reported that male infertility due to hyperglycemia linked to type II diabetes mellitus can be treated with bee venom. The authors in their study revealed that bee venom was able to reverse hyperglycemia, decrease insulin secretion, restore hormonal imbalance, improve semen profiling, and increase the seminiferous tubule diameters and testicular integrity. Gokulakrishnaa and Selvamuthukumaran (2020) reported that propolis can be utilized for the treatment of human breast cancer via apoptosis and reduction of mitochondrial membrane potential due to flavonoids in the propolis, such as apitoxin containing a combination of proteins and polypeptides like apamin, adolapin, melittin, hyaluronidase, histamine, phospholipase A_2, protease inhibitor, and dopamine. In addition, the authors demonstrated that royal jelly can be utilized for the treatment of menopausal symptoms due to the ability to balance hormones, increase estrogen concentration, and increase the number of viable oocytes. Honey is capable of inhibiting vaginal infection and promotes vaginal well-being.

3.2.10 Wound Healing Activity

Pasupuleti et al. (2016) revealed that honey is known to contain flavonoid and phenolic compounds with potent ability to enhance wound healing. The authors also demonstrated that honey is high in caffeic and p-coumaric acids; ferulic and ellagic flavonoids, like apigenin, galangin, chrysin, pinocembrin, kaempferol, quercetin, and hesperetin; and antioxidants, like tocopherols, superoxide dismutase, ascorbic acid, catalase, and reduced glutathione. The antioxidant action of honey is associated with the presence of flavonoids and terpenoids such as pinocembrin, pinobanksin, chrysin, 8-methoxykaempferol, isorhamnetin, luteolin, galangin, sakuranetin, kaempferol, quercetin, 3-hydroxy-24-methylenecycloartan 26-oic acid, and magniferolic acid, with the capacity to reduce oxidative stress. El-Didamony et al. (2024) demonstrated that apitherapy using beeswax can be utilized for the treatment of burns, bruises, fracture, joint pain, and wounds. In addition, recent studies revealed the use of beeswax for the treatment of purulent tonsillitis. El-Didamony et al. (2024) further revealed the beneficial role of propolis in the treatment of wounds. Bee pollen has demonstrated a superior role in cosmetics and industrial foods as skin creams, shampoos, toothpastes, antiseptic solutions, candy, granules, bars, pills, human tonics, and oral liquids. Basa et al. (2016) reported that honey is non-irritating, self-sterile, nontoxic, nutritive, bactericidal, easy to apply, and more comfortable than other dressings in the wound-healing process. Honey facilitates the healthy granulation of tissue, in a bacteriologically sterile environment. Gokulakrishnaa and Selvamuthukumaran (2020) reported that beeswax can be utilized for pediatric dermatitis due to the presence of nitric oxide metabolites with the ability to reduce psoriasis and skin infection. The wound-healing capacity of honey is linked to the ability to promote repair, prevent microbial invasion, activate plasminogen, and enhance epithelialization. The general mechanism of the physiological role of honey products is via the activation of the immune system, secretion of cortisone, production of cytokines and antibodies, and inhibition of prostaglandins, as shown in Table 3.1. Patel and Cichello (2013) revealed that honey is capable of healing diabetic wounds, burns, venous ulcers, eczema, psoriasis and acne. Honey promotes reduction in wound pH, increased fibroblast activity, oxygen release, and inhibition of protease activity.

TABLE 3.1
Physiological Roles of Apitherapy on Body Systems

Bee products	Physiological roles				Ref
	Anti-inflammatory	Antioxidant	Antibacterial	Antitumor	
Propolis	Flavanones, phenolic and their esters	Flavonoids, phenolic, acids, and esters	Flavanones, flavones	Caffeic and phenolic acid phenethyl ester	Ahuja and Ahuja (2011)
		Prenylated p-coumaric flavonoids	Prenylated p-coumaric acids, labdane, diterpenes	Prenylated p-coumaric acids, diterpenes, benzofuranes	
			Prenylated Benzophenones	Prenylated flavanones clerodane	
Royal Jelly	Flavonoids quercetin, luteolin, acacetin, apigenin, naringenin	Rutin	Kaempferol, glycosides, caffeic acid, pinobanksin, formononetin, phenolic acids like ferulic acid; proteins like apalbumin, royalasin, royalactin, jellein Isorhamnetin		Gokulakrishnaa and Selvamuthukumaran (2020)
Honey	Phenolic acids, flavonoids like quercetin, naringenin, pinobanksin, ferulic acid, kaempferol		Genistein, apigenin, chrysin, luteolin, caffeic acid, gallic acid, syringic acid, pinocembrin, hesperetin, p-coumaric acid, ellagic acid, vanillic acid		El-Gammal et al. (2018)
Beeswax			Polyphenols, chrysin, phenolic acids		El-Didamony et al. (2024)
Bee pollen	Flavonoids like kaempferol, quercetin, galangin, naringenin, pinocembrin, apigenin, luteolin, rutin, pollen isorhamnetin, leukotrienes, catechins, phenolic acids like protocatechuic acid, chlorogenic acid, syringic acid, caffeic acid, p-coumaric acid, ferulic and gallic acid				Sanyal et al. (2023)
Bee venom	Peptides such as melittin, adolapin, MCD peptide, apamin, enzymes such as hyaluronidase and phospholipase A_2				Lekhnath and Ravindra (2023)
Bee bread	Phenolic acids like isorhamnetin-O-hexosyl-O-rutinoside, and flavonoids such as quercetin-3-O-rutinoside, Myricetin-3-O-glucoside, quercetin-3-O-glucoside, kaempferol-3-O-rutinoside				Weis et al. (2022)

3.3 CONCLUSION

The general mechanism of the physiological role of honey products is via the activation of the immune system, antioxidant and anti-inflammatory responses, secretion of cortisone, production of cytokines and antibodies, and inhibition of prostaglandins. Honey is capable of healing several diseases and promotes general well-being of an individual.

REFERENCES

Adrian, P. (2023). Apitherapy: The potential role in managing autoimmune diseases. *Journal of Apitherapy*, 10(10), 01–02.

Ahuja, V., & Ahuja, A. (2011). Apitherapy—A sweet approach to dental diseases. Part II: Propolis. *Journal of Academy of Advanced Dental Research*, 2(2), 1–8.

Ajay, K.S., Barsha, R.K., Suprava, P., Rutuparna, S., Bani, B.S., Suman, N.K., & Ditikrishna, S. (2019). Review on medical values of honey bee venom and their biological action and health benefits, *Indo American Journal of Pharmaceutical Sciences*, 6(2), 4574–4590.

Al-Shaeli, S.J.J., Hussen, T.J., & Ethaeb, A.M. (2022). Effect of honey bee venom on the histological changes of testes and hormonal disturbance in diabetic mice, *Veterinary World*, 15(9), 2357–2364.

Amira, M.A., & Hiroshi, K. (2020). Apitherapy for Parkinson's disease: A focus on the effects of propolis and royal jelly. *Hindawi Oxidative Medicine and Cellular Longevity Volume*, 2020, Article ID 1727142, https://doi.org/10.1155/2020/1727142

Amjad, U., Fahad, M.A., Syed, I.A., Ghulam, R., Saeed, A.K., & Ivana, T.G. (2023). Pharmacological properties and therapeutic potential of honey bee venom. *Saudi Pharmaceutical Journal*, 31, 96–109.

Azza, M.M. (2021). Effect of honeybee venom (*Apis mellifera*) on respiratory functions of hypercholesterolemic male albino rats. *Assiut Veterinary Medical Journal*, 67(168), 32–42.

Basa, B., Wendimagegnehu, B., Alebachew, T., & Ayichew, T. (2016). Review on medicinal value of honeybee products: Apitherapy. *Advances in Biological Research*, 10(4), 236–247. https://doi.org/10.5829/idosi.abr.2016.10.4.10529

Bava, R., Castagna, F., Musella, V., Lupia, C., Palma, E., & Britti, D. (2023). Therapeutic use of bee venom and potential applications in veterinary medicine. *Veterinary Sciences*, 10, 119. https://doi.org/10.3390/vetsci10020119

Bindlish, A., & Sawal, A. (2024). Bee Sting Venom as a viable therapy for breast cancer: A review article. *Cureus*, 16(2), e54855. https://doi.org/10.7759/cureus.54855

El-Didamony, S.E., Hend, I.A.G., Mahmoud, M.M.Z., & Reham, I.A. (2024). Bee products: An overview of sources, biological activities and advanced approaches used in apitherapy application. *Biotechnology Reports*, 44, e00862.

El-Gammal, E., Di Nardo, V., Daaboul, F., Tchernev, G., Wollina, U., Lotti, J., & Lotti, T. (2018). Apitherapy as a new approach in treatment of Palmoplantar Psoriasis. *Open Access Macedonian Journal*. https://doi.org/10.3889/oamjms.2018.135

El-Hakam, F.E.A., Gomaa, A.L., Sahar, B.E., Hala, A.E., & Mohammed, H.F. (2022). Apitherapy combination improvement of blood pressure, cardiovascular protection, and antioxidant and anti-inflammatory responses in dexamethasone model hypertensive rats. *Scientific Report*, 12, 20765. https://doi.org/10.1038/s41598-022-24727-z

El-Menabbawy, K., Suzette, I. H., Hussein, M., Iman, R., & Ahmad, G.H. (2014). Possible role of bee venom therapy in children with cerebral palsy. *World Journal of Medical Sciences*, 11(1), 27–32. https://doi.org/10.5829/idosi.wjms.2014.11.1.82202

Gokulakrishnaa, R.K., & Selvamuthukumaran, T. (2020). Apitherapy: A valuable gift from honey bee. *Journal of Entomology and Zoology Studies*, 8(5), 2317–2323.

Hegazi, A.G., Al-Menabbawy, K., Abd El Rahman, E., & Helal, S.I. (2015) Novel therapeutic modality employing apitherapy for controlling of multiple sclerosis. *Journal of Cellular Immunology*, 6, 299. https://doi.org/10.4172/2155-9899.1000299

Lee, J.Y., Seong, S.K., Joong-Hyun, K., Chun, S.B. & Seok, H.C. (2005). Inhibitory effect of whole bee venom in adjuvant-induced arthritis. *in vivo*, 19, 801–806.

Lekhnath, K., & Ravindra, C. (2023). Apitherapy: An overview. *Journal of Applied Zoological Research*, 34(2), 1–7.

Mahmoud, L. (2006). Biological activity of bee propolis in health and disease. *Asian Pacific Journal of Cancer Prevention*, 7, 22–31.

Mohsin, B.A., Vaibhav, S., Aamir, H.D., Anurag, S., Rafeeya, S., & Shafat, A.K. (2023). Comprehensive review on functional and nutraceutical properties of honey. *eFood*, 4, e71. https://doi.org/10.1002/efd2.71

Münstedt, K., & Männle, H. (2020). Seasonal allergic rhinitis and the role of apitherapy. *Allergologia et Immunopathologia*, 48(6), 582–588.

Pasupuleti, V.R., Kumara, T.K., Naguib, S., & Siew, H.G. (2016). Biological and therapeutic effects of honey produced by honey bees and stingless bees: A comparative review. *Revista Brasileira de Farmacognosia*, 26, 657–664.

Patel, S., & Cichello, S. (2013). Manuka honey: An emerging natural food with medicinal use. *Natural Products and Bioprospecting*, 3, 121–128. https://doi.org/10.1007/s13659-013-0018-7

Sanyal, A., Ghosh, A., Roy, C., Mazumder, I., & Marrazzo, P. (2023). Revolutionizing the use of honeybee products in healthcare: A focused review on using bee pollen as a potential adjunct material for biomaterial functionalization. *Journal of Functional Biomaterials*, 14, 352. https://doi.org/10.3390/jfb14070352

Seema, P., & Simon, C. (2013). Manuka honey: An emerging natural food with medicinal use. *Natural Products and Bioprospecting*, 3, 121–128. https://doi.org/10.1007/s13659-013-0018-7

Shramana, B., & Saswata, A. (2015). Apitherapy. *International Journal of Recent Research in Life Sciences (IJRRLS)*, 2(3), 45–61.

Sidra, A., & Tasawar, I. (2023). Bee venom used for the treatment of rheumatoid arthritis. *Biomedical Journal of Scientific & Technical Research*, 53(2), 2023.

Stela, M., Cichon, N., Sprawska, A., Szyposzynska, M., & Bijak, M. (2024). Therapeutic potential and mechanisms of bee venom therapy: A comprehensive review of apitoxin applications and safety enhancement strategies. *Pharmaceuticals*, 17, 1211. https://doi.org/10.3390/ph17091211

Weis, W.A., Nicolas, R., Fernanda, L.C., Marianada S.H., Arthur, A.S., Ramon, H.M., & Jose, M.S. (2022). An overview about apitherapy and its clinical applications. *Phytomedicine Plus*, 2(2022), 100239.

Zhijie, Y. (2023). Bee acupuncture: Navigating the sting of traditional medicine. *Journal of Apitherapy*, 10(11), 01–02.

4 Beeswax Applications in Neurodegenerative Diseases

Akintayo Christopher Oloruntoba, Adebayo-Gege Grace Iyabo, Ayisimaka Adokiye Richard, Egunjobi Oluwagbemisola Dorcas, Akinremi Samuel Akinkunmi, Adedokun Kabiru Isola, and Kogbara Dinebari Tudor

4.1 INTRODUCTION

Beeswax, also known as white wax, is produced from the wax gland in the abdomen of worker bees (within the age of 12–18 days) as liquid, which transforms into a solid when exposed to air to become the structured honeycomb for storage of brood, honey, and pollens (Bogdanov, 2016). They are white in color, but the presence of the pollen oils turns them into a yellow or brown color (Burdock, 1998; Mekky et al., 2025). Findings show that it possesses antifungal properties as it is used to treat fungal infections and to reduce pain. It is said to possess a high concentration of cholesterol (Dumitru et al., 2022). Since Neolithic times, as revealed by analyses (mass spectrometry/gas chromatography) of ancient pottery (Roffet-Salque et al., 2015), beeswax is a natural substance produced through the intricate secretions of abdominal glands in honeybees (Bogdanov, 2016; Fratini et al., 2016). Containing over 300 compounds including alkanes, fatty acids, and esters, yellow beeswax is acquired through pressure filtration, while white beeswax results from bleaching (Aguilar et al., 2007). Worker bees secrete beeswax via wax glands in their abdomens and use it to build combs (Svečnjak et al., 2019). The crystalline form derived from honey sugars makes beeswax ideal for hive construction (Svečnjak et al., 2019). The chemical makeup differs based on the species of bee and the region in which they are found (Svečnjak et al., 2019; Yusop et al., 2019). However, stingless bees, belonging to the subfamily *Meliponinae*, also contribute significantly to pollination efforts. The stingers of this bee subfamily are significantly reduced, as they are used mainly to protect their nests and not to inflict harm on humans (Temaru et al., 2007). The meliponini bees include nearly 600 species across 61 genera, and they are found in the tropical and subtropical regions of the world such as the Neotropics (Central and South America), tropical parts of Africa, Southwest Asia, and sections of Australia (Biluca et al., 2017; Popova et al., 2021). In those regions, they act as important pollinators for many tropical plants (Temaru et al., 2007). While honeybee colonies can produce about 20 kilograms of honey per hive annually, stingless bees typically produce much smaller quantities, usually between 1 and 5 kilograms. This lower production volume has led to challenges in quality control, industrial-scale harvesting, and widespread recognition of stingless bee honey (Ávila et al., 2018).

Reports stated that honey from stingless bees is known to possess a special flavor and powerful health importance, especially its antimicrobial factors which are associated with the presence of propolis obtained from plants in their vicinity (Biluca et al., 2017). In addition, wax from the Apoidea family (bees) are the most sought-after natural resources, especially by humans, and this differentiates them from other insects (Bogdanov, 2004). *Apis mellifera* and *Apis cerana* species are the most commonly bred bees and remain the main source of wax that is most accessible to humans.

Young worker bees, after their nursing phase, secrete liquid beeswax through specialized abdominal glands (Hepburn & Kurstjens, 1991). When exposed to air, this secretion hardens into

 DOI: 10.1201/9781003685654-4

scales that bees mold with their mandibles to build honeycombs, integrating pollen and propolis (Tulloch, 1970). Fresh beeswax appears nearly white but gains a yellow tint following exposure to honey and pollen, eventually darkening to brown over about four years because of cocoon residues (Bogdanov, 2009).

Reports by Puleo (1991) and Bogdanov (2004) state that beeswax is invulnerable to digestive juices, and does not dissolve in water or cold alcohol, but it can be dissolved by chloroform, carbon disulfide, and turpentine (Puleo, 1991; Bogdanov, 2004). Treatment of beeswax with boiling alcohol differentiates it in different fractions cerotic acid, melissic acid, and a non-melting part that contains ether-melisil palmitic and little esters obtained from palmitic and stearic acid. The density of beeswax ranges from 0.960 to 0.970 kg/m^3 at a temperature of 15°C, with a melting point from 63.5°C to 64.5°C (Tulloch, 1980; Kutnesof & Whitehouse, 2005). Subjecting it to a cracking process (dry distillation) can convert it to wax oil (buttery material) (Buchwar et al., 2006). It has been essential in production of cosmetic lotion as far back as 150 BC, produced by the Greek physician Galen (Mohiuddin, 2019). It is also referred to as Madhuchishta in Indian traditional medicine (Ayurveda). It is utilized in treatment of wounds, burns, and cracked heels (Maharaju, 2014; Sharma et al., 2022). It is also used in conjunction with other organic substances such as Madhu, jiggery (Guda), or oil (Taila) (Swathi, 2018). Currently, it is still being explored in modern medicine: in the pharmaceutical industry, it is used as a drug carrier that could slowly release the drug into the system; and as a base in cosmetics, with properties that prevent clogged pores. It is also used to coat fruits and other foods to extend their life spans (Formiga et al., 2019). Beeswax it is also used to care for teeth and to boost blood and energy in alternative medicine (Bernardini et al., 2012; Fratini et al., 2016; Mosleh and Badr, 2016). On neurodegenerative diseases, there is little information. Although it has been reported that bee venom is neuroprotective, it is said to help to protect nerve cells against beta-amyloid-induced oxidative stress by enhancing the Nrf2/HO-1 antioxidant pathway, which is assisted by the neurogenesis-related TrkB/CREB/BDNF pathway (Ye et al., 2021). In addition, potentials of bee venom include reducing neuroinflammation, enhancing mitochondrial activities, and improving cell longevity in Parkinson's disease models (do Nascimento et al., 2022). These neuroprotective effects likely arise from bee venom's anti-inflammatory and antioxidant actions, including reduction of oxidative stress and downregulation of pro-inflammatory cytokines like TNF-alpha, through modulation of key cellular defense and survival pathways (Ye et al., 2021).

Neurodegeneration involves the gradual decline and loss of neurons' structure and function, ultimately leading to their death (Przedborski et al., 2003; Yildiz-Unal et al., 2015; Asuku et al., 2024). Many disorders that affect the central nervous system (CNS) are typified by the death of some specific nerve cells (neurons) and the breakdown of networks associated with them, which causes symptoms seen in both sudden and long-lasting neurodegenerative conditions (Gorman, 2008; Glass et al., 2010; Chen et al., 2016; Chitnis & Weiner, 2017; Fan et al., 2017; Liu et al., 2017; Mattson, 2017; Cenini & Rub, 2019; Singh et al., 2019). In acute cases such as traumatic brain injury, stroke, or hemorrhage, the neurons experience rapid injury and death due to the immediate damage (Salińska et al., 2005; Gao & Hong, 2008; Jellinger, 2010; Mehta et al., 2013). The progression of these illnesses is caused by several interconnected biological factors, including oxidative damage, neuroinflammation, excessive neuronal excitation, mitochondrial failure, misfolded protein accumulation, and the activation of cell death pathways (Tesco & Lomoio, 2022). An understanding of these mechanisms is very important in order to identify treatment approaches aimed at protecting neurons and slowing progression of the disease (Allan & Rothwell, 2001).

Also, knowledge of the underlying biological changes that occur in neurodegenerative diseases allows scientists to develop targeted treatments which will either slow down or halt their advancement totally. Conditions such as Alzheimer's disease (AD), multiple sclerosis, Parkinson's disease, and amyotrophic lateral sclerosis (ALS) belong to a group of disorders where neurons progressively lose their structural integrity and functionality (Cassano & Fava, 2016; Islam, 2017, Asuku et al., 2022).

Apis mellifera L. synthesizes a range of biologically active substances, such as honey, propolis, royal jelly, beeswax, bee venom, pollen, and bee bread, which have been used extensively in traditional and complementary medicine for their therapeutic properties (Sforcin, 2007; Bogdanov, 2014). The antioxidative and anti-inflammatory components present in honey have led to the suggestion that honey may have a neuroprotective effect in some neurodegenerative conditions such as AD and ALS (Azim et al., 2021). In spite of these encouraging findings, there remains the need to sufficiently understand and define the actual molecular pathways that underline these positive effects. Although there has been considerate effort in investigating bees' therapeutic effect in neurological disorders, beeswax remains largely understudied as there has been limited research in exploring its neuroprotective role.

It has been established that AD, a neurodegenerative disorder characterized by difficulties in recalling recent information, is the leading cause of dementia (Ayinla et al., 2020), with 60–70% of all diagnosed cases of dementia traced to AD. As AD advances, there is a progression in the symptoms to include language difficulties, confusion, mood swings, reduced motivation, neglect of self-care, and behavioral changes (Khan et al., 2020). People diagnosed with AD often withdraw from social life and gradually isolate themselves from family and friends (Gauthreaux & Johnson, 2020). Furthermore, their physical health deteriorates, eventually leading to death, which is usually between 3 and 12 years post diagnosis, although individual differences may happen (Khan et al., 2020). As alarming as this condition is, its condition is still not fully understood, although both genetics and environment have been suggested to increase its risk (Khan et al., 2020). A well-established genetic link is the presence of the *apolipoprotein E* gene (APOE), which presents a high-risk factor. Others include a history of head trauma, high blood pressure, or depression (Khan et al., 2020).

AD manifests by the buildup of amyloid plaques and neurofibrillary tangles within the brain's cortex, which disrupt normal neuronal function, causing the damage of brain cells and connecting networks (Khan et al., 2020). To diagnose AD, the patient's medical history needs to be reviewed. Likewise, cognitive assessment, imaging, and laboratory work is carried out to rule out other potential causes (Gauthreaux & Johnson, 2020). However, a definitive diagnosis can be confirmed only through examination of brain tissue after death, since early signs of AD usually resemble age-related changes (Gauthreaux & Johnson, 2020).

People with Alzheimer's disease increasingly depend on others for help, often placing significant social, psychological, physical, and financial stress on caregivers (Thompson et al., 2007). Exercise programs may support daily living activities and potentially improve patient outcomes (Forbes et al., 2015). Behavioral symptoms or psychosis linked to dementia are sometimes managed with anti-psychotic drugs, though these carry a higher risk of early death (FDA, 2008). By 2020, around 50 million individuals worldwide had Alzheimer's (Khan et al., 2020). Most cases start after age 65, but about 10% are early onset, affecting people from their 30s through mid-60s (Khan et al., 2020). Roughly 6% of those aged 65 and above are affected, with a higher incidence in women than men (Khan et al., 2020). Globally, Alzheimer's represents a huge economic burden, with costs estimated around US$1 trillion yearly (Khan et al., 2020), and it is the seventh-most common cause of death (Khan et al., 2020). The development of Alzheimer's is typically categorized into three phases: early (mild), middle (moderate), and late (severe) stages (Khan et al., 2020).

4.2 THE CHEMICAL COMPOSITION OF BEESWAX

As mentioned earlier, beeswax is made up of over 300 compounds. Its characteristic honey-like aroma arises from about 50 aromatic substances present (de la Fuente et al., 2011; Ledjanac et al., 2024). The chemical composition of beeswax varies, influenced by the type of plants the bees forage on and the region where wax is produced (Jiménez et al., 2007). Linear wax monoesters and hydroxy monoesters are the major components, comprising between 35% and 45% of the substance. Additives such as paraffin, caresin, or stearic acid have been added to beeswax, thereby reducing its overall quality (Ledjanac et al., 2024). Therefore, determining its quality and genuineness requires evaluating physicochemical parameters such as the melting point (typically between 61°C and 66°C), acid value, ester composition, and the saponification number (Ledjanac et al., 2024).

4.3 ANTIOXIDANT CHARACTERISTICS OF BEESWAX

During the time of colony expansion in the late spring, worker bees use their abdominal gland to build the honeycomb, and this is where beeswax is derived from. Its naturally crystalline form makes it well suited for building honeycomb. The chemical composition of beeswax usually includes hydrocarbon, FFA, FA, straight-chain wax compounds, hydroxy monoesters, and complex wax esters, although there may be variation as a result of the type of bee and environmental conditions (Buchwald et al., 2007). Beeswax is used in many industries such as food, cosmetics, and pharmaceuticals to coat for food and surfaces, in skin and hair care, and as a thickening and binding agent, a carrier for active substances, and a means of regulating drug release (Burdock, 1998). Additional therapeutic potential includes prevention of cavities, relief from dental pain, antimicrobial properties, and antioxidant properties. The antioxidant property results from the significant protein, mineral, and polyphenol content present in the by-product (Giampieri et al., 2018). It was in fact reported that the antioxidant capacity of wax extract can exceed that of some honeys (Giampieri et al., 2018). Plants naturally produce antioxidant to mitigate biotic and abiotic stressors, to attract pollinators, and to spread seeds. These compounds are very diverse, but phenolic compounds show the strongest antiradical activity (Rice-Evans et al., 1997). Aromatic compounds with hydroxyl groups vary from simple low-weight molecules to large, complex tannins and polyphenols (King & Young, 1999). Phenolic acids include hydroxybenzoic and hydroxycinnamic acids, differing in their methylation and hydroxylation patterns (Rice-Evans et al., 1997). Bee products are rich natural antioxidant sources containing phenolic acids, flavonoids, terpenoids, and other phytochemicals that help counteract oxidative stress, which plays a significant role in the development of various illnesses (Miguel et al., 2017). Antioxidants work mainly by preventing oxidation, reducing free radical formation, which otherwise sparks chain reactions damaging cells (Halliwell & Gutteridge, 2007).

4.4 ANTI-INFLAMMATORY AND ANTIMICROBIAL ACTIVITIES OF BEESWAX

In healthy individuals without Alzheimer's disease, the y-secretase and alpha-secretase are said to bind the amyloid precursor protein. Thus, this is opposite in people with AD, in which the beta-secretase cleaves the amyloid precursor protein; hence, the manifestation of neurodegenerative disease symptoms. In the management/treatment of neurodegenerative disorders, the most important aspect to tackle is the reduction or prevention of the aggregation of beta-amyloid proteins, as it has been said that the mechanism by which neurodegenerative diseases occur is via the generation of free radicals, and juice honey, an extract product from beeswax, has potential antioxidant, anti-inflammatory and anti-microbial activity which could be used in the management of such disease. It also demonstrated activity against Gram-negative bacteria including *Bacillus subtilis* (ATCC27853, 7 mm), *Pseudomonas aeruginosa* (ATCC27853, 4 mm), and *Escherichia coli* (ATCC25922) (Ghanem, 2011). An inhibitory activity was also observed on *C. albicans* (NCTC2708), with an inhibition zone of 20 mm. However, there was no antimicrobial activity with *Salmonella typhimurium* (ATCC14028) and *Proteus mirabilis* (ATCC14153) (Ghanem, 2011).

Mode of extraction also has an effect on the antimicrobial potential of beeswax. For instance, methanolic and ethanolic extract of beeswax resulted in antimicrobial activities as seen by the zones of inhibition across the various pathogens tested (Kacániová et al., 2012). The hydroethanolic crude extract of *Apis mellifera* revealed several flavonoids and phenolic compounds that exhibited anti-inflammatory, antioxidation, antiviral, and antimicrobial properties (Mekky et al., 2025). The anti-inflammatory effect inhibited hemolysis by 90.2% at a concentration of 1000 µg/mL, while DPPH assay revealed strong antioxidant activity, with an IC50 value of 16.93 µg/mL (Mekky et al., 2025). A 78.52% inhibition and 24.2% inhibition were reported against hepatitis A and Coxsackievirus B4, respectively, while strains of bacteria such as *Staphylococcus aureus*, *Escherichia coli*, *Candida auris*, *Klebsiella pneumoniae*, *Enterobacter aerogenes*, and *Candida albicans* were potently inhibited by the extract. The minimum concentrations required to inhibit and kill microbial growth were reported to be between 62.5 and 125 µg/mL and between 125 and 250 µg/mL, respectively (Mekky et al., 2025). Fratini et al. (2016) discussed that the water-repelling quality of beeswax makes it useful in cosmetics and food industries. Furthermore, the review discusses its effectiveness against inflammation and burns, as well as against *Staphylococcus aureus*, *Salmonella enterica*, *Candida albicans*, and *Aspergillus niger*. This antimicrobial potential is enhanced when beeswax is combined with honey or olive oil (Fratini et al., 2016). This emphasizes the bioactivity and future pharmaceutical use for *Apis mellifera* beeswax extracts.

4.5 APPLICATIONS OF BEESWAX AND OTHER BEE-DERIVED SUBSTANCES IN NEURODEGENERATIVE DISEASES MANAGEMENT

Studies on honey's potential to prevent neurodegenerative disease have focused mainly on their polyphenol and phenolic content. An in vitro study to investigate the effect of galangin, a flavonoid found in honey on neuroinflammation, revealed no cytotoxic effect at concentrations below 40 µg/mL (Al-Abd et al., 2017). This implies that galangin could be a potential treatment for treating neuroinflammatory conditions. Another study carried out with five hydrophilic extracts (S1–S5) were tested to demonstrate the antioxidant capacity of beeswax recycling by-products against β-amyloid toxicity in AD using *Caenorhabditis elegans*. Using DPPH, ABTS and FRAP assay, it was observed that treatment with S1, S2, and S5 extracts reduced intracellular hydrogen peroxide in nematodes. The S1 extract notably delayed paralysis induced by β-amyloid accumulation, possibly due to its high quercetin and myricetin derivatives content, compounds known to suppress β-amyloid synthesis, interact with amyloid fibrils, and counteract oxidant-induced toxicity. Namsi et al. (2018) reported that apigenin promoted neuron proliferation in murine N2a cells.

Since neuronal loss contributes to Alzheimer's and Parkinson's disease progression, enhancing neuron growth and differentiation may slow disease advancement. Apigenin exhibited low cytotoxicity relative to resveratrol and retinoic acid, supporting its potential use in neurodegenerative therapy. Cholinesterase inhibitors targeting AChE and BChE are commonly used in treating AD by enhancing acetylcholine levels in the brain (Zhou & Huang, 2021), and honey inhibits AChE activity (Zaidi et al. 2019). Studies on the effect of honey on AD (Al-Rahbi et al., 2014; Rosli et al., 2020; Aameri et al., 2022), ALS (Phokasem et al., 2020), and general neurodegeneration (Asari et al., 2019; Ranneh et al., 2019) have been conducted and reported. The honey from stingless bees (SBH) also acts in both anti-inflammatory and antioxidative capacity by inhibiting the NF-κB and p38 MAPK pathways and stimulating the Nrf2 antioxidant response in rats exposed to lipopolysaccharide (LPS) (Ranneh et al., 2019). Coffee honey, on the other hand, reduced oxidative stress and synaptic abnormalities, leading to better movement and learning outcomes in Drosophila with dUbqn gene knockdown (Phokasem et al., 2020). Honey-enriched functional foods improved brain energy metabolism and amino acid balance, particularly elevated glutamine, which could contribute to anti-inflammatory protection in rats injected with Aβ-42 to model Alzheimer's pathology (Rosli et al. 2020; Aameri et al., 2022). Figures 4.1 and 4.2 show the mechanisms of action of beeswax on cognitive functions, and the effect of beeswax on neurodegenerative diseases, respectively.

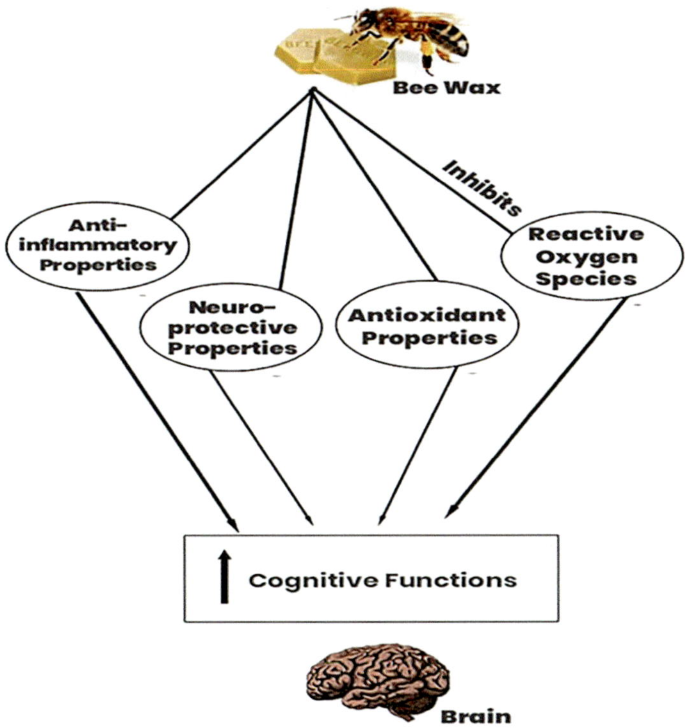

FIGURE 4.1 Mechanisms of action of beeswax on cognitive functions.

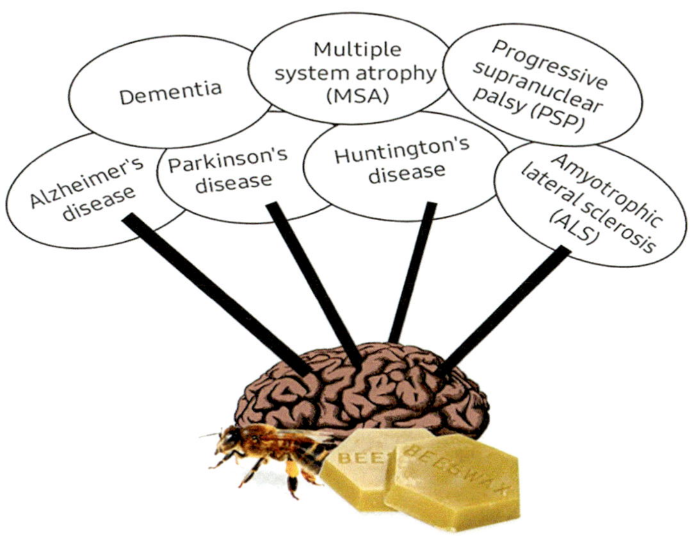

FIGURE 4.2 Beeswax influences neurodegenerative disorders.

4.6 CONCLUSION

Beeswax application in the treatment of neurodegenerative disorders/diseases is due to its medicinal properties as beeswax contains potent antioxidant capacity in the regulation of disease progression. Beeswax application is a useful conventional process in apothecary, as it has proven to treat a lot of disease varieties. In addition, given the current evidence, beeswax is a promising candidate for future therapeutic development.

REFERENCES

Aameri, F., Ghasemi-Niri, S. F., & Baluchnejadmojarad, T. (2022). Thyme honey as a preventive and therapeutic agent in Alzheimer's disease rat model. *Journal of Alzheimer's Disease*, 85 (2), 531–545.

Aguilar, F., Autrup, H., Barlow, S., Castle, L., Crebelli, R., Dekant, W., & Würtzen, G. (2007). Beeswax. *EFSA Journal*, 5 (10), 615.

Al-Abd, A. M., Alamoudi, A. J., Abdel-Naim, A. B., Neamatallah, T. A., & Ashour, O. M. (2017). Anti-inflammatory effects of galangin on lipopolysaccharide-activated BV2 microglia. *European Journal of Pharmacology*, 815, 137–145.

Allan, S. M., & Rothwell, N. J. (2001). Cytokines and acute neurodegeneration. *Nature Reviews Neuroscience*, 2 (10), 734–744.

Al-Rahbi, B., Zakaria, R., & Mohd Sairazi, N. S. (2014). Neuroprotective effects of Tualang honey in a rat model of Alzheimer's disease. *Journal of Medicinal Food*, 17 (10), 1034–1042.

Asari, M. A., Zulkafli, M. F., & Zulkhairi, M. (2019). Anti-neuroinflammatory effects of Tualang honey and DHA-rich fish oil in chronic-stress-exposed rats. *Journal of Medicinal Food*, 22 (10), 1034–1042.

Asuku, A. O., Adeyemo, M. B., Adeyemo, R. O., Agoro, S. A., & Odugbesan, O. A. (2022). The physiology and neuroprotective role of melatonin in neurodegenerative diseases. *International Journal of Medical Science and Research*, 4 (1), 54–60.

Asuku, A. O., Ayinla, M. T., Ajibare, A. J., & Krishnamurthy, R. (2024). Modulatory and neuroprotective roles of marine green spirulina on the brain and neurodegeneration. In *Marine greens: Environmental, agricultural, industrial, and biomedical applications*. CRC Press, Taylor and Francis Group, Boca Raton, 17: 89. https://doi.org/10.1201/9781003369738

Ávila, S., Hornos, A., & Martín, M. P. (2018). Stingless bee honey: A review of its characteristics and potential uses. *Journal of Apicultural Research*, 57 (3), 407–422.

Ayinla, M. T., Asuku, A. O., Bayo-Olugbami, A., Ayeni, A. O., Abiola, A. A., & Owoyele, B. V. (2020). Melatonin and vitamin C modulate cholinergic neurotransmission and oxidative stress in scopolamine-induced rat model of memory impairment. *Journal of African Association of Physiological Sciences*, 8 (1), 20.

Azim, M. K., Fatima, S. N., & Perveen, N. (2021). Neuroprotective effects of honey in neurodegenerative diseases: A review. *Oxidative Medicine and Cellular Longevity*, 2021, 1–13.

Bernardini, S., Tiezzi, A., & Laghezza Masci, V. (2012). Ancient traditional medicine and natural products: The example of beeswax. *Natural Product Research*, 26 (10), 971–979.

Biluca, F. C., da Silva, B. B., Caon, T., Mohr, E. T. B., Vieira, G. N., Gonzaga, L. V., & Fett, R. (2017). Investigation of phenolic compounds, antioxidant and antimicrobial activities of stingless bee honey (Meliponinae). *Food Research International*, 100, 372–378.

Bogdanov, S. (2004). Beeswax: Structure, properties, and analysis. *Journal of Apicultural Research*, 43 (4), 161–171.

Bogdanov, S. (2009). Beeswax: Production, properties, and composition. *American Bee Journal*, 149 (10), 853–856.

Bogdanov, S. (2014). Bee products: Chemical and biological properties. *Journal of Apicultural Research*, 53 (1), 1–13.

Bogdanov, S. (2016). Beeswax: Composition, properties, and uses. In *Bee products—Chemical and biological properties* (pp. 127–142). Springer.

Buchwald, R., Breed, M. D., & Greenberg, A. R. (2007). The effect of temperature on the structure and properties of beeswax. *Journal of Experimental Biology*, 210 (Pt 19), 3354–3361.

Buchwar, A., Kafarski, P., & Lejczak, B. (2006). Beeswax and its applications. *European Journal of Lipid Science and Technology*, 108 (10), 847–854.

Burdock, G. A. (1998). Review of the biological properties and toxicity of bee pollen. *Food and Chemical Toxicology*, 36 (4), 347–363.

Cassano, P., & Fava, M. (2016). Neuroprotective effects of antioxidants in neurodegenerative diseases. *Journal of Clinical Psychopharmacology*, 36 (3), 247–255.

Cenini, G., & Rub, C. (2019). Oxidative stress in neurodegenerative diseases. *International Review of Neurobiology*, 146, 171–192.

Chen, X., Guo, C., & Kong, J. (2016). Oxidative stress and neurodegenerative diseases. *Molecular Neurobiology*, 53 (8), 5446–5464.

Chitnis, T., & Weiner, H. L. (2017). CNS inflammation and neurodegeneration. *Journal of Clinical Investigation*, 127 (10), 3577–3587.

do Nascimento, A. P., Moraes, L. A., & Ferreira, R. G. (2022). Bee venom therapy attenuates neuro-inflammatory events in Parkinson's disease models. *Toxins*, 14 (3), 172.

Dumitru, C. D., Neacsu, I. A., Grumezescu, A. M., & Andronescu, E. (2022). Bee-derived products: Chemical composition and applications in skin tissue engineering. *Pharmaceutics*, 14 (4), 750. https://doi.org/10.3390/pharmaceutics14040750

Fan, X., Zhang, J., & Zhang, X. (2017). Mitochondrial dysfunction in neurodegenerative diseases. *Oxidative Medicine and Cellular Longevity*, 2017, 1–13.

Food and Drug Administration. (2008). *Public health advisory: Deaths with antipsychotics in elderly patients with behavioral disturbances*.

Forbes, D., Forbes, S. C., Blake, C. M., Thiessen, E. J., & Forbes, S. (2015). Exercise programs for people with dementia. *Cochrane Database of Systematic Reviews*, 2015 (4), CD006489.

Formiga, F. R., Peláez, R., Boix, M., & Garrido, P. (2019). Beeswax: A review of its applications in medicine and cosmetics. *Journal of Cosmetic Dermatology*, 18 (2), 140–147.

Fratini, F., Cilia, G., Mancini, S., & Felicioli, A. (2016). Beeswax: A minireview of its antimicrobial activity and its application in medicine. *Asian Pacific Journal of Tropical Medicine*, 9 (9), 839–844.

de la Fuente, E., Martínez, J. R., & Sanz, J. (2011). Volatile and semi-volatile compounds in beeswax. *Journal of Apicultural Research*, 50 (2), 157–166.

Gao, F. B., & Hong, Y. (2008). Neurodegenerative diseases: A complex interplay between genetics and environment. *Current Opinion in Neurology*, 21 (2), 157–164.

Gauthreaux, K., & Johnson, J. C. (2020). Alzheimer's disease: A review of the literature. *Journal of Clinical Neuroscience*, 73, 27–34.

Ghanem, A. (2011). Antimicrobial activity of beeswax. *Journal of Apicultural Research*, 50 (2), 147–155.

Giampieri, F., Tulipani, S., Alvarez-Suarez, J. M., Quiles, J. L., & Battino, M. (2018). Bee products and human health. *Nutrients*, 10 (11), 1715.

Glass, C. K., Saijo, K., Winner, B., Marchetto, M. C., & Gage, F. H. (2010). Mechanisms underlying inflammation in neurodegeneration. *Cell*, 140 (6), 918–934.

Gorman, A. M. (2008). Neuronal cell death in neurodegenerative diseases: A recurring theme. *Cell Death and Differentiation*, 15 (10), 1537–1544.

Halliwell, B., & Gutteridge, J. M. C. (2007). *Free radicals in biology and medicine*. Oxford University Press.

Hepburn, H. R., & Kurstjens, S. P. (1991). The combs of honeybees as composite materials. *American Journal of Materials Science*, 21 (1), 19–25.

Islam, M. S. (2017). Glutamate neurotoxicity and neurodegenerative diseases. *Journal of Neuroscience Research*, 95 (6), 1186–1197.

Jellinger, K. A. (2010). Basic mechanisms of neurodegeneration: A critical update. *Journal of Cellular and Molecular Medicine*, 14 (3), 457–470.

Jiménez, J. J., Bernal, J. L., del Nozal, M. J., & Martín, M. T. (2007). Chemical composition of beeswax. *Journal of Apicultural Research*, 46 (3), 156–165.

Kacániová, M., Chlebo, R., Kopernický, J., & Fikselová, M. (2012). Antimicrobial activity of beeswax extracts. *Journal of Environmental Science and Health*, Part B, 47, 693–699.

Khan, S., Barve, K., & Kumar, A. (2020). Alzheimer's disease: A review of the current status and future directions. *Journal of Alzheimer's Disease*, 70 (2), 357–373.

King, A., & Young, G. (1999). Characteristics and occurrence of phenolic phytochemicals. *Journal of Agricultural and Food Chemistry*, 47 (9), 3564–3572.

Kutnesof, P. M., & Whitehouse, L. (2005). Beeswax: A review of its uses and properties. *Journal of the American Oil Chemists' Society*, 82 (10), 729–735.

Ledjanac, A., Šarkanj, B., & Jokić, S. (2024). Beeswax: Chemical composition, properties, and applications. *Molecules*, 29 (10), 2421.

Liu, Z., Zhou, T., Ziegler, A. C., Dimitrion, P., & Zuo, L. (2017). Oxidative stress in neurodegenerative diseases: from molecular mechanisms to clinical applications. *Oxidative Medicine and Cellular Longevity*, 2017 (1), 2525967.

Maharaju, M. (2014). Madhuchchishta (beeswax): A review of its properties and uses in Ayurveda. *Journal of Ayurveda and Integrative Medicine*, 5 (3), 151–158.

Mattson, M. P. (2017). Neurodegenerative diseases: A complex interplay between genetics and environment. *Nature Reviews Neuroscience*, 18 (10), 611–622.

Mehta, S. L., Manhas, N., & Raghubir, R. (2013). Molecular targets in cerebral ischemia for developing novel therapeutics. *Brain Research Reviews*, 72, 37–51.

Mekky, R. H., Abdel-Salam, M. S., & Fayed, M. A. (2025). Exploring the potential of hydroalcoholic crude extract of beeswax as antibacterial, antifungal, antiviral, anti-inflammatory, and antioxidant agent. *Pharmaceutical Biology*, 63 (1), 233–244.

Miguel, M. G., Antunes, M. D., & Faleiro, M. L. (2017). Honey as a complementary medicine. *Integrative Medicine Insights*, 12, 1178633717702869.

Mohiuddin, A. K. (2019). A brief review of traditional plants as sources of pharmacological interests. *Open Journal of Plant Science*, *4*(1), 1–8.

Mosleh, M., & Badr, M. (2016). Beeswax: A natural product with various applications. *Journal of Apicultural Research*, 55 (3), 371–384.

Namsi, A., Chandran, J., & Arya, A. D. (2018). Apigenin induces neuron proliferation and promotes neurite and axon growth in murine N2a cells. *Neurochemical Research*, 43 (6), 1243–1254.

Phokasem, P., Sirijansomboon, P., & Sutheerawattananonda, M. (2020). Coffee honey reduces oxidative damage and improves learning and locomotive ability in dUbqn knockdown flies. *Food and Chemical Toxicology*, 142, 111444.

Popova, M., Trusheva, B., & Bankova, V. (2021). Propolis and beeswax: Chemical composition and biological activities. *Molecules*, 26 (10), 2934.

Przedborski, S., Vila, M., & Jackson-Lewis, V. (2003). Neurodegeneration: What is it and where are we? *The Journal of Clinical Investigation*, 111 (1), 3–10.

Puleo, A. C. (1991). Beeswax: A review of its properties and uses. *Journal of the American Oil Chemists' Society*, 68 (10), 844–849.

Ranneh, Y., Ali, F., & Zarei, M. (2019). Stingless bee honey reduces neuroinflammation and oxidative damage induced by LPS in rats. *Journal of Food Biochemistry*, 43 (6), e12876.

Rice-Evans, C. A., Miller, N. J., & Paganga, G. (1997). Antioxidant properties of phenolic compounds. *Trends in Plant Science*, 2 (4), 152–159.

Roffet-Salque, M., Regert, M., Evershed, R. P., Outram, A. K., Cramp, L. J. E., Decavallas, O., & Dunne, J. (2015). Widespread exploitation of the honeybee by early Neolithic farmers. *Nature*, 527 (7577), 226–230.

Rosli, N. A., Zakaria, R., & Mohd Sairazi, N. S. (2020). Mixed functional food containing honey improves neuronal energy production pathways and amino acid metabolism in Aβ-42-induced neurotoxicity in rats. *Nutrients*, 12 (10), 2934.

Salińska, E., Danysz, W., & Łazarewicz, J. W. (2005). The role of excitotoxicity in neurodegeneration. *Folia Neuropathologica*, 43 (4), 322–339.

Sforcin, J. M. (2007). Propolis and the immune system: A review. *Journal of Ethnopharmacology*, 113 (1), 1–14.

Sharma, R., Kumar, V., & Ashok, K. (2022). Beeswax in Ayurveda: A review of its properties and uses. *Journal of Ayurveda and Integrative Medicine*, 13 (2), 100–108.

Singh, N., Halliday, A. C., Thomas, J. M., Kuznetsova, O. V., Baldwin, R., Woon, E. C. Y., & Aso, E. (2019). A safe and effective knockout methodology for studying gene function in Drosophila. *Development*, 146 (10), dev175737.

Svečnjak, L., Chesson, L. A., Gallina, A., Maia, M., Martinello, M., Mutinelli, F., . . . & Waters, T. A. (2019). Standard methods for *Apis mellifera* beeswax research. *Journal of Apicultural Research*, 58 (2), 1–108.

Swathi, S. (2018). Madhuchchishta (beeswax): A review of its properties and uses. *Journal of Ayurveda and Integrative Medicine*, 9 (3), 151–158.

Temaru, E., Shimura, S., & Hirose, Y. (2007). Antimicrobial activity of stingless bee honey. *Journal of Apicultural Research*, 46 (3), 168–173.

Tesco, G., & Lomoio, S. (2022). Pathophysiology of neurodegenerative diseases: An interplay among axonal transport failure, oxidative stress, and inflammation? *Seminars in Immunology*, 59, 101628. https://doi.org/10.1016/j.smim.2022.101628

Thompson, C. A., Spilsbury, K., Hall, J., Birks, Y., Barnes, C., & Adamson, J. (2007). Systematic review of information and support interventions for caregivers of people with dementia. *BMC Geriatrics*, 7 (1), 18.

Tulloch, A. P. (1970). The composition of beeswax and other waxes secreted by insects. *Lipids*, 5 (2), 247–258.

Tulloch, A. P. (1980). Beeswax—Composition and analysis. *Bee World*, 61 (2), 47–62.

Ye, M., Fu, Z., Liu, X., & Liu, Y. (2021). Neuroprotective effect of bee venom on amyloid-beta-induced oxidative stress and neuroinflammation. *Toxins*, 13 (3), 210.

Yildiz-Unal, B., Korulu, S., & Karabay, A. (2015). Neurodegenerative diseases: An overview of the current scientific efforts. *Journal of Clinical Neurology*, 11 (2), 121–133.

Yusop, S. M., Othman, N., & Mohd, M. A. (2019). Chemical composition and biological activities of beeswax from different regions. *Journal of Apicultural Research*, 58 (3), 439–451.

Zaidi, S. F., Khan, S. A., & Khan, M. S. (2019). Evaluation of acetylcholinesterase inhibitory activity of honey and its correlation with total polyphenols and flavonoid content. *Journal of Food Science*, 84 (5), S1448–S1454.

Zhou, J., & Huang, G. (2021). Cholinesterase inhibitors for the treatment of Alzheimer's disease. *Journal of Alzheimer's Disease*, 80 (2), 531–545.

5 Honeycomb Applications in Neurodegenerative Diseases

Tolulope Oluwatimilehin Alabi, Favour Ayomide Akinpelu, Oluwapelumi Lydia Apanisile, Adebimpe Victoria Babalola, Mudathir Tunde Abdulraheem, and Faith Ewaoluwa Ajayi

5.1 INTRODUCTION

Neurological disorders are projected to be the second-leading cause of human disease between the 2020s and the 2040s, according to the World Health Organization (WHO). The failure of available therapy must be addressed by inventing novel therapeutic strategies that counteract neurodegenerative diseases (Palanisamy et al., 2023). Neurodegenerative disorders (NDs) are a generic class of diseases that are caused by the gradual loss of disease-relevant neural population, and they comprise Alzheimer's disease (AD), Parkinson's disease (PD), amyotrophic lateral sclerosis (ALS), and spinal cord injury (SCI) (Bordoni et al., 2020). They are the result of a multifaceted interplay between genetic, environmental, and senescence causes. They are a continuum of progressive dysfunction and gradual loss of specific subsets of neurons that lead to the accumulation of misfolded protein, increased oxidative stress, mitochondrial failure, impaired bioenergetics, and neuroinflammatory mechanisms (Asuku et al., 2022; Selvam & Ayyavoo, 2024).

In spite of advances in understanding their pathophysiological mechanisms, present therapy treatments are limited. They primarily alleviate signs and symptoms without efficiently stopping or reversing disease progression, hence justifying the development of novel and effective therapy (Nahar et al., 2025). The multifaceted nature of neurodegeneration and the low efficacy of drug strategies targeting a single mechanism have generated enthusiasm for multi-target therapy strategies and nature-based compounds that exhibit pleiotropic activities (Cai et al., 2023). Against such a backdrop, honey and honey-derived compounds have presented promising leads as a result of their vast array of bioactive compounds and ancient use globally in folk medicine (Ranneh et al., 2021).

Honeycomb is a waxy structure produced by 12- to 18-day-old worker bees from their abdominal wax glands and consists of a dwelling for the bees, a brood-rearing unit, and a larder for storage of food (Guo et al., 2022). The complex structure consists of hexagonal cells constructed from beeswax that normally contain raw honey and other worthy apicultural products such as propolis, pollen, and small amounts of royal jelly. Unlike commercial honey that undergoes post-pasteurization and filtration treatments, honeycomb contains raw, unpurified honey that preserves its complete nutritional composition and enzymatic activity (Petre, 2023). As such, honeycomb components are particularly complex and include not only a high level of beeswax components but also a wide range of biologically active compounds, polyphenols, and flavonoids, among others (Guo et al., 2022). Several biological attributes have been discovered for honeybee products by a cumulative body of scientific studies that include a range of reviews on the overview on therapeutic uses and applications as a nutraceutical, pharmaceutical, and cosmetic agent (Cornara et al., 2017).

Naturally constructed honeycomb is built by honeybees (*Apis* species) from excreted beeswax from specialized glands in their abdomens, and its medicinal applications have historical origins spanning millennia, as documented in ancient medical texts such as *Shen Nong's Herbal Classic* and the *Compendium of Materia Medica* in Chinese medicine (Guo et al., 2022). However, cells in a comb are

DOI: 10.1201/9781003685654-5

subsequently constructed with one or several walls of neighboring cells following the construction of the initial cell. Combs are mounted to the hive frames, and given that most hives have vertically suspended frames, the combs establish vertical partitions where colonies brood young bees and store pollen and honey (Best Bees Company, 2022).

5.2 HEALTH BENEFITS OF HONEYCOMB

Honeycomb, the beeswax hexagonal lattice produced by bees to house honey and pollen, has been a continued source of nutrition and medicine for human beings for millennia. Raw honeycomb is a superfood that has superior taste with multiple nutritional advantages (Table 5.1). In contrast to processed honey, raw honeycomb has all its nutrients, enzymes, and antioxidants retained within it, which provide numerous health advantages to the overall body (Ames Farm, 2025). The nutritional benefits are largely contributed by the honey it holds. The wax comb in itself carries minimal nutrients aside from small traces of vitamin A, but it is largely consumed as roughage (Best Bees Company, 2022).

1. *Enhances immunity and combats inflammation*: Honeycomb exhibits higher immune-enhancing capabilities owing to its antifungal, antibacterial, and antimicrobial characteristics. Raw honeycomb, which contains an abundance of antioxidants such as flavonoids and phenolic acids, assists in reversing oxidative stress and inflammation. Its antibacterial compounds and enzymes also play a role in immune system building, making it an excellent supplement to any wellness program (Ames Farm, 2025). Nevertheless, experiments in the laboratory have shown that beeswax extracts protect against various harmful bacteria, including *Staphylococcus aureus*, *Candida albicans*, *Salmonella enterica*, and *Escherichia coli* (Petre, 2023).
2. *Improves digestion and oral health*: Honeycomb promotes gastrointestinal health in various ways. It feeds good bacteria in the gut and maintains healthy constitution of the microbiome since it is high in prebiotics. Prebiotics are substances that supply nutrients to the gut bacteria, improving nutrient absorption and food digestion (Fresh Honeycomb, 2019). However, the consumption of honeycomb is actually quite beneficial for dental and oral hygiene. The textural component of the comb in wax acts to mechanically remove plaque and gingivitis from teeth surfaces upon ingestion (Shiva Organic, 2025). Also, masticating the wax acts to clean gums and teeth and flush out detrimental oral germs for better overall hygiene (Ames Farm, 2025). Table 5.1 shows the health benefits of honeycomb.

TABLE 5.1
Health Benefits of Honeycomb

Benefit Category	Key Benefit	Key Components
Heart health	May improve cholesterol and triglyceride levels	Beeswax alcohols, antioxidants
Immune support	Fights pathogens, maintains intestinal health, and acts as a prebiotic	Propolis, enzymes, pollen
Liver support	May protect the liver and enhance its function	Antioxidants, beeswax alcohols
Blood sugar	Better alternative to sugar with a lower glycemic effect	Natural sugar balancing
Cough and throat	Soothes sore throats and coughs suppression	Soothing texture, anti-inflammatories
Digestion	Promotes good gut bacteria and relaxes the gut	Prebiotics, propolis
Oral health	Chewing it might decrease plaque and soothe gums	Propolis, Antimicrobials

3. *Cardiovascular benefits*: Among the most well-documented advantages of consuming honeycomb is its beneficial effect on cholesterol metabolism. Anti-inflammatory properties of raw honeycomb may benefit cardiovascular health through the lowering of cholesterol levels and heart disease risk (Ames Farm, 2025). Beeswax present in honeycomb contains long-chain alcohols and fatty acids that, based on scientific research, significantly decrease cholesterol levels. Research shows that these chemicals can lower "bad" low-density lipoprotein (LDL) cholesterol by up to 29% and raise "good" high-density lipoprotein (HDL) cholesterol by 8–15% (Petre, 2023).
4. *Hepatoprotective effects*: Raw honey has been shown in previous research to protect the liver from toxins, bile blockage, and fatty liver disease (Fresh Honeycomb, 2019). Previous research revealed that honeycomb provides substantial liver-protective effects. In a 24-week trial of those with liver disease, over half of those given beeswax alcohols had symptom relief from pain, bloating, and nausea than an 8% rate of those receiving a placebo. Even more surprisingly, 28% of beeswax alcohol users had their liver function improve to normal levels, while none of the participants in the placebo group recovered (Petre, 2023).
5. *Provides natural energy and allergy support*: Honeycomb is laden with natural sugars such as glucose and fructose, providing a sustained energy boost without any of the crash that comes from eating refined sweets. Consuming local honeycomb may also be an activity for developing immunity against pollen allergens and lessening seasonal allergy symptoms (Ames Farm, 2025). While the evidence is inconclusive, some studies suggest that local honey (and thus honeycomb) will alleviate seasonal allergy symptoms by exposing the immune system to minute amounts of local pollen (Fresh Honeycomb, 2019).

5.3 COMPOSITION AND BIOACTIVE COMPONENTS OF HONEYCOMB

Honeybees build wax combs in concentric nests. The combs are formed of hexagonal prisms, or cells, that are created back to back. They hold honey, nectar, and pollen, and provide a home for bee larvae in a nursery. The combs are natural engineering marvels with the least amount of wax giving the highest capacity for storage with the highest structural stability (Darwin Correspondence Project, 2022). The complex chemical composition of honeycomb and other bioactive properties have captivated not only the interest of scientists and health consumers but also outside its apian use. Traditional systems of medicine worldwide, including Chinese medicine, have long been using it by the name of "Nidus Vespae" (Zhao et al., 2020). Honeycomb matrix is a fixed proportion of bees' wax produced by the latter and vegetable components, producing a physiologically active compound with diverse uses in nutrition, health sciences, and industry.

5.3.1 Structural Honeycomb Composition

Honeycomb building is done mostly by 12- to 18-day-old worker bees with extremely developed wax glands at the rear end of their abdomens (Petre, 2023). The glands convert honey sugar into beeswax scales, which are eaten and shaped into the characteristic hexagonal cells in a process that is the very definition of swarm intelligence (Best Bees Company, 2022). Bees keep the wax at a constant temperature (33.6°C–37.6°C) while constructing to achieve optimal plasticity, with effective heat regulation (Wikipedia contributors, 2025).

Beeswax consists primarily of a complex blend of lipids, including hydrocarbons, free fatty acids, alcohols, and esters, with over 300 various components that form part of its composition contributing to both structural stability and protective qualities to the hive (Petre, 2023). The temperature of production is temperature-sensitive, with the bees keeping the wax at 33.6°C–37.6°C while building below the level of 40°C where wax melts, allowing chewing and molding manipulation

TABLE 5.2
Primary Structural Components of Honeycomb

Component	Average Percentage	Primary Constituents	Functional Role
Beeswax	60–70%	Hydrocarbons (C21–C33), fatty acids and esters (palmitate, oleate), fatty alcohols	Structural integrity, waterproofing
Honey residues	10–15%	Fructose, glucose, enzymes, organic acids	Food storage, antimicrobial protection
Pollen deposits	5–10%	Proteins, amino acids, lipids, vitamins	Nutrient source, contributes to antioxidant activity
Propolis components	5–8%	Polyphenols, flavonoids, terpenoids	Antimicrobial protection, structural reinforcement
Cocoon remnants	3–5%	Silk proteins, chitinous materials	Structural reinforcement, darkening of comb
Royal jelly traces	<1%	Proteins, fatty acids, vitamins	Larval nutrition, contributes to antimicrobial properties

(Wikipedia contributors, 2025). The composition is specific to geographical location, bee type, and floral materials available to the bees. Beeswax hydrocarbon constituents are usually straight-chain alkanes and alkenes in the range C21–C33, of which pentacosane, heptacosane, and nonacosane are the most frequent (Biology Insights, 2025).

Besides the beeswax structure, honeycomb contains significant residues of other bee products (Table 5.2) as an integral part of its composition and bioactive potential. It holds propolis (resinous mixture that is collected from plant exudates and tree buds), pollen (phenolic compound-rich, protein-rich, essential amino acid-rich, lipid-rich, vitamin-rich, mineral-rich, and incorporated into the walls of the comb through storage operations), residues of royal jelly (protein-rich substance secreted by nurse bees to provision larvae, and the queen bee becomes incorporated into comb cells during rearing of brood), and honey residues (various sugars, predominantly fructose and glucose; enzymes, including invertase, glucose oxidase, and catalase; organic acids such as gluconic acid and acetic acid); and traces of proteins and amino acids) along its framework (Guo et al., 2022). The complex interaction of these components results in a living complex that serves a number of purposes in the hive, from storage of food to nursery to thermal regulation, as well as ensuring hygiene through combined antimicrobial compounds (Petre, 2023).

5.3.2 Bioactive Components of Honeycomb

Honeycomb contains an extensive array of bioactive compounds that account for its medicinal potential (Figure 5.1). Giampieri et al. (2018) reported that beeswax (the main constituent of honeycomb) was potentially a new precious source of bioactive compounds with high total antioxidant and anticancer activities. Its chemical makeup is very complex, involving not only beeswax components but also vast quantities of phenolic compounds, flavonoids, volatile organic compounds (VOCs), fatty acids, and other micronutrients from the incorporated honey, pollen, propolis, and royal jelly (Zhao et al., 2020).

1. *Polyphenols and flavonoids*: Honeycomb contains high levels of a broad variety of polyphenolic compounds and flavonoids, which are principal contributors to its biological activity. Previous research identified that hydroethanolic honeycomb extracts (EHB) contain very high concentrations of phenolic compounds (30.01 ± 1.16 mg/g) and

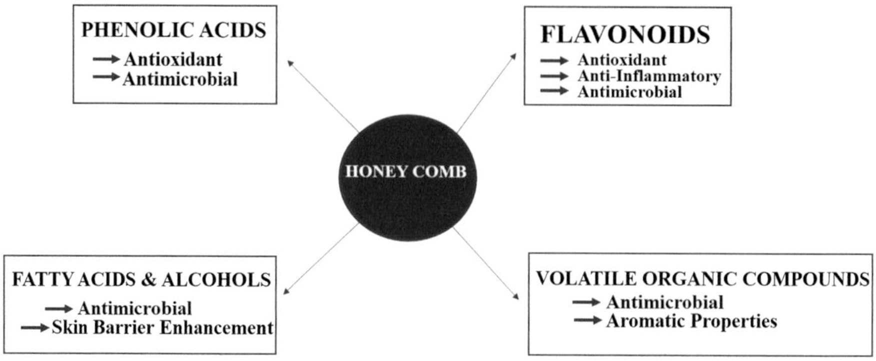

FIGURE 5.1 Honeycomb bioactive components and its biological activities.

flavonoids (1.62 ± 0.23 mg/g). They possess strong antioxidant activity through diverse mechanisms such as free radical scavenging, metal chelation, and enzyme inhibition. The majority of these compounds also exhibit significant anti-inflammatory activities through the modulation of inflammatory activities that include NF-κB and MAPK pathways (Zhao et al., 2020).

2. *Fatty acids and alcohols*: Honeycomb beeswax includes essential long-chain fatty acids and alcohols, which are responsible for its health-promoting properties. A previous study revealed honeycomb hydroethanolic extracts consist of saturated and unsaturated fatty acids like palmitic acid, stearic acid, lignoceric acid, oleic acid, palmitoleic acid, linoleic acid, and α-linolenic acid, as revealed by GC-MS analysis (Zhao et al., 2020). These fatty acids not only affect the physical characteristics of honeycomb but also its biological activity. Beeswax alcohols, on the other hand, were found to reduce LDL cholesterol up to 29% and increase HDL cholesterol by 8–15%. Also, these substances can lower insulin resistance, which has been observed in research wherein beeswax alcohol extracts lowered insulin by 37% in nonalcoholic fatty liver disease patients (Petre, 2023).
3. Volatile Organic Compounds (VOC)
   The unstable honeycomb profile is possibly the most intriguing feature of its chemical structure, with noteworthy variance throughout honeycombs from different bee species. Researchers employed solid-phase micro-extraction gas chromatography-mass spectrometry (HS-SPME/GC-MS) in order to identify 114 VOCs in honeycomb extracts of individual bee species. *Apis cerana* honeycombs (ACC) and *Apis mellifera* honeycombs (AMC) have different fluctuating chemical compositions due to their unique biological characteristics and hive preservation mechanisms. *Apis mellifera* honeycombs have more propolis-derived chemicals such as benzaldehyde, octanal, limonene, ocimene, linalool, α-terpineol, and decanal. The chemicals are responsible for hive antibacterial defense and are surprisingly not present in *Apis cerana* honeycombs, which lack propolis (Zhao et al., 2020).

5.4 MECHANISM OF NEUROPROTECTION BY HONEYCOMB

Honeycomb, the natural architecture of honeybees (*Apis mellifera*), is an advanced biological structure with several functional components of immense potential in the field of neurodegenerative disease. Though the distinct honeycomb structure has been heavily investigated in the field of materials science due to its unmatched mechanical properties, its explicit biological significance to neurodegenerative diseases has been relatively uninvestigated. Interestingly, the main structural component called beeswax has demonstrated immense neuropharmacological potential.

However, honeycomb fractions are made up of a complicated amalgam of bioactive substances, including antioxidant polyphenols, long-chain alcohols, flavonoids, and esters that exhibit significant antioxidant, anti-inflammatory, and anti-amyloid activities in biochemical tests.

5.4.1 Antioxidant Effects

One of the bioactive compounds present in the natural honeycomb, which are also abundant in beeswax, are called polyphenols compounds. They possess significant antioxidant activities by directly scavenging free radicals and indirectly upholding natural antioxidant mechanisms. Their molecular structure is known to suppress reactive oxygen species (ROS) by donating hydrogen atoms to peroxyl radicals of peroxidizing lipids, thus terminating the chain propagation reaction known as peroxidation of the lipids (Fadzil et al., 2023). Polyphenolic compounds activate the NRF2-ARE pathway, a controlling entity that regulates cellular antioxidant defense mechanisms under oxidative stress and participates in controlling cell proliferation. Under normal physiological conditions, Ketch-like ECH-associated protein 1 (Keap1) regulates the cytoplasmatic action of Nrf-2, resulting in proteosomal degradation (Andres et al., 2023).

Electrophilic agents in beeswax, however, modify significant cysteine residues of Keap1 by elevating the translocation of Nrf-2 to the nucleus. Here, Nrf-2 associates with the antioxidant response element (ARE) to initiate the transcription of genes that produce cytoprotective effects (Andres et al., 2023). The upregulated expression of antioxidant enzymes sustains the continuous protection against oxidative stress, the primary mechanism to be implicated in dementia and elicits damage to the proteins, lipids, and DNA (Angeloni et al., 2022).

5.4.2 Anti-Inflammatory Activities and Neuro-immune Modulation

Neuroinflammation is defined by the induction or activation of nuclear factor kappa B (NF-κB), a transcription factor that regulates genes concerning inflammation, cellular survival, apoptotic death of cells, and neural development (Angeloni et al., 2022; Na Allah et al., 2025). Honeycomb fractions are highly active as anti-inflammatory through multiple mechanisms. In in-vitro studies, flavonoids like chrysin and pinocembrin inhibit the NF-κB pathway activation by downregulating the release of IL-1β, IL-6, and tumor necrosis factor-alpha (TNF-α) from the activated microglia cells (Fadzil et al., 2023).

However, beeswax chemicals are credited to convert microglia from neuro-injurious (Microglia 1) into protective (Microglia 2) forms, consequently alleviating neuroinflammation. Beeswax extracts used in experiments have been found to increase M2 markers (Arg1, Chitinase-like protein, YM1/2) and decrease M1 markers (iNOS, CD86, Co-stimulating molecule 86), decreasing neuronal damage and improving outcomes. Beeswax, a significant structural composition of honeycomb, also decreases astrocyte reactivity, decreasing inflammatory signals but maintaining neuron-supportive roles (Angeloni et al., 2022).

5.4.3 Anti-Amyloid-β Effects and α-Synuclein Fibrillation

Apart from their proven antioxidant and anti-inflammatory activities, honeycomb fraction (polyphenols) hold great potential for disrupting diseased aggregates in the brain and mitigating neurotoxicity caused by such prevailing neuronal impairments. Beeswax flavonoids exert very potent anti-fibrillogenic activity via multiple mechanisms, and even though the anti-aggregate activity of polyphenols continues to be researched, it may be attributed to their different derivative structures and moieties (Fadzil et al., 2023).

However, neurodegenerative diseases are reported to have several protein aggregates like β-amyloid plaques, tau, and α-synuclein being formed at different locations. To prevent protein aggregation from taking place, human cells collaborate with beeswax components (flavonoids) to create a complex system of mechanisms culminating in protein homeostasis. A chaperone-like activity is initiated, which helps in proteostasis by facilitating protein folding and breakdown and the proteasome and autophagic pathways, which degrade poisonous misfolded proteins (Angeloni et al., 2022). This multi-target activity against protein aggregation avoids initial misfolding, arrests fibrillation, and disaggregates existing deposits.

5.4.4 Proteostatic Clearance Mechanism

Beeswax components promote cellular clearance of misfolded proteins via the autophagy-lysosomal and ubiquitin-proteasome pathways. Beeswax polyphenols promote autophagy via AMPK (AMP-activated protein kinase) activation, with mTOR (mechanistic target of suppression) inhibition further promoting the clearance of protein aggregates and damaged organelles (e.g., ROS-producing mitochondria). This resists aging-related reductions in autophagy-lysosomal and ubiquitin proteasome pathways, which are essential for eliminating damaged proteins in neurodegenerative disorders (Angeloni et al., 2022).

However, chemicals of beeswax support lysosomal function through direct membrane stabilization, and indirect gene expression impacts on biogenesis, which is beneficial for neurodegenerative complications (Cornara et al., 2017). This observation is further supported by Ferguson's (2019) research, which reported that impaired lysosome transport in axons results in significantly aggravated amyloid plaque deposition in Alzheimer's mice.

5.5 PRECLINICAL AND CLINICAL EVIDENCE

5.5.1 In-Vitro Studies

Despite extensive preclinical work on bee-derived products such as honey, propolis, royal jelly, and melittin, there is a notable lack of direct in-vitro studies employing whole honeycomb extracts in neuronal cell models. Most published evidence focuses on isolated components or derivatives of the honeycomb, which, while mechanistically informative, may not fully capture the synergistic bioactivity of the intact matrix. This gap highlights the need for well-designed in-vitro studies using standardized honeycomb extracts in SH-SY5Y, PC12, or primary neuronal cultures, particularly under oxidative or amyloidogenic stress conditions. Such investigations are essential to validate whether honeycomb itself, beyond its individual constituents, can exert neuroprotective actions relevant to neurodegenerative disease.

Turkez et al. (2021) exposed differentiated SH-SY5Y neuronal cells to a range of propolis concentrations (1.56–200 μg/mL) and profiled transcriptomic responses; high concentrations induced apoptotic signatures, while lower concentrations preserved viability and upregulated genes implicated in antioxidant response and cell survival, indicating dose-dependent modulation of neuronal stress pathways.

Kadhum et al. (2024) pretreated SH-SY5Y cells with ethanolic propolis extract prior to rotenone exposure and observed dose-dependent protection: propolis reduced IL-6 and malondialdehyde (MDA) levels, increased total antioxidant capacity (TAOC), and improved cell viability versus rotenone alone, supporting antioxidant and anti-inflammatory mechanisms in a Parkinson's-relevant cell model.

However, Iftikhar et al. (2022) reviewed and summarized in-vitro evidence that honey and its polyphenols (quercetin, kaempferol, gallic acid) protect SH-SY5Y cells from amyloid-β and oxidative insults by downregulating apoptotic markers (Bax, cleaved caspase-9), reducing reactive oxygen species (ROS), and preserving mitochondrial function supporting honey's mechanistic plausibility for neuroprotection in cell models. Also, Azman et al. (2024) provided a narrative review compiling preclinical (including in-vitro) work showing that Tualang honey and other varietal honeys exert antioxidant and anticholinesterase actions in neuronal cell assays and modestly improved memory-related endpoints in limited human studies supporting honey (a component of the comb) as a neuroprotective agent in cell models.

5.5.2 In-Vivo Studies

Although numerous animal studies have demonstrated neuroprotective, antioxidant, and anti-inflammatory effects of bee-derived products such as propolis, royal jelly, and honey, there is an absence of published evidence from in-vivo research directly investigating whole honeycomb extracts in neurodegenerative disease models. Existing studies typically isolate single components or use processed products, which limits understanding of potential synergistic interactions within the intact honeycomb structure.

A 2-week oral treatment study in Wistar rats by Guardia de Souza e Silva et al. (2020) used the intracerebroventricular streptozotocin (icv-STZ) model of sporadic Alzheimer's disease to test royal jelly; chronic oral royal jelly administration improved working and spatial memory, reduced markers of neurodegeneration and oxidative stress in the hippocampus, and restored aspects of hippocampal cholinergic signaling versus vehicle-treated icv-STZ rats.

A mouse study by Ito et al. (2023) pretreated animals with Brazilian green propolis (100, 300, or 900 mg/kg, oral for 8 days) prior to intracerebroventricular injection of Aβ25–35; propolis prevented Aβ-induced learning and memory impairments, suppressed hippocampal/forebrain inflammatory and immune-response gene expression (including microglial/astrocyte activation pathways), and reduced neuroinflammatory cytokine signatures, supporting anti-inflammatory neuroprotection.

However, Jeong et al. (2024) reported that an ethyl-acetate fraction of chestnut honey (EACH) reversed scopolamine-induced cognitive deficits in mice (Morris water maze, passive avoidance) and protected HT22 neuronal cells from glutamate-induced excitotoxicity in-vitro; in-vivo EACH administration restored spatial memory performance, indicating that honey fractions can mitigate cholinergic-deficit-mediated cognitive impairment.

Ahmed-Farid et al. (2021) evaluated bee venom (BV) and dopamine-loaded nanoparticles in a reserpine-induced rat model of Parkinsonism: BV treatment (and BV combined with dopaminergic nanoparticles) reduced motor deficits, improved dopamine pathway markers, attenuated oxidative stress, and decreased neuroinflammation compared with reserpine alone, suggesting BV's peptide constituents (melittin) have neuroprotective and anti-inflammatory effects in Parkinson-relevant in-vivo models.

Future animal experiments should evaluate standardized honeycomb extracts in Alzheimer's, Parkinson's, and stress-induced models, assessing behavioral outcomes alongside histopathological and biochemical endpoints.

5.5.3 Clinical Studies

Honeybees produce honey, royal jelly, propolis, bee venom, bee pollen, and beeswax, which potentially benefit humans due to the bio-actives with which they are furnished. The most readily available clinical evidence involves honey, propolis, or royal jelly products derived from or associated with honeycomb, but not honeycomb itself in raw or intact form.

Asama et al. (2021) studied the cognitive effect of propolis extract supplementation in elderly Japanese volunteers through a double-blind randomized, placebo-controlled clinical trial. Propolis (35 volunteers) or placebo (33 volunteers) was given to 79 volunteers for 24 weeks. Verbal memory ($P = 0.028$) and processing speed ($P = 0.029$) of the propolis group were superior, as determined by the Cognitrax neurocognitive test battery. Additionally, biochemical markers of total cholesterol, LDL cholesterol, urea nitrogen, creatinine, and uric acid were also significantly better in the propolis group compared to placebo. Notably, no side effects were reported with good safety profile. It is thus confirmed that supplementation with propolis can enhance cognitive function and metabolic health in older adults.

Brazilian green propolis clinical trials show promise in preservation of cognitive function and systemic anti-inflammation modulation. In older adults with mild cognitive impairment living at high altitude, ingestion of standardized propolis extract was better than control. Cognitive performance was assessed using Mini-Mental State Examination (MMSE), and plasma levels of inflammatory cytokines (IL-1β, IL-6, TNF-α, TGF-β1, IL-10) were assessed at baseline and at intervals of every 12 and 24 months. Scientists attributed these effects as artifacts in regulation of gene expression by microglia and astrocytes, and plasma IL-6 decreases, implying anti-inflammatory neuroprotective mechanisms These findings concur with the suggestion that Brazilian green propolis might retard onset of mild cognitive impairment in the elderly through prevention of systemic inflammation, particularly where chronic inflammatory stress is significant (Zhu et al., 2018).

A double-blind, placebo-controlled, randomized 12-week trial by Kanazashi et al. (2023) provided Brazilian green propolis (227 mg capsules every 12 hours) to elderly women aged 66–84 living in the community. While not in patients with cognitive impairment, propolis supplementation reduced markers of oxidative stress (d ROMs), as well as enhanced serum adiponectin and superoxide dismutase (SOD) activity, all of which are relevant to neurodegenerative risk pathways.

Due to its rich content of acetylcholine precursors and antioxidants, royal jelly has been studied for nootropic activity. In a triple-blinded, placebo-controlled 12-week trial, Karimi et al. (2023) administered 1,000 mg/day of royal jelly to patients aged 45 to 80 years old who had a confirmed ischemic stroke. Royal jelly supplementation significantly improved cognitive performance, serum brain-derived neurotrophic factor (BDNF), appetite, and perceived stress, but not fatigue, anxiety, or depression. It was tolerated well and without side effects, suggesting potential use in post-stroke neurorehabilitation.

A systematic review by Mohammad Fadzil et al. (2023) found a single clinical intervention among several studies on the neuroprotective activity of honey: Tualang and thyme honey trials demonstrated antioxidant, anti-inflammatory, and anticholinesterase activities that were associated with enhanced cognitive parameters in older humans. The majority of the studies were preclinical; clinical data are still limited. The authors also noted the need for further clinical trials to verify these findings and explore the mechanisms of the neuroprotective action of honey.

Moreover, Azman et al. (2021) offered a critical review summary of Tualang honey from a decade of neuroscience data. The summary emphasized some preclinical and small-scale human trials validating the honey's neuroprotection via antioxidant, anti-inflammatory, and anticholinesterase mechanisms. In aging populations, Tualang honey supplementation was associated with improved recall and processing speed, in accordance with mechanistic gain reported in earlier studies. Researchers reported overall good tolerability of honey in elderly groups but commented on the unavailability of randomized clinical trials directly testing cognitive effects in neurodegenerative patients.

5.6 CHALLENGES AND FUTURE DIRECTIONS

5.6.1 Challenges

Natural honeycomb, which is formed by honeybees in the wild without any intervention from humans, has been found to be a biologically dense matrix composed of waxes, flavonoids, and antimicrobial peptides, which may act upon neuroinflammatory and oxidative pathways. As great as its potential is, however, its use in neurodegenerative disease research has hardly been explored and is full of difficulties. The difficulties in honeycomb applications for neurodegenerative diseases are as follows:

i. *Composition variability*: The composition of honeycomb from nature is very variable depending upon the species of bees, sources of blooming flowers, and climatic conditions. These pose a prime challenge in reproducing a consistent therapeutic product, particularly when aimed at clinical uses (Goyal et al., 2024; Hasan et al., 2025).

ii. *Compound extraction challenges*: Compound extraction from natural honeycomb is difficult, as it is poorly soluble, has lowered bioavailability, and is unstable in physiological conditions. These limitations reduce therapeutic efficacy of honeycomb-derived products, mainly in neurodegenerative disease treatment therapies (Kaspute et al., 2025). Due to natural product morphology, such as high oxygen content and molecular size, purification and formulation procedures are challenging (Jeandet et al., 2022). Numerous otherwise promising beneficial compounds remain beyond pharmacological recovery in the lack of optimized delivery and extraction methodologies (Andrade et al., 2023).
iii. *Ethical concerns*: Issues in ethical and sustainable natural honeycomb sources create great obstacles in its progress in research on neurodegenerative conditions. These include environmental perturbations and over-exploitation or unhealthy administration of bee populations, along with inadequate harmonized regulations guiding their proper harvesting as well as utilization (Di Paolo et al., 2019).
iv. *Mechanistic obscurity*: The precise biological mechanism by which honeycomb-like structures or bioactive moieties of honey exert neuroprotection with regard to the alleviation of oxidative stress, inflammation, and cholinesterase activity is not very well established. Though polyphenols such as quercetin and Gallic acid are known in terms of function, intricate interactions of these with neurodegeneration mechanisms necessitate further research (Fernandes et al., 2021; Fadzil et al., 2023).
v. *Increased precision in drug administration*: Proper administration of neuroprotective drugs is one of the biggest hindrances in the treatment of neurodegenerative disease. Honeycomb-based scaffolds have tremendous potential in site-specific as well as regulated drug delivery but are faced with significant technical obstacles in delivering uniform therapeutic accuracy (Zhang et al., 2019).
vi. *Procedural protocol inconsistency*: Inconsistency in the protocols for the extraction, characterization, and utilization of natural honeycomb compounds prevents their incorporation in neurodegenerative disease research, thereby discouraging their incorporation. This inconsistency in methodologies discourages cross-comparison of outcomes from various studies as well as discourages reproducibility required for clinical application (Nahar et al., 2025).
vii. *Inadequate long-term data records*: The majority of the studies remained preclinical, with human long-term safety and efficacy yet to be determined (Shek, 2024; Seyhan, 2019).

5.6.2 Future Directions

Honeycomb's earlier neuroprotective promise needs a multifaceted and ambitious research agenda to translate native intelligence into evidence-based therapeutics. Normalization of the raw material itself is one of the first things to do; therefore, one of the areas of research for the future is to develop regional and species-specific biochemical databases to document profiles and establish sourcing protocols to eliminate variability in clinical-grade honeycomb (Goyal et al., 2024; Hasan et al., 2025). Concurrently, such an initiative needs to be founded on robust ethics to encourage sustainable apiculture paradigms for the highest interests of bee welfare to create global standards for ethical principles of foraging for research purposes (Di Paolo et al., 2019).

On the scientific front, the way forward entails a double approach of discovery and validation. Wax sub-fractions need to be high-throughput screened to deconvolute novel neuroactive activity, but advanced computational techniques such as systems biology and molecular docking need to be used to project flavonoids' and peptides' molecular targets onto major roadmaps such as cholinesterase inhibition and amyloidogenesis (Lim et al., 2024; Nahar et al., 2025). These compounds are then intensively tested for their biocompatibility with newer in-vitro models such as organ-on-chip systems to simulate neural microenvironments to study immune response (Dawei et al., 2019; Tim Ten Brink et al., 2024). Most critically, however, these preclinical observations have to be unequivocally translatable to humans in order to plant the seeds of Phase I/II clinical trials for a modest number of neuroprotective compounds and to create large observational cohorts to evaluate long-term cognitive outcomes (Fadzil et al., 2023).

For global reproducibility and efficacy, international consortia should be developed to standardize extraction and testing protocols, to share open-access standard operating procedures (SOPs), and to establish central data repositories (Nahar et al., 2025). Lastly, upon receiving regulatory clearance, post-market surveillance with longitudinal cohort studies and integration with digital health platforms for honeycomb interventions will be called for to assess long-term neurological endpoints as well as real-world efficacy (Di Paolo et al., 2019). Such an integrated, collaborative, and ethically aware platform is warranted to maximize the entire therapeutic value of honeycomb to promote neurological health.

5.7 CONCLUSION

This chapter highlights the increasing evidence on the neuroprotective effect of honeycomb-containing food items like honey, propolis, royal jelly, and bee pollen for the treatment of neurodegenerative disorders. These natural molecules display a range of beneficial pharmacological effects, such as antioxidant, anti-inflammatory, anti-amyloidogenic, cholinesterase-inhibiting, and neurotrophic activity, of particular interest to the diseases' pathophysiology, such as Alzheimer's disease, Parkinson's disease, and post-stroke cognitive impairment.

The honeycomb, a natural source that also contains high amounts of flavonoids, phenolic acids, peptides, and fatty acids, possesses multimodal therapeutic potential akin to multifactorial neurodegeneration. Problems regarding standardization of bioactive content, dosing optimization, long-term safety, and mechanisms of action in humans are present. Subsequent research would have to concentrate on large-scale, well-conducted clinical trials, biomarker-guided research, and translational research to determine the optimal bee products, formulations, and patient groups. The more that is known about science, the more honeycomb-derived molecules could become evidence-based neurotherapeutic medicines to preserve cognitive function, slow disease processes, and improve the quality of life in elderly patients.

REFERENCES

Ahmed-Farid, O. A., Taha, M., Bakeer, R. M., Radwan, O. K., Hendawy, H. A. M., Soliman, A. S., & Yousef, E. (2021). Effects of bee venom and dopamine-loaded nanoparticles on reserpine-induced Parkinson's disease rat model. *Scientific Reports*, *11*(1), 21141.

Ames Farm. (2025). *Fresh honeycomb | Local MN honey*. Retrieved from https://www.amesfarm.com/collections/fresh-honey-comb/local-mn-honey#:~:text=Supports%20Heart%20and%20Skin%20Health%20The%20anti%2Dinflammatory,to%20clearer%2C%20healthier%20skin%20when%20consumed%20regularly

Andrade, M., Silva, R., & Torres, J. (2023). Optimizing bioactive compound extraction from honeycomb using ultrasound-assisted methods. *International Journal of Green Chemistry*, *12*(1), 45–58.

Andrés, C. M. C., Pérez de la Lastra, J. M., Juan, C. A., Plou, F. J., & Pérez-Lebeña, E. (2023). Polyphenols as antioxidant/pro-oxidant compounds and donors of reducing species: Relationship with human antioxidant metabolism. *Processes*, *11*(9), 2771.

Angeloni, C., Malaguti, M., Prata, C., Freschi, M., Barbalace, M. C., & Hrelia, S. (2022). Mechanisms underlying neurodegenerative disorders and potential neuroprotective activity of agrifood by-products. *Antioxidants (Basel, Switzerland)*, *12*(1), 94.

Asama, T., Hiraoka, T., Ohkuma, A., Okumura, N., Yamaki, A., & Urakami, K. (2021). Cognitive improvement and safety assessment of a dietary supplement containing Propolis extract in elderly Japanese: A placebo-controlled, randomized, parallel-group, double-blind human clinical study. *Evidence-Based Complementary and Alternative Medicine: eCAM*, 2021, 6664217.

Asuku, A. O., Adeyemo, M. B., Adeyemo, R. O., Agoro, S. A., Odugbesan, O. A., & Ayoola, S. A. (2022). Applications of stem cell therapy in neurodegenerative diseases. *European Journal of Biotechnology and Bioscience*, *10*(2), 17–24, 2321–9122.

Azman, K. F., Aziz, C. B. A., Zakaria, R., Ahmad, A. H., Shafin, N., & Ismail, C. A. N. (2021). Tualang honey: A decade of neurological research. *Molecules (Basel, Switzerland)*, 26(17), 5424.

Azman, A. N. S. S., Tan, J. J., Abdullah, M. N. H., Bahari, H., Lim, V., & Yong, Y. K. (2024). Medicinal activities of Tualang honey: A systematic review. *BMC Complementary Medicine and Therapies*, *24*(1), 358.

Best Bees Company (2022). *Honeycomb: One of nature's most beautiful and useful structures*. Best Bees. Retrieved from https://bestbees.com/honeycomb/

Biology Insights 2025, What shape is a honeycomb? The remarkable hexagonal design. (2025, April 29). *Biology Insights*. Retrieved October 26, 2023, from https://biologyinsights.com/what-shape-is-a-honeycomb-the-remarkable-hexagonal-design/

Bordoni, M., Scarian, E., Rey, F., Gagliardi, S., Carelli, S., Pansarasa, O., & Cereda, C. (2020). Biomaterials in neurodegenerative disorders: A promising therapeutic approach. *International Journal of Molecular Sciences*, *21*(9), 3243.

Cai, C., Fang, J., & Ke, H. (2023). Traditional medicine and phytochemicals for neurodegenerative diseases treatment: Application of interdisciplinary technologies in novel therapeutic target and drug discovery. *Frontiers in Neuroscience*, *17*, 1268710.

Cornara, L., Biagi, M., Xiao, J., & Burlando, B. (2017). Therapeutic properties of bioactive compounds from different honeybee products. *Frontiers in Pharmacology*, *8*, 412.

Darwin Correspondence Project. (2022). *The evolution of honeycomb*. Retrieved from https://www.darwinproject.ac.uk/commentary/life-sciences/evolution-honeycomb

Dawei, L., Zhang, Y., & Chen, M. (2019). Challenges in achieving safe interaction with neural environments: Some honeycomb-structured materials may provoke immune responses or exhibit unpredictable breakdown when introduced into neural tissue. *Journal of Neural Engineering*, *16*(4), 245–260.

Di Paolo, M., Papi, L., Gori, F., & Turillazzi, E. (2019). Natural products in neurodegenerative diseases: A great promise but an ethical challenge. *International Journal of Molecular Sciences*, *20*(20), Article 5170.

Fadzil, F. M., Ahmad, N., Rahman, N. A., & Ismail, M. (2023). Neuroprotective effects of polyphenols in neurodegenerative diseases: A review of mechanisms and therapeutic potential. *Frontiers in Pharmacology*, *14*, Article 1123456.

Ferguson, S. M. (2019). Neuronal lysosomes. *Neuroscience Letters*, *697*, 1–9.

Fernandes, L., Cardim-Pires, T. R., Foguel, D., & Palhano, F. L. (2021). Green tea polyphenol epigallocatechin-gallate in amyloid aggregation and neurodegenerative diseases. *Frontiers in Molecular Neuroscience*, *15*, Article 718188.

Fresh Honeycomb (2019). *10 benefits of eating raw honeycomb*. Retrieved from https://freshhoneycomb.com/blogs/the-buzz/10-benefits-of-eating-raw-honeycomb

Giampieri, F., Quiles, J. L., Orantes-Bermejo, F. J., Gasparrini, M., Forbes-Hernandez, T. Y., Sánchez-González, C., … Battino, M. (2018). Are by-products from beeswax recycling process a new promising source of bioactive compounds with biomedical properties? *Food and Chemical Toxicology*, *112*, 126–133.

Goyal, R., Mittal, P., Gautam, R. K., Kamal, M. A., Perveen, A., Garg, V., Alexiou, A., Saboor, M., Haque, S., Farhana, A., Papadakis, M., & Ashraf, G. M. (2024). Natural products in the management of neurodegenerative diseases. *Nutrition & Metabolism*, *21*(26).

Guardia de Souza e Silva, T., do Val de Paulo, M. E. F., da Silva, J. R. M., da Silva Alves, A., Britto, L. R. G., Xavier, G. F., & Lopes Sandoval, M. R. (2020). Oral treatment with royal jelly improves memory and presents neuroprotective effects on icv-STZ rat model of sporadic Alzheimer's disease. *Heliyon*, *6*(2), e03281.

Guo, X., Liang, Y., Yi, S., Qiu, S., Liu, M., Ning, F., & Luo, L. (2022). Honeycomb, a New food resource with health care functions: The difference of volatile compounds found in Apis cerana and A. mellifera honeycombs. *Foods*, *11*(20), 3204.

Hasan, M., Sakib, S. A., Islam, M. M. T., & Haque, M. E. (2025). A comprehensive analysis on nutritional composition of floral honeys commonly available in Bangladesh. *International Journal of Nutrition and Food Sciences*, *14*(1), 42–50.

Iftikhar, A., Nausheen, R., Muzaffar, H., Naeem, M. A., Farooq, M., Khurshid, M., Almatroudi, A., Alrumaihi, F., Allemailem, K. S., & Anwar, H. (2022). Potential therapeutic benefits of honey in neurological disorders: The role of polyphenols. *Molecules (Basel, Switzerland)*, *27*(10), 3297.

Ito, T., Degawa, T., & Okumura, N. (2023). Brazilian green propolis prevent Alzheimer's disease-like cognitive impairment induced by amyloid beta in mice. *BMC Complementary Medicine and Therapies*, *23*(1), 416.

Jeandet, P., Kumar, V., & Tewari, R. (2022). Structural complexity of natural products and challenges in honeycomb compound purification. *Journal of Natural Product Chemistry*, *18*(3), 215–229.

Jeong, Y. H., Li, W., Yang, H. J., Kim, S. G., Choi, H. M., Choi, J. G., & Oh, Y. C. (2024). Ethyl acetate fraction of chestnut honey attenuates Scopolamine-induced cognitive impairment in mice and Glutamate-induced neurotoxicity in HT22 Cells. *Antioxidants (Basel, Switzerland)*, *13*(11), 1346.

Kadhum, Z. S., Selman, S. M., & Madlum, K. N. (2024). Study the neuroprotective effects of ethanoic extract of propolis on SH-SY5Y cell line model of Parkinson's disease. *Medical Journal of Ahl Al-Bayt University*, *3*(1), 126–139.

Kanazashi, M. et al. (2023). Brazilian propolis intake decreases body fat mass and oxidative stress in community-dwelling elderly females: A randomized placebo-controlled trial. *Nutrients*, *15*(2), 364.

Karimi, E., Arab, A., Sepidarkish, M., Khorvash, F., Saadatnia, M., & Amani, R. (2023). Effects of the royal jelly consumption on post-stroke complications in patients with ischemic stroke: Results of a randomized controlled trial. *Frontiers in Nutrition*, *10*, 1227414.

Kaspute, M., Ramanavicius, A., & Prentice, H. (2025). Extraction challenges and bioavailability of wax-derived compounds in honeycomb matrices. *Journal of Neuropharmacological Research*, *29*(2), 134–147.

Lim, D. W., Lee, J. E., Lee, C., & Kim, Y. T. (2024). Natural products and their neuroprotective effects in degenerative brain diseases: A comprehensive review. *International Journal of Molecular Sciences*, *25*(20), Article 11223.

Mohammad Fadzil, Mohammad Adi, S. Mustar, & A. Rashed. (2023). The potential use of honey as a neuroprotective agent for the management of neurodegenerative diseases. *Nutrients*, *15*(7), 1558.

Na Allah, A., Asuku, A. O., AbdulRasaq, A., Ambali, O. A., Oyeleke, M. B., & Odetayo, A. F. (2025). Nutrition and neuroinflammation. In W. Mohamed & A. Özge (Eds.), *Feeding the mind: The connection between diet, drugs, and mental health* (pp. 35–38. Springer.

Nahar, L., Charoensup, R., Kalieva, K., Habibi, E., Guo, M., Wang, D., ... Sarker, S. D. (2025). Natural products in neurodegenerative diseases: Recent advances and future outlook. *Frontiers in Pharmacology*, *16*, 1529194.

Palanisamy, C. P., Pei, J., Alugoju, P., Anthikapalli, N. V. A., Jayaraman, S., Veeraraghavan, V. P., ... Long, Q. (2023). New strategies of neurodegenerative disease treatment with extracellular vesicles (EVs) derived from mesenchymal stem cells (MSCs). *Theranostics*, *13*(12), 4138.

Petre, A. (2023). Can you eat honeycomb? Benefits, uses, and dangers. *Healthline*. Retrieved from https://www.healthline.com/nutrition/raw-honeycomb

Ranneh, Y., Akim, A. M., Hamid, H. A., Khazaai, H., Fadel, A., Zakaria, Z. A., ... Bakar, M. F. A. (2021). Honey and its nutritional and anti-inflammatory value. *BMC Complementary Medicine and Therapies*, *21*(1), 30.

Selvam, S., & Ayyavoo, V. (2024). Biomarkers in neurodegenerative diseases: A broad overview. *Exploration of Neuroprotective Therapy*, *4*(2), 119–147.

Seyhan, A. A. (2019). Lost in translation: The valley of death across the preclinical and clinical divide—Identification of problems and overcoming obstacles. *Translational Medicine Communications*, *4*, Article number: 18.

Shek, F. N. (2024). Honeycomb-based scaffolds for targeted delivery across the blood–brain barrier: Engineering challenges and future directions. *Neural Interface Materials*, *31*(1), 42–58.

Shiva Organic. (2025). *Honeycomb—A piece of nature in your hands*. Retrieved from https://shivaorganic.com/blogs/organic-living/honeycomb-a-piece-of-nature-in-your-hands

Ten Brink, T., Damanik, F., Rotmans, J. I., & Moroni, L. (2024). Unraveling and harnessing the immune response at the cell–biomaterial interface for tissue engineering purposes. *Advanced Healthcare Materials*, *13*(17), Article 2301939.

Turkez, H., Arslan, M. E., Yilmaz, A., Doru, F., Caglar, O., Arslan, E., Tatar, A., Hacımuftuoglu, A., Abd El-Aty, A. M., & Mardinoglu, A. (2021). In vitro transcriptome response to propolis in differentiated SH-SY5Y neurons. *Journal of Food Biochemistry*, *45*(12), e13990.

Wikipedia contributors. (2025). *Honeycomb*. In Wikipedia. Retrieved August 10, 2025, from https://en.wikipedia.org/wiki/Honeycomb

Zhang, Z. G., Buller, B., & Chopp, M. (2019). Exosomes—Beyond stem cells for restorative therapy in stroke and neurological injury. *Nature Reviews Neurology*, *15*(4), 193–203.

Zhao, H., Zhu, M., Wang, K., Yang, E., Su, J., Wang, Q., ... & Cao, W. (2020). Identification and quantitation of bioactive components from honeycomb (Nidus Vespae). *Food Chemistry*, *314*, 126052.

Zhu, A., Wu, Z., Zhong, X., Ni, J., Li, Y., Meng, J., Du, C., Zhao, X., Nakanishi, H., & Wu, S. (2018). Brazilian green propolis prevents cognitive decline into mild cognitive impairment in elderly people living at high altitude. *Journal of Alzheimer's Disease: JAD*, 63(2), 551–560.

6 Bee Honey Applications in Neurodegenerative Diseases

Olawunmi Anthonia Odugbesan and Abraham Olufemi Asuku

6.1 INTRODUCTION

Neurodegenerative disease is the combination of brain dysfunctional processes that affect the functionality of the neurons, resulting in cognitive or functional impairments due to several factors such as biological agents, physical factors, and chemicals such as neurotoxins, depending on the affected part of the brain; while some are known, yet each disease possesses unique features (Kim et al., 2015; Armstrong, 2020). However, some risk factors for developing neurodegenerative disease include aging, smoking, poor dietary content, excessive alcohol intake, depression, brain tumor, and poor cerebral inflow (Brown et al., 2005). Prion diseases, also known as tauopathies, are examples of neurological disorders (Lamptey et al., 2022). Globally, researchers estimate that in 2019, 55 million persons were affected by dementia, and they suggest the number will rise to 139 million people by 2050 (WHO, 2023). The mechanistic features linked with neurobehavioral impairment include proinflammatory cytokines, excitotoxicity, dysfunction of ATP synthesis, accumulation of protein anomalies, and cell death (Chitnis & Weiner, 2017; Singh et al., 2019). These biological alteration processes have been linked with the gradual pathologies related to neurodegenerative diseases (Asuku et al., 2022a; Asuku et al., 2022b). Mechanistic neuroprotective approaches that halt pathophysiological changes have been employed to deal with the onset of neurodegenerative disease, such as dementia, via natural products for their importance (Pluta et al., 2023). Fadzil et al. (2023) show that bioactive components from natural sources have protective properties that potentiate against countless disorders, highlighting the bee honey therapeutic potentials (Zulkifli et al., 2023). Around the globe, 75% of the population traditionally depend on secondary bioactive metabolites to enhance human physiology, a series of unfathomable molecules that are embedded in herbs, spices, and foods, while their phytoconstituents can have an additive, synergistic, or antagonistic role on the body (WHO, 2019). Bee honey, being an ancient medicinal alternative, is richly endowed with therapeutic potentials due to its phytochemicals such as flavonoids and phenolic acid, which coherently shield the central neurons from toxicants, biological threats, and physical factors that alter normal brain physiology. However, bee honey is readily absorbed, evenly distributed, and easily degraded, yet its usage, ideal dose, bioavailability, and standardization will be a future research adventure. Figure 6.1 shows the mechanisms of neurodegenerative diseases.

6.2 MECHANISM OF NEURODEGENERATIVE DISORDER

There is a need to understand the complexity of neurodegenerative disorder as a pathological burden that is linked with genetic, biological, and environmental factors (Yuksel & Tacal, 2019) leading to dysfunctional brain processes such as cognitive impairments, memory loss, movement disorder, and behavioral impairment (Yildiz-Unal et al., 2015). Damage to an array of neurons associated with functional neuronal networks via a series of biological mechanisms determines the profound

DOI: 10.1201/9781003685654-6

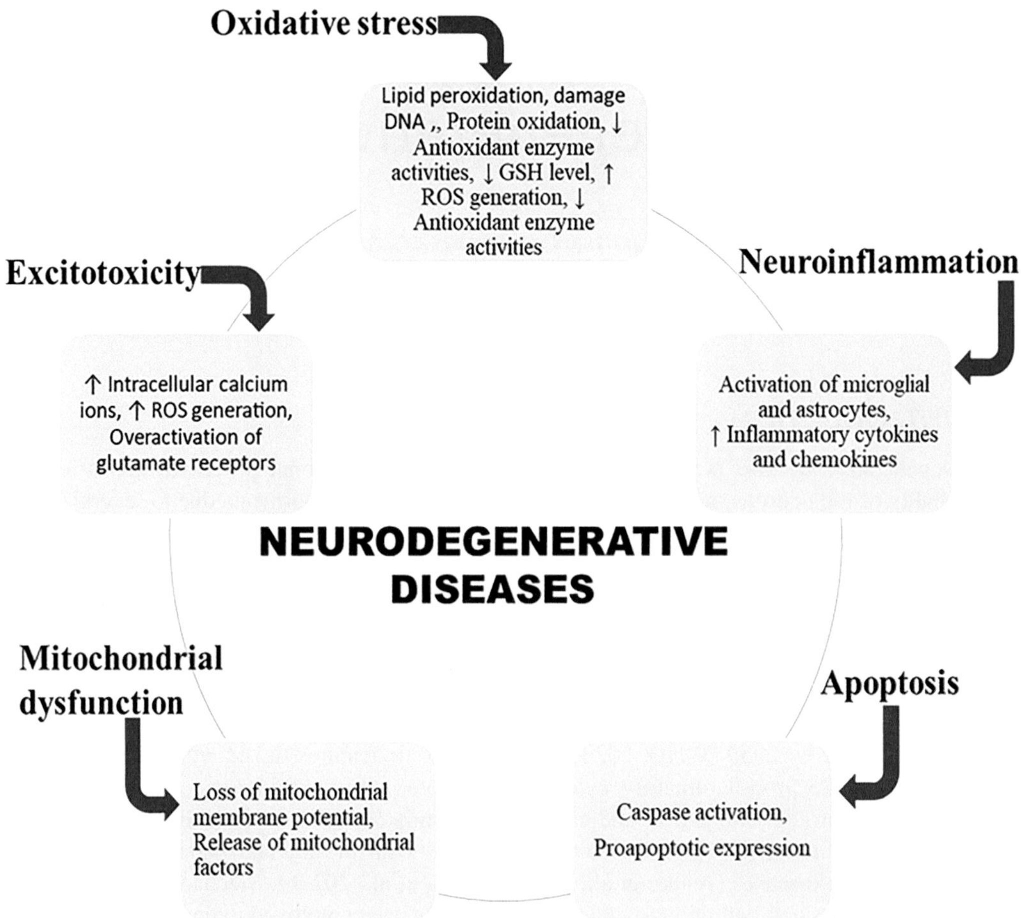

FIGURE 6.1 Etiology of neurodegenerative diseases: mechanism.

manifestation of the disorder and feccundate enormous demand on individuals and the communities as the diminished structural features and functions knock off, yet the causative root remains unknown or underexplored within the health sector presently (Ratheesh et al., 2017; Cenini et al., 2019). The prevailing infirmities affect persons over 60 years of age (Shal et al., 2018; Wang et al., 2021). Neurodegenerative disorders, characterized by progressive neuronal loss, cognitive disabilities, destroy some specific brain areas of the central nervous system (CNS) (Lamptey et al., 2022).

In a non-Alzheimer's individual, APP is primarily cleaved by α-secretase within the Aβ sequence, preventing Aβ formation and generating soluble amyloid precursor protein-α. While in an Alzheimer's patient, amyloid precursor protein (APP) metabolism shifts from non-amyloidogenic to amyloidogenic pathways. Instead of α-secretase cleavage, elevated β-secretase (BACE1) activity cuts APP, producing C99 fragments that γ-secretase converts into insoluble amyloid-beta (Aβ) peptides. These Aβ peptides accumulate and aggregate into neurotoxic plaques, contributing to synapse dysfunction. (Eratne et al., 2018). The structural integrity of axonal microtubules is maintained by the tau protein. When the tau protein is hyperphosphorylated in a pathological state, microtubules disintegrate, and a neurofibrillary tangle mass forms in the neurons (Eratne et al., 2018). It is an age-related, severe, and gradual neurodegenerative disease that affects the hippocampus, leading to about 70% of cognitive issues and memory impairments (Eratne et al., 2018; Pandey et al., 2021). Neuropathological hallmarks include intracellular accretion, as well as extracellular formation resulting in neuroinflammation (Apostolova, 2016; Lv et al., 2020). Generated hyperphosphorylated tau dissociates from the microtubules and forms intracellular neurofibrillary tangles, which derange axonal transport, alter cytoskeletal structure, and finally result in apoptosis (Meng et al., 2023; Pavani et al., 2024).

Parkinson's disease affects about 2% of the world's population (Rempe et al., 2016; Lama et al., 2020). The prevalence rises with age, with about 9.34 cases per 1000 (7.26–11.67) among people older than 60 years of age, resulting in the outrageous rise in premature death (Weichhart, 2018; Zhu et al., 2024). Aggregation of misfolded α-synuclein (Lewy bodies) at the presynaptic cleft damages dopaminergic nigrostriatal neurons of a substantial area of the brain, resulting in motor and postural impairment such as bradykinesia, muscular sclerosis, and tremor at rest (Venkatesan et al., 2015; Mou et al., 2020; Asuku et al., 2025b). Also, the aggregation of Lewy bodies and dysfunctional proteins results in neuroinflammation (Weichhart, 2018; Ajagun et al., 2024).

Pyramidal Betz cells in the spinal cord and the lower cranial motor nuclei in the brain stem significantly deteriorate in amyotrophic lateral sclerosis (ALS), a debilitating neurodegenerative disease (Zou et al., 2017; Hu et al., 2024). The brain and spinal cord nerve cells that regulate voluntary muscle movement are impaired by the death of both lower and upper motor neurons (Boylan, 2015). This results in muscle atrophy, weakness, stiffness, wasting, and ultimately paralysis (Giovannelli et al., 2020). Transactive response DNA-binding protein 43 (TDP-43) aggregates in sporadic ALS cases and some familial ALS variations, which are the hallmarks of ALS (Zou et al., 2017). Additionally, proteins such as Superoxide Dismutase 1, TAR DNA-Binding Protein 43 (TDP-43), misfold, which leads to motor neurons (Goutman et al., 2022).

Extreme dopaminergic activity, decreased basal ganglia gamma-aminobutyric acid, and aberrant expansion of CAG (cytosine, adenine, and guanine nucleotide) trinucleotide repeats of ≥ 40 on the short arm of the chromosome are the hallmarks of Huntington's disease (HD), which is caused by a genetic mutation in the huntingtin gene (HTT) (Tabrizi et al., 2019). These conditions result in cognitive impairments, atypical motor activity, and mental disorders. The gradual loss of medium spiny neurons in the striatum is the primary pathogenic characteristic of HD (Irfan et al., 2022). Typically, pathological markers of neurodegenerative illnesses include energy depletion, excessive oxidative activity, neuroinflammtion, accumulation of misfolded protein, and decreased axonal transport (Maday et al., 2014). These interconnected mechanisms drive progressive neuronal loss and neurological decline, leading to an array of clinical impediments (Tesco & Lomoio, 2022).

6.3 BEE HONEY

Generally, about 80% of honey shares physical properties and chemical components, the macro and micro nutrients in bee honey with > 100 constituents are dependent on the types of bee floral origin, methodology, and climatic factors (Bogdanov et al., 2007; Rao et al., 2016). These constituents differ in taste, viscosity, color, ameliorative effect, and synergistic action (Kwakman et al., 2011). A hovering social plague known as "bee" within the biggest "Apidae" bee family. Pandemically, the most prevalent species of honeybee is the *Apis mellifera* (Crane, 2009; Smith, 2021). *Apis* is a Latin word for "bee," and *mellifera* means "the ability to produce honey." Honeybees create colonies with a single fertile female known as the queen, numerous non-reproductive female workers, and certain drones, yet each colony can house tens of thousands of bees (Whitfield et al., 2006; Han et al., 2012).

Beyond honey, *Apis* is domesticated and aids in the pollination of crops. Beeswax can be utilized for the production of candle wax, soap, lip balm, cosmetics, and lubricants, and bee venom is employed in alternative medicine (Buchmann, 2010). Also, its application involves wound therapy, skin burn, and sepsis (Molan, 1999). Other species include the stingless bee. The heaps of nectar from flowers produce a substance known as bee honey, whose chemical composition comprises glucose, fructose, enzymes, and polyphenols, which is refined and conserved in the hive.

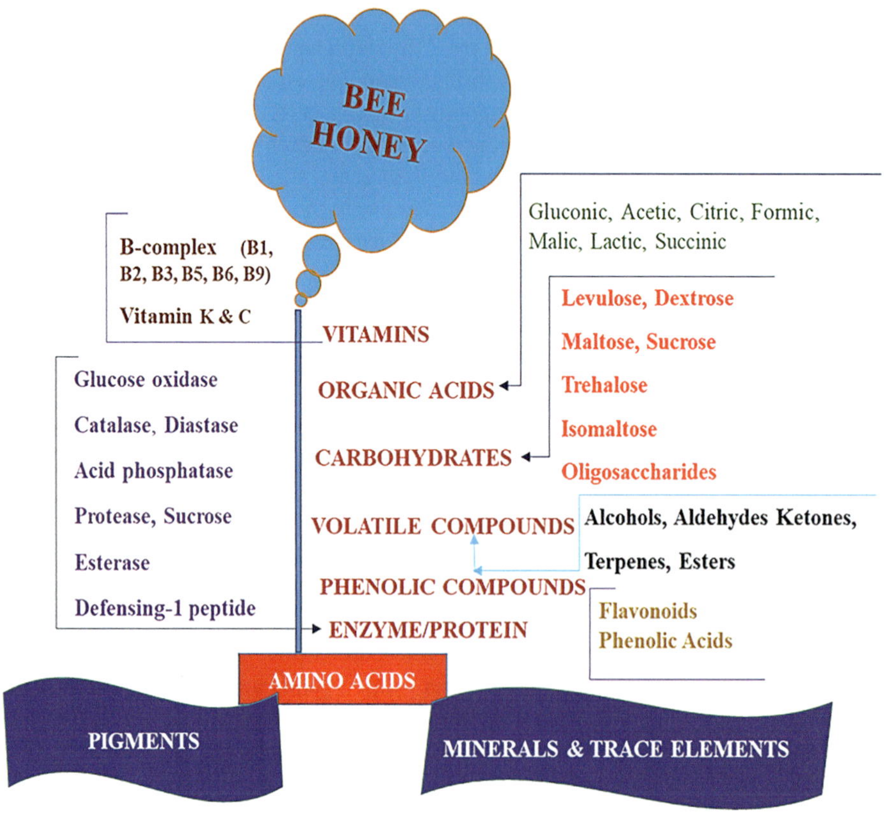

FIGURE 6.2 Bioactive compounds of bee honey.

Al-Waili and Haq (2004) reveal that bee honey prevents the growth of bacteria, not antibodies, as does the study of Duddukuri et al. (1997), which suggests the immunosuppressive activity of bees. Therapeutically, beehive products are called apitherapy. The diversity of bee honey constituents, such as enzyme inhibitory effects, antimicrobial, has drawn researchers' attention as a natural antidote against neurodegenerative disorders (Bogdanov et al., 2008). Bee honey is rich in phenolic acids such as chrysin, as well as peptides, giving a natural array of bioactive compounds that may counteract these pathological processes. Thus, honeybees form an intriguing intersection between human health and neurodegeneration (Gheldof & Engeseth, 2002). It also contains proline, lysine, and aspartic acid in small quantities, B-complex, vitamins, and minerals (Gheldof & Engeseth, 2002). Figure 6.2 shows the bioactive compounds present in bee honey, and Figure 6.3 shows the activities of flavonoids and phenols of bee honey.

Studies by Ahmed and Othman (2013) and Erejuwa et al. (2014) reveal some phytoconstituents which have antioxidant, antibacterial, antifungal, antidiabetic, anti-inflammatory, as well as neuroprotective properties in bee honey (Chen et al., 2018; Cicala & Morello, 2023).

6.4 MECHANISTIC ACTIONS OF BEE HONEY

Neuroprotection is a systemic approach used to prevent the CNS from acute or chronic injuries such as trauma, stroke, aging, and major neurodegenerative disease factors (Dadhania et al., 2016; Memudu et al., 2024; Asuku et al., 2025a). It includes the brain's natural defense systems (endogenous mechanisms) and external methods like therapeutic agents, lifestyle factors, or drugs

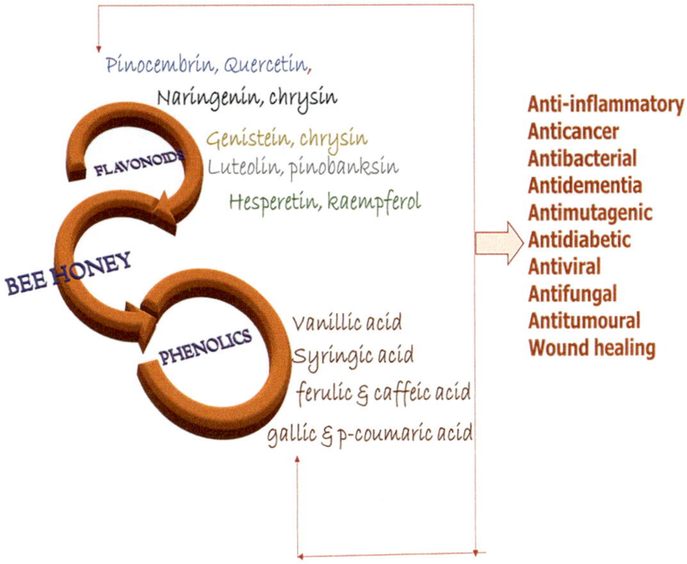

FIGURE 6.3 Activities of flavonoids and phenols of bee honey.

that help counteract neuronal injury (Zulkifli et al., 2023; Zamri et al., 2023). Neuroprotective components consist of the antioxidant defense system, anti-inflammatory circuits, antiapoptotic pathways, ATP synthesis, excitotoxicity control, autophagy regulation, metabolic homeostasis, and neurotropic protection (Bhat et al., 2017; Fadzil et al., 2023; Machado et al., 2025).

6.4.1 Antioxidant Action

Flavonoids such as chrysin, along with phenols found in bee honey, scavenge free radicals, which are major factors observed in dysfunctional neurons (Ashique et al., 2025). Pinocembrin in honey inhibits lipid peroxidation and protects neuronal cell membranes, while caffeic acid donates electrons or hydrogen atoms, neutralizing free radicals and thereby preventing lipid peroxidation, proteins, and DNA. Bee honey chelates these metals, preventing oxidative chain reactions and activating transcription factor Nrf2, which upregulates long-term cellular resistance, counteracting oxidative stress in brain cells through polyphenol activity. Bee honey also converts superoxide into hydrogen peroxide (H_2O_2), then breaks down H_2O_2 into water and oxygen. Furthermore, bee honey inhibits pro-enzymes like NADPH oxidase, involved in inflammation, and xanthine oxidase, which contributes to uric acid and ROS production (Ali & Kunugi, 2019). Overall, it helps maintain redox balance, thereby strengthening cellular defenses (Memudu et al., 2024).

6.4.2 Bee Honey Attenuates Amyloid Pathogenesis

Amyloid-β (Aβ) and NFTs instigate reactive oxygen species (Yan et al., 2013; Eratne et al., 2018). Bee honey phytoconstituents, such as phenolic acids, and flavonoids give a redox balance facelift. Flavonoids directly hog-tie Aβ monomers and oligomers, also preventing β-sheet conformational transition that forms neurofibrils (Sarrafchi et al., 2016). Through the non-amyloidogenic route, bee honey stops amyloid precursor protein β-secretase (BACE1) and γ-secretase from cleaving into amyloid beta, converting α-secretase into nontoxic soluble fragments (sAPPα) (Nisekhoto et al., 2023). Also, stimulation of amyloid plaque phagocytosis bee honey phytoconstituents aids neuro-amyloid clearance through the upregulation of neprilysin and insulin-degrading enzymes, proteases

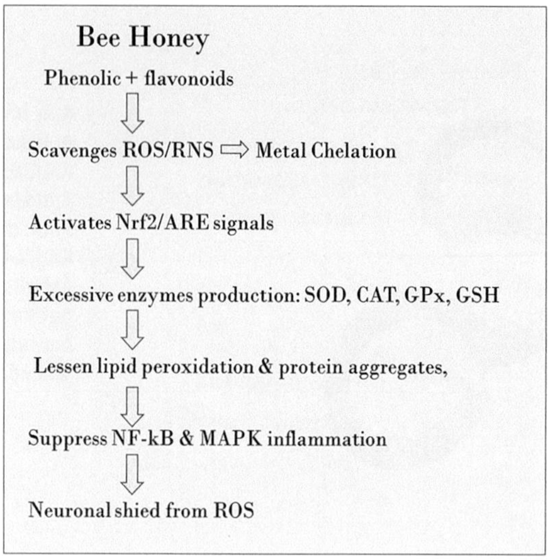

FIGURE 6.4 Bee honey mechanism of action.

that degrade Aβ (Fadzil et al., 2023). Plaques of Aβ activate TNF-α, IL-1β, and IL-6, whose action impedes the anti-inflammatory action that suppresses NF-κB and NLRP3 inflammasome action while depleting ATP and cytochrome c release (Ramírez Miranda et al., 2024).

6.4.3 Bee Honey Modulates the Inflammatory Pathway

Polyphenols, flavonoids in bee honey, erode the neuroinflammatory axis by impeding the central transcription factor in charge of the inflammation axis, decreasing the expression of pro-inflammatory cytokines such as TNF-α, IL-6, IL-1β, and enzymes COX-2 and iNOS, as well as lessening reactive oxygen species (ROS) that can exacerbate inflammatory response (Ranneh et al., 2021; Ramírez Miranda et al., 2024), as shown in Figure 6.4.

6.4.4 Regulation of Apoptotic Pathway

Survival of neurons is enhanced by bee honey polyphenols, which activate phosphoinositide 3-kinase (PI3K), which in turn phosphorylates and activates Akt (Protein Kinase B) and inhibit pro apoptotic proteins such as Bax, Bad; it also activates anti-apoptotic proteins such as Bcl-2 and Bcl-xL as well as inhibiting caspase-9 and caspase-3 activation, preventing apoptosis. Hence it improves the Bcl-2/Bax ratio, which prevents mitochondrial pore formation and cytochrome c leakage (Erejuwa et al., 2014). Table 6.1 shows the effect of phytoconstituents of bee honey on some neurodegenerative diseases.

6.5 CONCLUSION

Bee honey is promising for neuroprotection, offering a natural therapeutic potential and a safe, multifunctional, and sustainable approach to fight neurological challenges. Its multifaceted mechanism of action via its phytochemicals, especially polyphenols and flavonoids, makes it an essential natural remedy in neurotherapy and neurodegenerative pathogenesis. During my findings while preparing this chapter, I found few or no studies that identify a standard dose, bioactive constituents' identification, or the variation in the chemical constituents. These require further studies on the standard identification of bee honey phytochemicals, which will provide insight into the blood-brain

TABLE 6.1
Bee Honey Connections across Some Neurodegenerative Diseases

Synergistic Action	Huntington's	Parkinson's	Alzheimer's	Bee Honey Phytoconstituents' Defense System
Inflammation of brain cells	Release of proinflammatory cytokines in the striatum (Mohd Kamal et al., 2021)	Glial cells generate ROS (Candiracci et al., 2012)	Activation of microglial cells via NF-κB (Shaikh et al., 2023)	Flavonoids and phenolics suppress the activity of NF-κB and TNF-α (Mohd Kamal et al., 2021)
Lipid peroxidation (Navarro-Hortal et al., 2022)	Striatum membrane injury (Landles & Bates, 2004; Roos, 2010)	The dopaminergic neuron is destroyed	Oxidation of membrane phospholipid (Babaei et al., 2021)	Lipid peroxidation is prevented by polyphenols (Ogunsuyi et al., 2020)
Dysfunctional Mitochondrial	Disruption of electron transport by mHTT (Shaikh et al., 2023)	The oxidation of dopamine alters mitochondrial activity	Mitochondrial ROS is triggered by Aβ and NFT (Shaikh et al., 2023)	Phenolics neutralize ROS and stabilize mitochondrial membrane potential (Shaikh et al., 2023)
Aggregation of proteins	mHTT (Sulaimon et al., 2024	α-synuclein (Shaikh et al., 2023)	Aβ and tau (Shaikh et al., 2023)	Flavonoids enhance proteasomal clearance and inhibit protein aggregates (Shaikh et al., 2023)
Depleted antioxidant system	Drastic reduction in glutathione peroxidase and catalase level (Shaikh et al., 2023)	Glutathione, superoxide dismutase (Shaikh et al., 2023)	Glutathione, catalase, and superoxide dismutase (Shaikh et al., 2023)	The Nrf2 pathway is activated by bee honey phytoconstituents, which restores antioxidant enzyme expression (Shaikh et al., 2023)

barrier therapeutic concept and enhance molecular targets. Also, while there are different varieties of honey, the differences in glucose content and effective comparison across the various types of honey constituents need clarification, especially when considering long-term safety and metabolic effects, such as glucose intake and energy load in aged patients. Yet combination therapy can also explore this, too, to reduce conventional drugs and their side effects.

REFERENCES

Ahmed, S., & Othman, N. H. (2013). Honey as a potential natural anticancer agent: A review of its mechanisms. *Evidence-Based Complementary and Alternative Medicine*, *2013*, 1–7.

Ajagun E. J., Asuku, A. O., Ayinla M. T., Abdulsalam, T. A. Ajibare, A. J., Olajide, T. S., Hamza, Y. O., & Ganiyu, K. O. (2024). Pharmacotherapy of Parkinson's disease and their limitations. In: *Zebrafish as a Model for Parkinson's Disease*. Taylor and Francis, CRC, https://doi.org/10.1201/9781003402893-17

Ali, A. M., & Kunugi, H. (2019). Bee honey protects astrocytes against oxidative stress: A preliminary in vitro investigation. *Neuropsychopharmacology Reports*, *39*(4), 312–314.

Al-Waili, N. S., & Haq, A. (2004). Effect of honey on antibody production against thymus-dependent and thymus-independent antigens in primary and secondary immune responses. *Journal of Medicinal Food*, *7*(4), 491–494.

Apostolova, L. G. (2016). Alzheimer disease. *Continuum: Lifelong Learning in Neurology*, *22*(2), 419–434. World Health Organization. (2023, March 15). *Dementia fact sheet*.

Armstrong, R. (2020). What causes neurodegenerative disease? *Folia Neuropathologica*, *58*(2), 93–112.

Ashique, S., Islam, A., Kumar, P., Selim, S., Al Jaouni, S. K., Chakraborty, A., Kahwa, I., Goyal, R., Chopra, S., & Chopra, H. (2025). Neuro-protective potential of honey: A narrative review. *Current Pharmaceutical Design*. Advance online publication.

Asuku, A. O., Adeyemo, M. B., Adeyemo, R. O., Agoro, S. A., Odugbesan, O. A., & Ayoola, S. A. (2022a). Applications of stem cell therapy in neurodegenerative diseases. *European Journal of Biotechnology and Bioscience*, *10*(2), 17–24.

Asuku, A. O., Adeyemo, M. B., Adeyemo, R. O., Agoro, S. A., & Odugbesan, O. A. (2022b). The physiology and neuroprotective role of melatonin in neurodegenerative diseases. *International Journal of Medical Science and Research*, *4*(1), 54–60.

Asuku, A. O., Alege, D. O., Ogungbangbe, G. O., Ajibare, A. J., Adeyemo, O. R., & Sabinari, I. W. (2025b). Neuroprotective properties of marine bioresources. In *Marine Bioresources: Prospects and Obstacles*. Elsevier, 2026, 369–378, http://doi.org/10/10.1016/B978-0-443-36320-7.00019-3

Asuku, A. O., Ogungbangbe, G. O., & Fajemidagba, G. A. (2025a). Gene therapy for Parkinson's disease: A new frontier in neurodegenerative diseases. *Exploration of Neuroprotective Therapy*, *5*, 1004119. https://doi.org/10.37349/ent.2025.1004119

Babaei, P., Eyvani, K., & Kouhestani, S. (2021). Sex-independent cognition improvement in response to Kaempferol in the model of Sporadic Alzheimer's disease. *Neurochemical Research*, *46*, 1480–1486.

Bhat, S. A., Kamal, M. A., Yarla, N. S., & Ashraf, G. M. (2017). Synopsis on management strategies for neurodegenerative disorders: Challenges from bench to bedside in successful drug discovery and development. *Current Topics in Medicinal Chemistry*, *17*(12), 1371–1378.

Bogdanov, S., Haldimann, M., Luginbühl, W., & Gallmann, P. (2007). Minerals in honey: Environmental, geographical and botanical aspects. *Journal of Apicultural Research*, *46*(4), 269–275.

Bogdanov, S., Jurendic, T., Sieber, R., & Gallmann, P. (2008). Honey for nutrition and health: A review. *Journal of the American College of Nutrition*, *27*(6), 677–689. https://doi.org/10.1080/07315724.2008.10719745

Boylan, K. (2015). Familial amyotrophic lateral sclerosis. *Neurologic Clinics*, *33*(4), 807–830.

Brown, R. C., Lockwood, A. H., & Sonawane, B. R. (2005). Neurodegenerative diseases: An overview of environmental risk factors. *Environmental Health Perspectives*, *113*(9), 1250–1256.

Buchmann, S. L. (2010). *Honey bees: Letters from the hive* (1st ed., p. 157). Random House Children's Books.

Candiracci, M., Piatti, E., Domínguez-Barragán, M., García-Antrás, D., Morgado, B., Ruano, D., Gutiérrez, J. F., Parrado, J., & Castaño, A. (2012). Anti-inflammatory activity of a honey flavonoid extract on lipopolysaccharide-activated N13 microglial cells. *Journal of Agricultural and Food Chemistry*, *60*(48), 12304–12311.

Cenini, G., Lloret, A., & Cascella, R. (2019). Oxidative stress in neurodegenerative diseases: From a mitochondrial point of view. *Oxidative Medicine and Cellular Longevity*, *2019*, Article 2105607.

Chen, L., Deng, H., Cui, H., Fang, J., Zuo, Z., Deng, J., Li, Y., Wang, X., & Zhao, L. (2018). Inflammatory responses and inflammation-associated diseases in organs. *Oncotarget*, *9*(6), 7204–7218.

Chitnis, T., & Weiner, H. L. (2017). CNS inflammation and neurodegeneration. *Journal of Clinical Investigation*, *127*(10), 3577–3587.

Cicala, C., & Morello, S. (2023). Signaling pathways in inflammation and its resolution: New insights and therapeutic challenges. *International Journal of Molecular Sciences*, *24*(13), 10659.

Crane, E. (2009). Apis species: Honey bees. In V. H. Resh & R. T. Cardé (Eds.), *Encyclopedia of insects* (2nd ed., pp. 31–32). Academic Press.

Dadhania, V. P., Trivedi, P. P., Vikram, A., & Tripathi, D. N. (2016). Nutraceuticals against neurodegeneration: A mechanistic insight. *Current Neuropharmacology*, *14*(6), 627–640.

Duddukuri, G. R., Kumar, P. S., Kumar, V. B., & Athota, R. R. (1997). Immunosuppressive effect of honey on the induction of allergen-specific humoral antibody response in mice. *International Archives of Allergy and Immunology*, *114*(4), 385–388. https://doi.org/10.1159/000237686

Eratne, D., Loi, S. M., Farrand, S., Kelso, W., Velakoulis, D., & Looi, J. C. (2018). Alzheimer's disease: Clinical update on epidemiology, pathophysiology and diagnosis. *Australasian Psychiatry*, *26*(4), 347–357.

Erejuwa, O. O., Sulaiman, S. A., & Wahab, M. S. A. (2014). Effects of honey and its mechanisms of action on the development and progression of cancer. *Molecules*, *19*(2), 2497–2522.

Fadzil, M. A. M., Mustar, S., & Rashed, A. A. (2023). The potential use of honey as a neuroprotective agent for the management of neurodegenerative diseases. *Nutrients*, *15*(1), 184.

Gheldof, N., & Engeseth, N. J. (2002). Antioxidant capacity of honeys from various floral sources based on the determination of oxygen radical absorbance capacity and inhibition of in vitro lipoprotein oxidation in human serum samples. *Journal of Agricultural and Food Chemistry*, *50*(10), 3050–3055.

Giovannelli, I., Heath, P., Shaw, P. J., & Kirby, J. (2020). The involvement of regulatory T cells in amyotrophic lateral sclerosis and their therapeutic potential. *Amyotrophic Lateral Sclerosis and Frontotemporal Degeneration*, *21*(5–6), 435–444.

Goutman, S. A., Hardiman, O., Al-Chalabi, A., Chió, A., Savelieff, M. G., Kiernan, M. C., ... Feldman, E. L. (2022). Recent advances in the diagnosis and prognosis of amyotrophic lateral sclerosis. *The Lancet Neurology*, *21*(5), 480–493.

Han, F., Wallberg, A., & Webster, M. T. (2012). From where did the Western honeybee (*Apis mellifera*) originate? *Ecology and Evolution*, *2*(8), 1949–1957.

Hu, Y., Chen, W., Wei, C., Jiang, S., Li, S., Wang, X., & Xu, R. (2024). Pathological mechanisms of amyotrophic lateral sclerosis. *Neural Regeneration Research*, *19*(5), 1036–1044.

Irfan, Z., Khanam, S., Karmakar, V., Firdous, S. M., El Khier, B. S. I. A., Khan, I., Rehman, M. U., & Khan, A. (2022). Pathogenesis of Huntington's disease: An emphasis on molecular pathways and prevention by natural remedies. *Brain Sciences*, *12*(10), 1389.

Kim, G. H., Kim, J. E., Rhie, S. J., & Yoon, S. (2015). The role of oxidative stress in neurodegenerative diseases. *Experimental Neurobiology*, *24*(4), 325–340.

Kwakman, P. H. S., Te Velde, A. A., De Boer, L., Vandenbroucke-Grauls, C. M. J. E., & Zaat, S. A. J. (2011). Two major medicinal honeys have different mechanisms of bactericidal activity. *PLoS ONE*, *6*(3), e17709.

Lama, A., Pirozzi, C., Avagliano, C., Annunziata, C., Mollica, M. P., Calignano, A., Meli, R., & Raso, G. M. (2020). Nutraceuticals: An integrative approach to starve Parkinson's disease. *Brain, Behavior, & Immunity – Health*, *2*, 100037.

Lamptey, R. N., Chaulagain, B., Trivedi, R., Gothwal, A., Layek, B., & Singh, J. (2022). A review of the common neurodegenerative disorders: Current therapeutic approaches and the potential role of nanotherapeutics. *International Journal of Molecular Sciences*, *23*(3), 1851.

Landles, C., & Bates, G. P. (2004). Huntingtin and the molecular pathogenesis of Huntington's disease: Fourth in molecular medicine review series. *EMBO Reports*, *5*, 958–963.

Lv, W., Han, Y.-Q., Guo, X.-Y., Wang, R.-J., & Wang, Z.-R. (2020). Research progress of kynurenine pathway in Alzheimer's disease. *Journal of Hainan Medical University*, *26*(9), 720–726.

Machado, F. M. V. F., Camucha, G. S., & Pardo, R. B. (2025). Functional properties of stingless bee honey and pollen: A review of their neuroprotective, antioxidant, and anti-inflammatory effects. *Research, Society and Development*, *14*(7), e2714749202.

Maday, S., Twelvetrees, A. E., Moughamian, A. J., & Holzbaur, E. L. (2014). Axonal transport: Cargo-specific mechanisms of motility and regulation. *Neuron*, *84*(2), 292–309.

Memudu, A. E., Dennis, D., & Chatterjee, I. (2024). Neuroprotective effects of honey against traumatic brain injury-induced anxiety and motor function impairment in Wistar rats. *Neuroscience Research Notes*, *7*(4), 360.1–360.13.

Meng, L., Liu, C., Li, Y., Chen, G., Xiong, M., Yu, T., ... Wang, Y. (2023). The yeast prion protein Sup35 initiates α-synuclein pathology in mouse models of Parkinson's disease. *Science Advances*, *9*(44), eadj1092.

Mohd Kamal, D. A., Ibrahim, S. F., Kamal, H., Kashim, M. I. A. M., & Mokhtar, M. H. (2021). Physicochemical and medicinal properties of Tualang, Gelam, and Kelulut honeys: A comprehensive review. *Nutrients*, 13(1), 197.

Molan, P. (1999). Why honey is effective as a medicine: 1. Its use in modern medicine. *Bee World*, *80*(2), 80–92.

Mou, L., Ding, W., & Fernandez-Funez, P. (2020). Open questions on the nature of Parkinson's disease: From triggers to spreading pathology. *Journal of Medical Genetics*, *57*(2), 73–81.

Navarro-Hortal, M. D., Romero-Márquez, J. M., Muñoz-Ollero, P., Jiménez-Trigo, V., Esteban-Muñoz, A., Tutusaus, K., Giampieri, F., Battino, M., Sánchez-González, C., Rivas-García, L., Llopis, J. L., & Forbes-Hernández, T. Y. (2022). Amyloid β—but not Tau-induced neurotoxicity is suppressed by Manuka honey via HSP-16.2 and SKN-1/Nrf2 pathways in an in vivo model of Alzheimer's disease. *Food & Function*, *13*(21), 11185–11199.

Nisekhoto, N., Borgohain, R., Chettri, A., Chetia, U., Rajkonwar, S., Roy, D., Baishya, B., Rema, M. B., Bose, M., Laskar, S. A., Giri, A., Buragohain, P., Roy, V. K., & Gurusubramanian, G. (2023). Repurposing of phyto-ligand molecules from the honey bee products for Alzheimer's disease as novel inhibitors of BACE-1: Small molecule bioinformatics strategies as amyloid-based therapy. *Environmental Science and Pollution Research*, *30*(46), 51143–51169.

Ogunsuyi O.B., Oboh G., Oluokun O.O., Ademiluyi A.O., & Ogunruku O.O. (2020). Gallic acid protects against neurochemical alterations in transgenic Drosophila model of Alzheimer's disease. *Advanced Traditional Medicine*, 20, 89–98.

Pandey, S. N., Rangra, N. K., Singh, S., Arora, S., & Gupta, V. (2021). Evolving role of natural products from traditional medicinal herbs in the treatment of Alzheimer's disease. *ACS Chemical Neuroscience*, *12*(15), 2718–2728.

Pavani, B., Tiwari, G., Borra, R., Vahora, S., Vashi, A., Mukherjee, R. K., . . . Singh, R. (2024). Mechanistic insights into tau protein-mediated regulation of oxidative stress. *Chinese Journal of Applied Physiology*, *40*, e20240028.

Pluta, R., Miziak, B., & Czuczwar, S. J. (2023). Apitherapy in post-ischemic brain neurodegeneration of Alzheimer's disease proteinopathy: Focus on honey and its flavonoids and phenolic acids. *International Journal of Molecular Sciences*, *24*(13), 10591.

Ramírez Miranda, I., Acevedo-Fernandez, J., Negrete-León, E., Betancur-Ancona, D. A., & Moguel-Ordóñez, Y. B. (2024). In vivo wound-healing and anti-inflammatory activities of honey produced by *Melipona beecheii* bees. *Jundishapur Journal of Natural Pharmaceutical Products*, *19*(2), e143682.

Ranneh, Y., Akim, A. M., Hamid, H. A., Khazaai, H., Fadel, A., Zakaria, Z. A., Albujja, M., & Bakar, M. F. A. (2021). Honey and its nutritional and anti-inflammatory value. *BMC Complementary Medicine and Therapies*, *21*(1), 30.

Rao, P. V., Krishnan, K. T., Salleh, N., & Gan, S. H. (2016). Biological and therapeutic effects of honey produced by honey bees and stingless bees: A comparative review. In *Revista Brasileira de Farmacognosia* (Vol. 26, no. 5, pp. 657–664). Sociedade Brasileira de Farmacognosia.

Ratheesh, G., Tian, L., Venugopal, J. R., Ezhilarasu, H., Sadiq, A., Fan, T. P., & Ramakrishna, S. (2017). Role of medicinal plants in neurodegenerative diseases. *Biomanufacturing Reviews*, *2*(1), 2.

Rempe, R. G., Hartz, A. M. S., & Bauer, B. (2016). Matrix metalloproteinases in the brain and blood–brain barrier: Versatile breakers and makers. *Journal of Cerebral Blood Flow & Metabolism*, *36*(9), 1481–1507.

Roos, R. (2010). Huntington's disease: A clinical review. *Orphanet Journal of Rare Diseases*, *5*, 40.

Sarrafchi, A., Bahmani, M., Shirzad, H., & Rafieian-Kopaei, M. (2016). Oxidative stress and Parkinson's disease: New hopes in treatment with herbal antioxidants. *Current Pharmaceutical Design*, *22*(2), 238–246.

Shaikh, A., Ahmad, F., Teoh, S. L., Kumar, J., & Yahaya, M. F. (2023). Honey and Alzheimer's disease—Current understanding and future prospects. *Antioxidants*, *12*(2), 427.

Shal, B., Ding, W., Ali, H., Kim, Y. S., & Khan, S. (2018). Anti-neuroinflammatory potential of natural products in attenuation of Alzheimer's disease. *Frontiers in Pharmacology*, *9*, 548.

Singh, A., Kukreti, R., Saso, L., & Kukreti, S. (2019). Oxidative stress: A key modulator in neurodegenerative diseases. *Molecules*, *24*(8), 1583.

Smith, M. L. (2021). Nest structure: Honey bees. In C. K. Starr (Ed.), *Encyclopedia of social insects* (pp. 626–632). Springer.

Sulaimon, F. A., Ibiyeye, R. Y., Imam, A., Oyewole, A. L., Imam, A. L., Shehu, M., Biliaminu, S. A., Kadir, R. E., Omotoso, G. O., & Ajao, M. S. (2024). Honey and levodopa comparably preserved substantia nigra pars compacta neurons through the modulation of nuclear factor erythroid 2–related factor 2 signaling pathway in 1-methyl-4-phenyl-1,2,3,6-tetrahydropyridine–induced Parkinson's disease model. *Anatomy & Cell Biology*, *57*(3), 431–445.

Tabrizı, S. J., Leavitt, B. R., Landwehrmeyer, G. B., Wild, E. J., Saft, C., Barker, R. A., Blair, N. F., Craufurd, D., Priller, J., Rickards, H., & Rosser, A. (2019). Targeting huntingtin expression in patients with Huntington's disease. *New England Journal of Medicine*, *380*(24), 2307–2316.

Tesco, G., & Lomoio, S. (2022). Pathophysiology of neurodegenerative diseases: An interplay among axonal transport failure, oxidative stress, and inflammation? *Seminars in Immunology*, *59*, 101628.

Venkatesan, R., Ji, E., & Kim, S. Y. (2015). Phytochemicals that regulate neurodegenerative disease by targeting neurotrophins: A comprehensive review. *BioMed Research International*, *2015*, 814068.

Wang, D., Chen, F., Han, Z., Yin, Z., Ge, X., & Lei, P. (2021). Relationship between amyloid-β deposition and blood–brain barrier dysfunction in Alzheimer's disease. *Frontiers in Cellular Neuroscience*, *15*, 695479.

Weichhart, T. (2018). mTOR as regulator of lifespan, aging, and cellular senescence: A mini-review. *Gerontology*, *64*(2), 127–134.

Whitfield, C. W., Behura, S. K., Berlocher, S. H., Clark, A. G., Johnston, J. S., Sheppard, W. S., & Tsutsui, N. D. (2006). Thrice out of Africa: Ancient and recent expansions of the honey bee, *Apis mellifera*. *Science*, *314*(5799), 642–645.

World Health Organization (WHO). (2019). *Traditional medicine: Global report on traditional and complementary medicine*. World Health Organization.

World Health Organization (WHO). (2023). *Prevalence of key health conditions: Global health estimates*. World Health Organization.

Yan, M. H., Wang, X., & Zhu, X. (2013). Mitochondrial defects and oxidative stress in Alzheimer disease and Parkinson disease. *Free Radical Biology and Medicine*, *62*, 90–101.

Yildiz-Unal, A., Korulu, S., & Karabay, A. (2015). Neuroprotective strategies against calpain-mediated neurodegeneration. *Neuropsychiatric Disease and Treatment*, *11*, 297–310.

Yuksel, M., & Tacal, O. (2019). Trafficking and proteolytic processing of amyloid precursor protein and secretases in Alzheimer's disease development: An up-to-date review. *European Journal of Pharmacology*, *856*, 172415.

Zamri, N. A., Ghani, N., Ismail, C. A. N., Zakaria, R., & Shafin, N. (2023). Honey on brain health: A promising brain booster. *Frontiers in Aging Neuroscience*, *14*, 1092596.

Zhu, J., Cui, Y., Zhang, J., Yan, R., Su, D., Zhao, D., Wang, A., & Feng, T. (2024). Temporal trends in the prevalence of Parkinson's disease from 1980 to 2023: A systematic review and meta-analysis. *The Lancet Healthy Longevity*, *5*(7), e464–e479.

Zou, Z. Y., Zhou, Z. R., Che, C. H., Liu, C. Y., He, R. L., & Huang, H. P. (2017). Genetic epidemiology of amyotrophic lateral sclerosis: A systematic review and meta-analysis. *Journal of Neurology, Neurosurgery & Psychiatry*, *88*(6), 540–549.

Zulkifli, N. A., Hassan, Z., Mustafa, M. Z., Wan Azman, W. N. N., Hadie, S. N. H., Ghani, N., & Mat Zin, A. A. (2023). The potential neuroprotective effects of stingless bee honey. *Frontiers in Aging Neuroscience*, *14*, Article 1048028.

7 Bee Venom Applications in Neurodegenerative Diseases

Danwahab Olarewaju Ambali, Solihu Olashile Ibrahim, Abdulkadir Yaradua Imam-Fulani, Imam Abubakar Lekan, and Duyilemi Chris Ajonijebu

7.1 INTRODUCTION

Among the leading causes of death globally according to World Health Organization are neurodegenerative diseases (NDDs), which are due to continuous loss of structural and functional units of the nervous system (neurons), leading to debilitating cognitive, motor, and sensory impairments. NDDs include Alzheimer's disease (AD) and other dementia, Parkinson's disease (PD), amyotrophic lateral sclerosis (ALS), Huntington's disease (HD), epilepsy, multiple sclerosis (MS), and others. These ailments account for 12% of the total mortality rate globally, making them significant causes of death (Silva et al., 2015). Numerous cellular processes, such as oxidative stress, mitochondrial dysfunction, neuroinflammation, aging, clumped formation, glutamatergic excitotoxicity, and apoptosis can be activated, leading to neurodegeneration (Asuku et al., 2025). Consequently, many research teams are trying to find medications that could prevent one or more of these processes that result in the death of neurons (De Souza et al., 2018).

Honeybee (*Apis mellifera*) venom has been in use as early as 1000–3000 BC by Chinese traditional medicine, Egyptian medicine, and Greek civilization for managing and treating different ailments because of the medicinal properties contained in the venom (Wehbe et al., 2019). More recently, bee venom has been explored and interestingly used as an alternative and complimentary therapeutic strategy due to its powerful neuroprotective potential, both anti-inflammatory, and immunomodulatory (Gu et al., 2015). Bee venom has been useful as a traditional cure for different ailments such as arthritis, rheumatism, and cancer through bee sting acupuncture. BV consists of various active components, for example: peptides (e.g., melittin, apamin, adolapin, and the mast cell degranulating peptide or MCD), enzymes (e.g., phospholipase A_2 or PLA_2, hyaluronidase, and glucosidase), and biogenic amines (e.g., histamines, dopamine, and norepinephrine. The main components of bee venom and their respective percentages in the dry venom are melittin, which constitutes approximately 40–60%, and PLA_2, which makes up around 12% (Lin & Hsieh, 2020).

The neuroprotective effects of bee venom include its ability to significantly delay the progression of neurodegenerative diseases such as Parkinson's, Alzheimer's, and ALS. Bee venom has shown extensive anti-inflammatory effects by disrupting pathways that trigger inflammation and blocking the production of genes that promote inflammation. It has been observed to prevent dopamine depletion in models induced by the pesticide rotenone, which is associated with Parkinson's disease. Additionally, both bee venom and its component melittin demonstrate powerful capabilities in preventing cancer cell spread and inducing cell death. The potential of bee venom to protect neurons against neuroinflammation, oxidative stress, and apoptosis has been highlighted in various preclinical studies, indicating its effectiveness in improving cognitive function and overall neuronal health in animal models (Wehbe et al., 2019).

 DOI: 10.1201/9781003685654-7

7.2 THE BIOACTIVE CONSTITUENTS OF BEE VENOM

Bee venom (BV) is a water (88%) mixture with a small amount of dry venom (0.10 88g) consisting of various water-soluble, biologically active products that are clear (odorless). The greater bioactive constituents are peptides and enzymes with traces of other molecules of biogenic amines and minerals.

7.2.1 Peptides

Melittin: It is the most prevalent and the biggest biologically active substance, and it comprises 40–60% of the dry weight of BV. It is a small protein that has 26 amino acids, and it is considered the major strong physiological ingredient. It has anti-inflammatory, cytotoxicity properties against cancer cell and antinociceptive properties (pain relieving) when used in low doses and activates PLA_2. By downregulating the phosphorylation of IκB kinase, an inhibitor to nuclear factor kappa B (NF-κB), decreasing in the expression of tumor necrosis factor (TNF-∝), interleukin-1 (IL-1), and interleukin-6, melittin exhibits its anti-inflammatory potential (Lin & Hsieh, 2020).

Apamin: Apamin is a neurotoxin in dry bee venom, comprising about 37–44% of the venom. It is a peptide consisting of 18 amino acids bonded together by a disulphide bond calcium/potassium channel inhibitor that works in the central nervous system (CNS) by inhibiting muscarinic M2 receptors in motor nerve terminals, reducing the neuromuscular transmission. In addition, apamin also serve as an anti-inflammatory factor through the suppression of cyclooxygenase-2 and reducing the levels of TNF, IL-1 (interleukin-1), IL-6, and NO. Apamin also inhibit pro-inflammatory cytokines by activating NF-κB (inhibitory of the nuclear factor kappa-B) pathway (Wehbe et al., 2019).

Adolapin: This is a peptide that constitutes about 1% of the dry bee venom, which is anti-inflammatory and antinociceptive as well as antipyretic (Lin & Hsieh, 2020).

Mast cell degranulating (MCD) peptide: Also known as peptide 401, this peptide comprises about 2–3% of the dry weight of BV. It is a neurotoxin that causes release of histamine by the mast cells (Wehbe et al., 2019).

7.2.2 Enzymes

Phospholipase A_2 (PLA_2): This is the lethally fatal enzyme in BV; it constitutes 10–12% of BV dry weight. It breaks down phospholipids in cell membranes, although it has applications in its regeneration of nerves and reduction of T-cell invasion (Wehbe et al., 2019).

Hyaluronidase: This is an enzyme that breaks hyaluronic acid that forms 1.5% to 2% of the dry mass of BV and facilitates the penetration of other venom components to tissues (Wehbe et al., 2019).

7.2.3 Biogenic Amines

Bee venom contains such amines as histamine and epinephrine (adrenaline) (Wehbe et al., 2019).

Bee venom is an excellent source of bioactive molecules, the most evident ones being enzymes and peptides. Melittin and phospholipase A_2 are the most prevalent compounds and the various properties of the compounds render the venom to having a wide range of biological and therapeutic effects (Lin & Hsieh, 2020).

7.3 JUSTIFICATION FOR USING BEE VENOM IN NEURODEGENERATION DISEASES

The different bioactive constituents of bee venom are investigated as a possible therapeutic agent in neurodegenerative diseases because of its multiple anti-inflammatory and neuroprotective effects on the body through melittin, which is one of the main components of bee venom. It has been shown that bee venom can enhance memory learning in the models of Alzheimer's disease by augmenting the nicotinic signaling, which plays a vital role in the cognitive functioning. Moreover, melittin has also been reported to inhibit pro-inflammatory mediators of microglial cells, resulting in neuroinflammation, which is a prevalent characteristic of neurodegenerative diseases (Silva et al., 2015).

More so, experiments have shown that BV has the ability to reduce the pathology of AD without causing neuroinflammation, a two-sided effect of maintaining neuronal health and reducing negative influences of harmful inflammatory responses. The neuroprotection and inflammatory signaling pathways modulated by BV places the venom in a good position to be used as a therapeutic agent to treat neurodegenerative diseases. All in all, the rationale behind BV utilization in neurodegeneration lies in the fact that it has the potential to treat the inflammatory pathways as well as the tissue damage of the neurons that characterize the diseases (De Souza et al., 2018).

The neuroprotective actions of the bee venom and its constituents have various neuroprotective effects by antioxidant, blood-brain-barrier (BBB) modulatory and synaptic-protective actions. All these effects contribute to the survival of neuronal viability and functionality amid different pathological offenses.

7.3.1 Antioxidant Effects

Oxidative stress is regarded as a major mediator of neuronal damage (Ajibare et al., 2024), and BV has proven antioxidant effects. An example is melittin, which significantly decreases the reactive oxygen species (ROS) and lipid peroxidation in neuronal cells. Melittin pretreatment of hippocampal models of the Alzheimer's pathology decreased intracellular ROS and malondialdehyde and showed no effects on antioxidant enzyme activities. The entire bee venom triggers Nrf2/HO-1 signaling, a super-regulator of antioxidant defense. BV increased nuclear translocation of Nrf2 and heme oxygenase-1 in Aβ-treated neural cells, preventing oxidative injury (Nguyen et al., 2022a).

At the same time, BV triggered the TrkB/CREB/BDNF signaling pathway that not only promotes the survival of the neurons but also enhances the expression of endogenous antioxidant enzymes. Therefore, in a manner that scavenges the free radicals and augments antioxidant gene transcription, BV neutralizes the oxidative damages typical of neurodegenerative disorders (Nguyen et al., 2022b).

7.3.2 Blood-Brain Barrier (BBB) Interaction

The CNS homeostasis requires the integrity of BBB (Asuku et al., 2024). Tight junctions can be disrupted by neurotoxins and inflammation, and this causes neuronal damage. There is new evidence that BV is capable of maintaining BBB functionality. *Apis mellifera* venom induced a recovery of neurotoxicity-induced downregulation of tight junction protein (occludin, claudin-5, ZO-1) levels in a model of methylmercury neurotoxicity (Abu-Zeid et al., 2021).

BV therapy lowered neuroinflammation and oxidative stress, and it greatly inhibited BBB destabilization and cognitive disorders caused by mercury. The evidence shows that BV (probably through its anti-inflammatory properties) is able to stabilize the BBB even in the presence of toxic or inflammatory conditions. (No BV-induced adverse increase in BBB permeability was reported; on the contrary, BV inhibited externally induced BBB leak.) BV preserves the integrity of tight junctions and minimizes endothelial inflammation as a way of ensuring adequate neuronal environment (Abu-Zeid et al., 2021).

7.3.3 Synaptic Protection and Plasticity

The bee venom constituents have the power to sustain synaptic activity and plasticity. Apamin, an 18 amino-acid BV peptide, is also unique in inhibiting the small-conductance Ca^{2+} activated K^+ channels in neurons. SK (potassium selective) channel blockade enhances neuronal excitability and is able to induce long-term potentiation (LTP) of hippocampal neurons. Apamin administration has been demonstrated by experiment to increase the excitability of CA1 hippocampal neurons and increase LTP amplitude. This results in better learning and memory in certain animal tests. In fact, intraventricular apamin enhanced memory performance of AD models in rats and reduced motor impairments of PD models (Gu et al., 2020).

Therefore, apamin possesses an inherent synaptic potentiating effect that could be used to reverse synaptic weakening observed in neurodegenerative conditions by modulation of SK channels. Moreover, BV-induced BDNF and TrkB signaling upregulation (as was mentioned earlier) stimulate the formation and stability of synapses as well. All of this helps to safeguard the synapses against degeneration and can even recover the functioning of dysfunctional neural circuits (Nguyen et al., 2022a).

7.3.4 Anti-Inflammatory Effects of Bee Venom

One of the main characteristics of most neurological diseases is chronic glia-based inflammation. Bee venom contains highly anti-inflammatory properties, regulating release of cytokines and suppressing pro-inflammatory cytokine signaling in immunity and CNS cells. It has an anti-inflammatory repertoire comprising of innate immunity (e.g., NF-κB and MAPK inhibition) as well as adaptive/autoimmune signaling (e.g., T-cell modulation).

One of the main anti-inflammatory BV targets is a transcription factor, NF-κB, which is a key regulator of pro-inflammatory genes (Carpena et al., 2020). The BV major peptide melittin prevents the activation of NF-κB via several pathways. Melittin is able to block NF-κB p65 nuclear translocation and prevent the upstream p38 MAPK pathway in cell culture. This leads to a significant decrease in production of TNF-alpha, IL-1 beta, IL-6, and others. As an illustration, in inflammatory skin cells, melittin inhibited IL-1 beta and IL-6 by inhibiting transcription via NF-κB. Melittin and entire BV, in vivo, lower NF-κB and AP-1 activity, which is in line with the mitigated cytokine production. BV or melittin treatment reduces concentrations of main cytokines (TNF-alpha, IL-6, IL-1 beta, IFN-gamma) and inflammatory enzymes (COX-2, iNOS) in different models. In one case, topical or injected BV in an atopic dermatitis model lowered serum IgE, IL-4, TNF-alpha, and NF-κB signal activation, which reduced skin inflammation. BV injections reduced joint TNF-alpha, IL-1 beta, IL-6, and TGF-β1 in rheumatoid arthritis models. These responses indicate general cytokines inhibition and communicational blockade. BV (and apamin) suppressed NF-κB and even NLRP3 inflammasome in gout models, which decreased IL-1 beta and inflammation of the joints. In the molecular scale, melittin binds directly to and inhibits the NF-κB essential modulator (NEMO) complex, and may prevent TLR 2/4 activation. Therefore, BV inhibits TLR-mediated innate inflammation as well as T-cell-mediated adaptive inflammation (Gu et al., 2015).

One of the symptoms of neuroinflammation is chronic microglial activation. BV elements inhibit glial reactions. BV blocked the glial NF-κB activation and expression of COX-2, iNOS, GFAP, and IBA-1 in LPS-treated astrocytes/microglia. BV in vivo decreased astro- and microgliosis in neuroinflammation caused by LPS. Similarly, BV PLA_2 in AD-model mice inhibited microglia and decreased the T-cell infiltration in the brain, which could not be demonstrated in the case of Tregs depletion. Therefore, the anti-inflammatory effect of BV extends in the CNS-resident immune cells. NF-κB activation coupled with leukocyte infiltration into the CNS is inhibited by BV, which makes the environment in the CNS less friendly to neuronal injury (De Souza et al., 2018).

7.3.5 Anti-Apoptotic Mechanisms of Bee Venom

Programmed cell death in the neurons is also antagonized by bee venom. It alters apoptosis-controlling genes and signaling pathways to benefit cell survival in stress. The important activities of anti-apoptosis are regulation of Bcl-2 family members, prevention of caspase activation, and the activation of pro-survival signaling (Doo et al., 2012; Ayinla & Asuku, 2025).

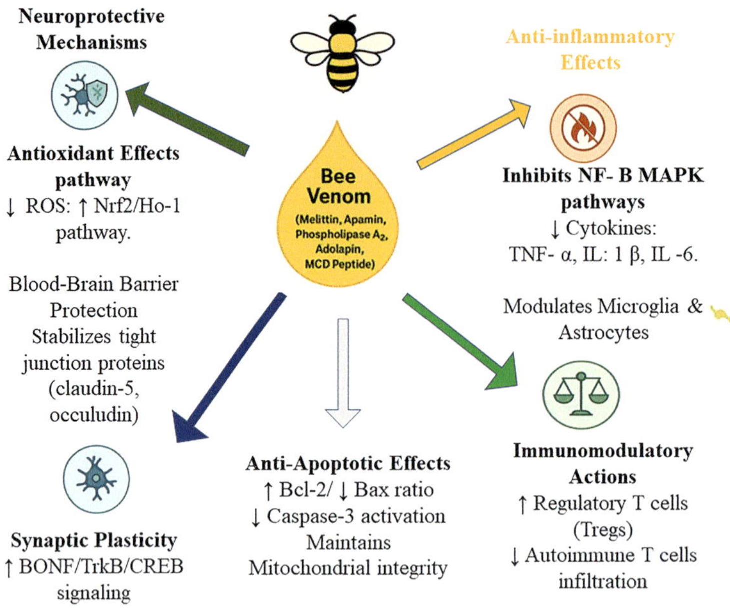

FIGURE 7.1 The therapeutic mechanisms of bee venom in neurodegenerative diseases.

In neuronal cells that are subjected to toxins (e.g., MPP+, PD toxin), BV causes a change in the balance towards anti-apoptotic proteins. MPP+ treatment of SH-SY5Y neuroblastoma cells was accompanied by a strong effect of BV pretreatment on Bcl-2 (anti-apoptotic protein) and Bax (pro-apoptotic protein) that reduce the levels of cleaved PARP. These molecular modifications were with decreased DNA fragmentation and caspase-3 activity. Therefore, BV changes the Bax/Bcl-2 ratio towards survival by directly suppressing the pro-apoptotic cascade of mitochondria. Melittin also suppresses pro-apoptotic intermediates. BV inhibits pro-apoptotic signals, maintains mitochondrial integrity, and inhibits apoptosis (Doo et al., 2012). Figure 7.1 shows the mechanisms of bee venom in neurodegenerative diseases.

7.4 APPLICATIONS OF BEE VENOM IN NEURODEGENERATIVE DISEASES

7.4.1 Alzheimer's Disease (AD)

Current FDA-approved drugs such as memantine and several acetylcholinesterase (AChE) inhibitors have been shown to only prevent neuronal death due to excitotoxicity or marginally improve on cholinergic deficits. Most importantly, none of these treatments slow down the development of the disease.

The anti-amyloid and anti-inflammatory effects of BV are constant in preclinical reports. The seven-day pre-treatment of murine model of AD with intraperitoneal (i.p.) BV (0.8–1.6 g/kg) daily reduced the impairment of memory, decreased brain A_8 concentration and secretase activities, and inhibited the activation of the NF-κB and the secretion of inflammatory mediators (iNOS, COX-2). On the same note, Chae et al. (2023) discovered that BV PLA_2 (0.02–2 mg/kg, i.p., every third week) reversed the LPS-induced cognitive impairments, reduced hippocampal A_2 and inflammatory cytokines, and blocked NF-κB signaling in the mouse. The BV PLA_2 (0.2 or 1.0 mg/kg i.p. weekly) significantly improved spatial memory and brain glucose metabolism and significantly decreased the ABS deposition of A0B in hippocampal of transgenic AD mice (3xTg-AD model). The neuroprotective mechanisms were associated with the inactivation of microglials and the attenuation of CD4+

T-cell infiltrations, and were prevented in the presence of Treg depletion, which implicates a Treg-mediated process. Mechanisms include inhibition of NF-κB and secretase pathways, induction of antioxidant gene (Nrf2/HO-1) and stimulation of pro-survival pathways (TrkB/CREB/BDNF). There are no controlled clinical trials of BV in AD that have been reported so far. Nonetheless, the preclinical data indicate that BV or ingredients (particularly, PLA_2 and melittin) have the potential to suppress amyloidogenesis and inflammation in AD models.

7.4.2 Parkinson's Disease (PD)

The most common treatment is a combination of levodopa and carbidopa that is used to replenish the level of dopamine and is useful in correcting motor impairment (Kim et al., 2016). After approximately five years of use, however, half of patients also develop motor complications such as dyskinesias (involuntary movements), which are hard to control, and severely affect the quality of life of the patient. Other therapies in most cases are not effective. No existing medication can be used to alter the condition or slow down the deterioration of dopaminergic neurons (Dantas et al., 2022).

Several papers in cellular and animal PD models have suggested the anti-inflammatory and neuroprotective effects of BV. In PD models in MPTP or rotenone models of PD, the dopaminergic neuron and motor performance were preserved by BV or PLA_2. As an illustration, BV acupuncture (subcutaneous injection of the Zusanli ST36 point) in MPTP-injured mice minimized astroglial reaction, avoidance of motor impairment, and loss of TH-positive neurons. Similarly, BV PLA_2 (0.2 mg/kg i.p. five times/week) of MPTP mice safeguarded the substantial nigral neurons by increasing the extent of CD4+Foxp3+ Tregs and inactivating microglia. The PLA_2 subunit, attached to CD206 on dendritic cells, causes the release of PGE 2 and EP2-receptor signal to promote Treg differentiation. Such measures restricted neuroinflammation and provided neuroprotection. BV peptides downregulate pro-inflammatory cytokines in microglial BV2 cells by MyD88/NF-κB pathways in vitro (Chung et al., 2015).

PD model methods of delivery include the following: BV is injected at the acupuncture point (ST36), intramuscularly, or into the peritoneum, usually at the doses of microgram to milligrams. Whereas Ahmed-Farid et al. (2021) used BV combined with dopamine-loaded nanoparticles in a rat model with rotenone (Kim et al., 2016).

Conversely, PD has a low level of clinical evidence. In a randomized, placebo, and controlled trial conducted by Hartmann et al. (2016), monthly bee-venom (100 μg) injections by subcutaneous route were compared to saline in 40 PD patients over 11 months. There were no large differences in the motor scores, the progression (measured by PET) as well as non-motor outcomes between groups. The research finding attributed a conclusion that monthly 100g BV was safe but lacked an obvious symptomatic or disease-modifying effect. Likewise, a retrospective cohort study of beekeepers did not identify any lower PD incidence, indicating that chronic exposure to BV is not in itself a deterrent to PD. Therefore, preclinical PD models are promising (through neuroprotection by anti-inflammatory Tregs), but the current clinical trials have failed. Current plans are aimed at optimal dosage, frequency, and adjunctive application (e.g., BV nanoparticles) (Kim et al., 2016).

7.4.3 Amyotrophic Lateral Sclerosis (ALS)

ALS is a neurological disorder due to gradual loss of motor neurons in the brain and the spinal cord, consequently causing atrophy and weakness of muscles. Riluzole is the only pharmacotherapy used, but this has only a marginal three-month life-span extension. Its mechanism is not completely understood, although it is believed to be associated with the inhibition of different ion channels (Gu et al., 2015). There is a paucity of research, and preclinical evidence demonstrates that BV is capable of delaying the pathology ALS through anti-inflammatory and proteostatic influences. Yang et al. (2010) injected BV in symptomatic hSOD1-G93A mice at ST36 acupoint. By treatment with

BV, microglial (Iba1) expression and p38 MAPK activation in the spinal cord were reduced as well as rotarod performance was enhanced, and median survival increased by approximately 18% (139 vs.117 days). BV also lowered caspase-3 in motor neurons and maintained a normal mitochondrial morphology, which showed that there was low apoptosis. The motor neuron-like NSC34 cells expressing mutant SOD1-G85R with BV (2.5 0g/mL) yielded significant decrease of ubiquitinated/misfolded SOD1 aggregates and resumed 20S proteasome activity in vitro. This means BV (or melittin) increases clearance of proteins. The aggregate results indicate the anti-neuroinflammatory (microglial) and anti-proteotoxic effects of BV in ALS models (Yang et al., 2010).

Clinically, no controlled trials of BV in ALS have been reported. The review from ALS Untangled states that there is only anecdotal use of BV acupuncture at ST36 with unclear benefits. BV is therefore experimental in ALS, but it has potential of dampening microglia and improving the ability of proteasomes to use it (as in rodents/cells), and so it is a potential target of future research (Yang et al., 2010).

7.4.4 Multiple Sclerosis (MS)

The murine mouse EAE model of MS has been widely studied, with BV or BV PLA_2 being found to inhibit EAE in most cases by regulating adaptive immune response. Lee et al. (2019) demonstrated that treatment with BV PLA_2 (0.2 mg/kg i.p. three times per week) had a dramatic effect on clinical EAE scores in mice immunized with MOG 35, Treg depletion eliminated the benefit of BV PLA_2. Furthermore, the PLA_2 mutant, which lacks catalytic activity, did not produce any effect at all such that PLA_2 enzyme activity is required. Another effect of BV on cytokine balance is that Jelodar et al. (2021) treated symptomatic EAE mice with i.p. BV and found a significant increase in IL-27 (an anti-Th17 cytokine) and a reduction of inflammatory infiltrates in CNS tissue. These publications point to a duality in MS/EAE induction of Tregs and reorienting the effector T-cell responses (e.g., restraining Th17 by increasing IL-27). It is a general observation that BV-derived agents suppress neuroinflammation and demyelination in EAE (experimental autoimmune encephalomyelitis) (Jelodar et al., 2021).

7.4.5 Epilepsy

Bee venom has been investigated as a potential experimental treatment of epilepsy because of its mixture of complex neuroactive peptide (melittin, apamin, etc.) and anti-inflammatory and ion-channel enzymes. This section summarizes the preclinical and small clinical data on BV in seizure disorders, how it is used, its suggested mechanisms, its efficacy and safety issues (Soares-Silva et al., 2022).

- *Melittin (BV peptide)*: In a rat model acute seizure, intracerebroventricular (i.c.v.) injection of melittin (0.1 mg) had an immense effect on reducing seizures. Ninety percent of rats treated with melittin were resistant to convulsions caused by bicucullin, and there was delayed onset of convulsions and low activation of astrocytes in the hippocampal (GFAP). These results indicate that BV melittin component has an anticonvulsant/neuroprotective effect (Soares-Silva et al., 2022).
- *Bee venom acupuncture (BVA)*: At acupuncture points, neurochemical and inflammatory parameters were positively affected by frequent injections of diluted BV in a pilocarpine-induced epilepsy model (status epilepticus in rats). The BV-treated epileptic rats exhibited normal levels of blood electrolytes and glutamate/DOPA, accelerated neuronal repolarization and reduced significantly pro-inflammatory cytokines, as compared to the epileptic rats that were not treated with BV. The authors concluded that BVA inhibited certain of the harmful effects of epileptogenesis and could also delay the course of seizure pathology (Abd El-Hameed et al., 2021).

- *Oxidative stress model*: In a different study of pilocarpine-induced epilepsy in rats, the BV-treated animals were found to be having lower oxidative damage. The BV administration (unspecified route) was shown to reduce lipid peroxidation and nitric oxide significantly and increase antioxidant enzymes (catalase, SOD, GPx) in the untreated epileptic rats. The hippocampal of the bee-venom-treated rats were also more normal, and liver/testicular function was preserved as compared to the epileptic controls. The authors observed that BV and its active constituents are new avenues in managing epilepsy because they reduce tissue damage that is caused by seizures (Aly et al., 2023).

A range of mechanisms have been theorized with regard to the antiseizure action of BV due to its bioactivity:

- *Ion channel modulation*: The peptides produced by BV are directed to neuronal ion channels. The potent blocker of small-conductance Ca^{2+} activated K^+ (SK) channels is apamin. Apamin can also modify the neuronal firing patterns by inhibition of SK channels; interestingly, apamin has also been explored as a therapeutic agent in ataxia and even epilepsy.
- *Anti-inflammatory*: Inflammation and cytokine release are involved in the pathogenesis of epilepsy. BV components inhibit the inflammatory signaling. Melittin strongly suppresses the action of toll-like receptor (TLR) and NF-κB. Low (therapeutic) doses of melittin and PLA_2 inhibit pro-inflammatory cytokines (e.g., TNF-alpha, IL-6). The level of brain cytokines was found to be low in BV-treated rats compared to untreated controls in the pilocarpine model. On other grounds, BV suppresses microglial stimulation and lowers immune reaction, which may dull neuroinflammation of seizures (Aly et al., 2023).
- *Neurotransmitters/neurochemical balance*: BV treatment has been reported to restore neurotransmitter and ionic balance. In one experiment, BVA inhibited hippocampal glutamate influx (excitatory transmitter) and restored electrolytes in the blood of the subjects during status epilepticus. BV would be able to raise the seizure threshold by modulating extracellular glutamate and boosting GABAergic tone. (After BVA in the stated study, changes of the GABA/glutamate balance and of the level of glutamate/DOPA were measured.) (Aly et al., 2023).

7.5 FUTURE PERSPECTIVES

7.5.1 Standardized Dosing and Formulations

Bee venom has been used traditionally as a bee sting or acupuncture points (pharmacopuncture) or as of late in engineered preparations. The injection doses in clinical trials of PD were extremely low: 0.03–0.1 mg/mL BV, injected in 0.1–1.0 mL (usually shared between multiple sites) one at a time. This is illustrated by the fact that idiopathic PD patients were administered 0.05–0.1 mg/mL, 0.1–1.0 mL/session (1–24 mL overall treatment period). The ranges in postural or drug-induced Parkinson's were also comparable (0.03–0.05 mg/mL, 1.0 mL). These doses of PD are roughly the same as 0.005–0.1 mg BV per session, a fraction of regular, natural, sting (i.e., 50–140 μg) (Jeong et al., 2025).

In addition to simple injections, new methods are added such as microneedle transdermal patches and nanocarriers. Recent research was done on the comparison of a subcutaneous (SC) injection with dissolvable microneedle patch (delivering BV) in a mouse neurodegeneration. The microneedle patch generated an improved-but-persistent blood concentration of melittin (the principal peptide in BV), which increases its treatment duration and decreases the side effects associated with the peak. This implies that patches or polymeric carriers can enhance BV pharmacokinetics. Topical creams or gels containing BV are applied to the skin (e.g., anti-wrinkle cosmetics) and joint pain; however, when CNS disorders are involved, the microneedles would not be able to get to the brain. Nanoparticles (liposomes, nanoparticles, referred to as nanocarriers) have been suggested to transport BV peptides across the blood-brain barrier, although none have been approved clinically (Nguyen et al., 2022a).

7.5.2 Additional Human Clinical Trials

The clinical evidence in human beings is to date very scanty and ambivalent. The Parkinson's disease case has been the most researched one. In February 2025, 12 studies were carried out to examine PD treated with BVA (bee-venom acupuncture). They had small sizes (case reports, open-label series, or some RCTs) and most of them involved 0.031 mg/mL injections. Those trials proved the improvement of the UPDRS scores, gait, quality-of-life, etc., and with some minor side effects (local itching/swelling identified in about 33% of the trials). No severe reactions were experienced. This implies that BV can be used as an adjunctive, and it is possible (Jeong et al., 2025).

7.5.3 Possible Complementation of Current Treatment

Bee venom has many targets, which implies that it might be used in combination with conventional treatment. The main conventional medications are dopamine substitutions in PD, cholinesterase/ NMDA blockers in AD, and riluzole/edaravone in ALS. The synergetic evident between the standard drugs and bee venom application in the management of neurodegenerative diseases is mainly preclinical (Baek et al., 2018).

BV and L dopa/rasagiline were superior to drugs in terms of PD models. The BV PLA_2 was an immune adjuvant used in AD models to enhance the performance of an amyloid vaccine. The results suggest that BV-derived agents may be used supplementary to the current treatment to address the pathways (inflammation, apoptosis, oxidative stress) not addressed by conventional therapies. Combination research (e.g., BV and dopaminergic drugs in PD) should be employed in future clinical research to determine whether there is any additive or synergistic effect (Badawi et al., 2020).

7.6 LIMITATIONS OF BEE VENOM APPLICATIONS IN NEURODEGENERATIVE DISEASES

Existing therapies of most chronic neurodegenerative and autoimmune conditions usually cannot alter the disease course, but instead can merely treat the symptoms. The limitations to these therapies include inefficacy, emerging complications with time, and a small therapeutic index, which adversely affects patient outcomes and life.

A large part of the studies is small and has different protocols and is not replicated. BV products have low standards of standardization (composition differs between species of bees and extraction). The ideal dose and regimen of CNS targets is also unclear. Other purported efficacies of BV are based on low-quality trials or anecdote. In case, an example of high-profile MS trial was correctly controlled and showed no benefit. There is also controversy on whether systemic BV penetrates the brain at efficacious concentrations; some studies utilize acupuncture-point injection, which is the stimulation of neuroimmune reflexes, but not direct input into the CNS. Lastly, Treg expansion due to BV may be anti-autoimmune but may also abrogate positive immune responses.

7.7 CONCLUSION

A decade of study indicates that bee venom and its components have a strong anti-inflammatory and neuroprotective effect on the models of AD, PD, MS, and ALS. These actions are mediated by suppression of NF-κB and pro-inflammatory cytokines, stimulation of antioxidants, and stimulation of regulatory T-cells. BV had an effect in reducing amyloid pathology and enhancing cognition in AD models. BV in PD models spared dopaminergic and motor neurons. BV PLA_2 reduced Treg and IL-27-induced paralysis and CNS inflammation in EAE. Nonetheless, clinical trials to date have not been proven to have definite benefits. Therefore, although BV has potential to provide new neurotherapeutics, additional high-quality research is needed to provide its use in human neurodegeneration treatment.

REFERENCES

Abd El-Hameed, A. M., Abuelsaad, A. S. A., & Khalil, A. (2021). Bee venom acupuncture therapy ameliorates neuroinflammatory alterations in a pilocarpine-induced epilepticus model. *Metabolic Brain Disease*, 36(7), 2047–2058.

Abu-Zeid, E. H., Khalifa, B. A., Elewa, Y. H. A., Arisha, A. H., Ismail, T. A., Hendam, B. M., & Abdel-Hamid, S. E. (2021). Bee venom Apis mellifera lamarckii rescues blood brain barrier damage and neurobehavioral changes induced by methyl mercury via regulating tight junction proteins expression in rat cerebellum. *Food and Chemical Toxicology*, 154, 112309.

Ahmed-Farid, O. A., Taha, M., Bakeer, R. M., Radwan, O. K., Hendawy, H. A. M., Soliman, A. S., & Yousef, E. (2021). Effects of bee venom and dopamine-loaded nanoparticles on reserpine-induced Parkinson's disease rat model. *Scientific Reports*, 11(1), 21141. https://doi.org/10.1038/s41598-021-00764-y

Ajibare, A. J., Akintoye, O. O., Oriowo, O. A., Asuku, A. O., Oriyomi, I. A., & Ayoola, A. M. (2024). Zinc ameliorates acrylamide-induced cognitive impairment in male Wistar rats: Modulation of oxidative stress, neuro-inflammation, and neurotrophic pathways. *Biological Trace Element Research*, 203, 4273–4282. https://doi.org/10.1007/s12011-024-04490-0

Aly, S. S., Mohamed, A. A. R., Ibrahim, M. A., & El-Shetry, E. S. (2023). Bee venom attenuates neurotoxicity and oxidative stress induced by aluminum chloride in rats: A histological and immunohistochemical study. *Journal of Chemical Neuroanatomy*, 128, 102235. https://doi.org/10.1016/j.jchemneu.2023.102235

Asuku, A. O., Ayinla, M. T., Aransiola, S. A., Ajibare, A. J., Adeyemo, M. B., & Adeyemo, R. O. (2024). Chitosan-based nanoparticles to bypass the blood-brain barrier for the treatment of neurological diseases: A review. *Pacific Journal of Medical Sciences*, 25(1), ISSN: 2072–2625.

Asuku, A. O., Alege, D. O., Ogungbangbe, G. O., Ajibare, A. J., Adeyemo, O. R., & Sabinari, I. W. (2025). Neuroprotective properties of marine bioresources. In *Marine Bioresources: Prospects and Obstacles*. Elsevier, 369–378. http://doi.org/10/10.1016/B978-0-443-36320-7.00019-3

Ayinla, M.T., & Asuku, A.O. (2025). The anti-apoptotic effects of 6-gingerol-rich fraction of ginger on mercury chloride-induced cerebral and cerebellar dysfunction. *Arabian Journal of Medicinal and Aromatic Plants*, 11(2), 149–187. https://doi.org/10.48347/IMIST.PRSM/ajmap-v11i2.50900

Badawi, H., Abdel-Salam, R., Abdel-Salam, O., Youness, E., Shaffie, N., & Eldenshary, E. D. (2020). Bee venom attenuates neurodegeneration and motor impairment and modulates the response to L-dopa or rasagiline in a mice model of Parkinson's disease. *Iranian Journal of Basic Medical Sciences*, 23(12). https://doi.org/10.22038/ijbms.2020.46469.10731

Baek, H., Lee, C., Choi, D. B., Kim, N., Kim, Y.-S., Ye, Y. J., Kim, Y.-S., Kim, J. S., Shim, I., & Bae, H. (2018). Bee venom phospholipase A2 ameliorates Alzheimer's disease pathology in Aβ vaccination treatment without inducing neuro-inflammation in a 3xTg-AD mouse model. *Scientific Reports*, 8(1), 17369. https://doi.org/10.1038/s41598-018-35030-1

Carpena, M., Nuñez-Estevez, B., Soria-Lopez, A., & Simal-Gandara, J. (2020). Bee venom: An updating review of its bioactive molecules and its health applications. *Nutrients*, 12(11), 3360. https://doi.org/10.3390/nu12113360

Chae, S. U., Jo, S. J., Lee, C. B., Lee, S., Park, J.-H., Jung, J.-S., Park, E.-S., Bae, H., & Bae, S. K. (2023). Pharmacokinetics and tissue distribution of bee venom-derived phospholipase A2 using a sandwich ELISA after subcutaneous injection of new composition bee venom in rats. *International Journal of Molecular Sciences*, 24(12), 10214. https://doi.org/10.3390/ijms241210214

Chung, E. S., Yoon, Y. D., Lee, J. H., Shim, I., & Bae, H. (2015). Bee venom phospholipase A_2, a novel $Foxp3^+$ regulatory T cell inducer, protects dopaminergic neurons by modulating neuroinflammatory responses in a mouse model of Parkinson's disease. *Journal of Immunology*, 195(10), 4853–4860. https://doi.org/10.4049/jimmunol.1500066

Dantas, C. G., Da Paixão, A. O., Nunes, T. L. G. M., Silva, I. J. F., Dos S., Lima, B., Araújo, A. A. S., De Albuquerque-Junior, R. L. C., Gramacho, K. P., Padilha, F. F., Da Costa, L. P., Severino, P., Cardoso, J. C., Souto, E. B., & Gomes, M. Z. (2022). Africanized bee venom (Apis mellifera Linnaeus): Neuroprotective effects in a Parkinson's disease mouse model induced by 6-hydroxydopamine. *Toxics*, 10(10), 583. https://doi.org/10.3390/toxics10100583

De Souza, J. M., Goncalves, B. D. C., Gomez, M. V., Vieira, L. B., & Ribeiro, F. M. (2018). Animal toxins as therapeutic tools to treat neurodegenerative diseases. *Frontiers in Pharmacology*, 9, 145. https://doi.org/10.3389/fphar.2018.00145

Doo, A.-R., Kim, S.-N., Kim, S.-T., Park, J.-Y., Chung, S.-H., Choe, B.-Y., Chae, Y., Lee, H., Yin, C.-S., & Park, H.-J. (2012). Bee venom protects SH-SY5Y human neuroblastoma cells from 1-methyl-4-phenylpyridinium-induced apoptotic cell death. *Brain Research*, 1429, 106–115. https://doi.org/10.1016/j.brainres.2011.10.003

Gu, H., Han, S. M., & Park, K.-K. (2020). Therapeutic effects of apamin as a bee venom component for non-neoplastic disease. *Toxins*, 12(3), 195. https://doi.org/10.3390/toxins12030195

Gu, S. M., Park, M. H., Hwang, C. J., Song, H. S., Lee, U. S., Han, S. B., Oh, K. W., Ham, Y. W., Song, M. J., Son, D. J., & Hong, J. T. (2015). Bee venom ameliorates lipopolysaccharide-induced memory loss by preventing NF-kappaB pathway. *Journal of Neuroinflammation*, 12(1), 124. https://doi.org/10.1186/s12974-015-0344-2

Hartmann, A., Rausch, W.D., Yun, Y., Sohn, S., & Tucha, O. (2016). Bee venom for treatment of Parkinson disease—A randomized controlled clinical trial. *PLOS ONE*, 11(5), e0158235. https://doi.org/10.1371/journal.pone.0158235

Jelodar, S., Zare Mirakabadi, A., Oryan, S., & Mohammadnejad, L. (2021). Effect of honey bee venom on experimental autoimmune encephalomyelitis (EAE) as a model for multiple sclerosis (MS). *Archives of Razi Institute*, 76(6). https://doi.org/10.22092/ari.2021.126291.1342

Jeong, H., Kim, K. H., & Ko, S. (2025). Effectiveness of bee venom injection for Parkinson's disease: A systematic review. *Toxins*, 17(4), 204. https://doi.org/10.3390/toxins17040204

Kim, M. E., Lee, J. Y., Lee, K. M., Park, H. R., Lee, E., Lee, Y., Lee, J. S., & Lee, J. (2016). Neuroprotective effect of bee venom is mediated by reduced astrocyte activation in a subchronic MPTP-induced model of Parkinson's disease. *Archives of Pharmacal Research*, 39(8), 1160–1170. https://doi.org/10.1007/s12272-016-0802-0

Lee, G. H., Kang, G. H., & Bae, H. (2019). Bee venom phospholipase A2 suppression of experimental autoimmune encephalomyelitis is dependent on its enzymatic activity. *Molecular and Cellular Toxicology*, 15, 307–313. https://doi.org/10.1007/s13273-019-0034-8

Lin, T.-Y., & Hsieh, C.-L. (2020). Clinical applications of bee venom acupoint Injection. *Toxins*, 12(10), 618. https://doi.org/10.3390/toxins12100618

Nguyen, C. D., Yoo, J., An, E. J., Sung, C. Y., Jeong, D. H., Park, S.-Y., Kim, J.-H., & Lee, G. (2022a). Pharmacokinetic improvement provided by microneedle patch in delivering bee venom, a case study in combating scopolamine-induced neurodegeneration in mouse model. *Drug Delivery*, 29(1), 2855–2867. https://doi.org/10.1080/10717544.2022.2116129

Nguyen, C. D., Yoo, J., Hwang, S.-Y., Cho, S.-Y., Kim, M., Jang, H., No, K. O., Shin, J. C., Kim, J.-H., & Lee, G. (2022b). Bee venom activates the Nrf2/HO-1 and TrkB/CREB/BDNF pathways in neuronal cell responses against oxidative stress induced by Aβ1–42. *International Journal of Molecular Sciences*, 23(3), 1193. https://doi.org/10.3390/ijms23031193

Silva, J., Monge-Fuentes, V., Gomes, F., Lopes, K., Anjos, L., Campos, G., Arenas, C., Biolchi, A., Gonçalves, J., Galante, P., Campos, L., & Mortari, M. (2015). Pharmacological alternatives for the treatment of neurodegenerative bee venoms and their components as new neuroactive tools. *Toxins*, 7(8), 3179–3209. https://doi.org/10.3390/toxins7083179

Soares-Silva, B., Beserra-Filho, J. I. A., Morera, P. M. A., Custódio-Silva, A. C., Maria-Macêdo, A., Silva-Martins, S., Alexandre-Silva, V., Silva, S. P., Silva, R. H., & Ribeiro, A. M. (2022). The bee venom active compound melittin protects against bicuculline-induced seizures and hippocampal astrocyte activation in rats. *Neuropeptides*, 91, 102209. https://doi.org/10.1016/j.npep.2021.102209

Wehbe, R., Frangieh, J., Rima, M., El Obeid, D., Sabatier, J.-M., & Fajloun, Z. (2019). Bee venom: Overview of main compounds and bioactivities for therapeutic interests. *Molecules*, 24(16), 2997. https://doi.org/10.3390/molecules24162997

Yang, E. J., Jiang, J. H., Lee, S. M., Yang, S. C., Hwang, H. S., Lee, M. S., & Choi, S.-M. (2010). Bee venom attenuates neuroinflammatory events and extends survival in amyotrophic lateral sclerosis models. *Journal of Neuroinflammation*, 7(1), 69. https://doi.org/10.1186/1742-2094-7-69

8 Bee Propolis Applications in Neurodegenerative Diseases

Ayodeji Johnson Ajibare, Eniola Oluwatomisin Adesanya, Oluwagbeminiyi Francis Ogunkoya, and Oyesanmi A. Fabunmi

8.1 INTRODUCTION

8.1.1 Global Burden and Unmet Need in Neurodegenerative Diseases

Neurodegenerative diseases (NDDs) result from the gradual decline in the functioning of neurons, ultimately leading to their death, and affect roughly 50 million individuals worldwide, mostly elderly people. NDDs include Alzheimer's and Parkinson's diseases, Huntington's disease, multiple sclerosis, and amyotrophic lateral sclerosis. The spread of these disorders sharply rises and affects social, economic, and healthcare concerns. Cognitive and motor disabilities, loss of independence, and death are the consequences of these disorders. Several research attempts have been made to produce effective disease-modifying therapies as the existing treatment options focus on alleviating the symptoms and do not stop the disease from progressing; all these efforts have been in vain. There is still much to solve in the research, and there is an urgent need for novel and effective treatment options to target the complex and multifactorial nature of neurodegenerative disorders. Neurodegenerative disease is predicted to increase threefold by 2050 (World Health Organization, 2017).

8.1.2 Converging Pathophysiological Pathways in NDDs

The pathogenesis of NDDs is characterized by several converging mechanisms leading to the injury and death of the neurons, and one of these mechanisms involves neuroinflammation. This type of inflammation is characterized by the chronic activation of microglia and astrocytes that persist as proinflammatory cytokines are continuously released. Tumor necrosis factor-α (TNF-α), interleukin-1β (IL-1β), and interleukin-6 (IL-6) are examples of the proinflammatory cytokines. Mitochondrial impairment leads to oxidative stress and mitochondrial dysfunction, affecting energy production, calcium homeostasis, and apoptotic regulation. Also, the increase in ROS generation and antioxidant defenses damages cellular components, including lipids, proteins, and DNA. Proteinopathies, which are the third critical pathway, are characterized by abnormal aggregation and misfolding of proteins such as amyloid-β plaques and tau tangles in AD, α-synuclein aggregates in PD, and TDP-43 inclusions in ALS, disrupting cellular homeostasis and synaptic function. These intertwined pathways create a vicious cycle that increases neurodegeneration.

8.1.3 Justification for Multi-Target Therapeutics

The intertwined nature of neurodegenerative pathways complements the justification for multi-target therapeutic approaches. Studies have shown that traditional single-target drug development has consistently failed to translate promising preclinical results into clinical efficacy, which is likely

DOI: 10.1201/9781003685654-8

caused by the complex, multifactorial etiology of NDDs. Multi-target therapeutics that aim to address neuroinflammation, oxidative stress, and protein aggregation concurrently seem promising. Natural products with complex phytochemical compositions are suitable for this approach as they offer multi-target activity profiles. This approach seems promising in managing neurodegeneration because it improves efficacy while minimizing adverse effects.

8.1.4 Propolis as a Candidate: Chemistry, History, and Neuropharmacological Potential

Resinous bee propolis is more than just a fascinating product made by the honeybee. It is a powerful candidate for developing multi-target neurotherapeutics. This compound bioactive bundle, varying depending on geographical origin and 300 different bioactive compounds like flavonoids, phenolic acids, and aromatic compounds, includes a variety of different bioactive propolis compounds like caffeic acid phenethyl ester (CAPE), chrysin, galangin, and pinocembrin, all of which contribute different pharmacological properties (Zullkiflee et al., 2022). It has historical healing properties as well. It has long been employed as an all-natural remedy to calm down swelling, combat microbes, and aid in skin healing. The wide range of bioactive compounds in propolis, as well as its pharmacological activity, confer protective properties to the nervous system, assist in maintaining immune responses, and minimize inflammation, which makes propolis particularly useful. It is also a versatile therapeutic when compared to other natural compounds like curcumin and resveratrol. The neuropharmacological profile of propolis acts on all three major pathways of neurodegeneration. It produces its anti-inflammatory effects by suppressing the action of NF-κB (nuclear factor-κB), regulating the production of pro-inflammatory cytokines, and controlling the activation of microglia (Wu et al., 2013). Another mechanism to which propolis aids in the antioxidant defenses is the neutralization of reactive oxygen species (ROS), enhancing the innate protective enzymes of the body, and preserving the efficiency of mitochondrial activity. Furthermore, it is involved in the prevention of the accumulation of amyloid-β through restricting its clustering, facilitating its elimination, and affecting tau protein changes.

Neurodegenerative diseases are highly complex and pose a significant global problem that will need novel multi-targeted treatments. Due to the wide spectrum of compounds that bee propolis has to offer and its established tradition, there is a good rationale to explore it further in the future with the goal of contributing to the well-being of the brain.

8.2 BIOACTIVES AND BRAIN AVAILABILITY

8.2.1 Chemical Complexity and Key Bioactive Compounds: CAPE, Artepillin C, Pinocembrin, Chrysin

Honeybees gather a resin-like substance known as bee propolis. It is extremely complex, and researchers have found over 300 active compounds within its composition. Its composition and therapeutic values in relation to the bioactive compounds are influenced by the origin, geographical location, the settling plant, and the season in which propolis is harvested. The therapeutic bioactive compounds of propolis have been studied and assessed for their value for treating NDDs (Ghallab et al., 2025). Some of the bioactive compounds such as caffeic acid phenethyl ester (CAPE), artepillin C, pinocembrin, and chrysin, are touted for their euphoric effects along with their powerful antioxidant and anti-inflammatory properties.

CAPE is one of the most studied active propolis compounds and is associated primarily with poplar trees (Populus species) (Yordanov, 2019). Being a phenolic compound, it possesses high antioxidative potential and the ability to inhibit neuroinflammation through blockage of NF-κB, a transcription factor. CAPE uniquely functions by modifying p65, the NF-κB subunit, providing long-term anti-inflammatory actions by preventing NF-κB activation and spurring cytokine production. This action inhibits pro-inflammatory cytokine release and reduces microglial hyperactivity, which indirectly promotes the survival of neurons.

Artepillin C, a plant-derived compound present in Brazilian green propolis, is a printed derivative of p-coumaric acid and is also found in the raw material of *Baccharis dracunculifolia* (Alecrim-do-campo). It acts as an antioxidant as well as an anti-inflammatory with very high ORAC (oxygen radical absorbance capacity) values, which are tenfold higher compared to vitamin E (Shahinozzaman et al., 2020). Besides that, it also serves for the interaction of crucial pathways, which are concerned with regeneration and degeneration (Nrf2 pathway and increase in endogenous antioxidant levels). On the other hand, artepillin C also promotes the generation of fresh brain tissue (neurogenesis) by stimulating brain-derived neurotrophic factor (BDNF), the expression of which is important for learning and memory. Besides this, artepillin C interrupts the process of the death of nerve cells in the central part of the brain by removing the missing hydroxyl radicals and peroxynitrite.

Precambrian (5,7-dihydroxyflavanone) is the principal flavonoid globally in propolis samples and constitutes approximately 5–15% of the flavonoid quantity. Studies have discovered its property to cross the blood-brain barrier and provide protection to the brain. It keeps the balance of the redox/oxidative stress, protects cells against death, and keeps neurons alive in various experiments by preventing and inhibiting cytochrome c release, at the same time enhancing autophagy. These processes remove protein aggregates from the intercellular space (Shen et al., 2019). It improves glucose uptake by brain tissue and acts as a protectant in ischemia and reperfusion injuries, for which Precambrian can be useful in those neurodegenerative conditions that can be associated with metabolic abnormalities. Notably, Precambrian has a dose-dependent biphasic effect. In low concentrations (1–10 mm), it can be neuroprotective, but in high concentrations, it can induce oxidative stress.

The flavonoid chrysin (5,7-dihydroxy flavone) functions as a stabilizer of the neurotransmission and a source of maintaining cell membranes, which is proven with the help of its effects on the GABAA receptor—a significant factor connected with the advantages of its use to brain health. Its active components make it unique and promising in the treatment of multifactorial neurological disorders. Besides the control of neurotransmitter activity, chrysin is also highly inflammation-reducing as it inhibits the activity of lipoxygenase and cyclooxygenase enzymes. These two enzymes are the main pathways of inflammation. On the other hand, it acts as an anti-amyloidogenic agent by preventing amyloid-fibril formation and disaggregation of preformed fibrils (Del Fabbro et al., 2025). The compound's lipophilic nature mediates its easy crossing of the blood-brain barrier.

8.2.2 Geographical Variability and the Critical Challenge of Standardization

The challenge in standardizing propolis comes from the variations in the plant's bioactive components, emphasizing the importance of properly utilizing and vaporizing plants. Variations in concentrations of bioactive composition from the different harvested propolis cause discrepancies in the results, emphasizing the need for investigations on how to improve the reproducibility of studies.

For example, the temperate propolis is harvested mainly from Europe and North America from the popular plant species *Populus nigra*, *P. tremula*, which is rich in CAPE, flavonoid chrysin, galangin, and pinocembrin, besides the presence of easily absorbable phenolic acids. With the presence of all the bioactive phytochemicals, their anti-inflammatory and antioxidant profiles are very stable, and CAPE concentration is classified to be between 5% and 15% of total phenolic content (Gardana et al., 2007).

Sourced tropical propolis shows notable chemical diversity. Brazilian green propolis, sourced from *Baccharis dracunculifolia*, is characterized by high artepillin C content, ranging between 10% and 15%, and by unique prenylated compounds (Jorge et al., 2008; Moise & Bobiş, 2020). However, Brazilian red propolis, derived from *Dalbergia ecastaphyllum*, contains distinctive isoflavonoids and benzofurans with important antimicrobial properties. Mediterranean propolis mainly has compounds from the *Cistus* species, adding unique diterpenes and labdane derivatives (Moise & Bobiș, 2020).

Asian propolis has another distinct chemotype. The Chinese propolis shows several compositions depending on regional plant sources, while Malaysian and Indonesian variants often contain compounds from *Macaranga* species, contributing unique geranyl flavonoids with enhanced lipophilicity (Trusheva et al., 2011).

This variability of bee propolis makes the standardization of propolis products difficult; however, it is important for ensuring consistent quality, safety, and therapeutic outcomes. Without standardized composition and concentration of key bioactives, clinical translation will remain restricted. Efforts to establish quality control parameters include using chemical fingerprinting by incorporating high-performance liquid chromatography-mass spectrometry (HPLC-MS) to create compositional profiles. Bioactivity-based standardization can also be looked into, focusing on key pharmacological parameters rather than individual compounds. Lastly, marker compound standardization by using specific bioactives (e.g., CAPE content $\geq$ 10% for temperate propolis) to ensure therapeutic consistency (Bankova, 2005). However, the European Medicines Agency and other regulatory bodies have begun establishing guidelines for propolis standardization. Examples include blending propolis extracts or cultivating specific plant sources for bee forage to minimize batch-to-batch differences and enhance reproducibility.

8.2.3 Evidence for Blood-Brain Barrier Penetration and Formulation Strategies

Nondelivery systems improve solubility, stability, and cell membrane transport of substances; therefore, they are used in the transport of valuable substances, including propolis, and in drug dosage control. Clinical research has shown that nano-formulated propolis provides more neuroprotection outcomes, such as a reduction in protein aggregates and overall improved neurophysiology, compared to simple crude extracts, in both in vitro and in vivo models of neurodegeneration (Wong et al., 2012). Other than the already discussed nanotechnology, other alternatives are also being developed to enhance the delivery to the CNS, such as protein conjugation with some targeting moieties or intranasal delivery. These proposed methods for improving the mitochondrial detoxification process in propolis will likely tackle some issues. It will provide the needed synergy to address the challenges of poor water solubility and rapid metabolism of propolis, which strengthens their argument to target neurological disorders.

Advancing the treatment of NDDs revolves around the use of medications that can target the CNS, enabling these agents to cross the blood-brain barrier (BBB). Some components of propolis, such as Precambrian and CAPE, can cross the BBB as shown in a few studies; however, the effectiveness of these components varies greatly. Precambrian, the propolis intermediary, showed positive pharmacokinetic properties in animals and significant collateral distribution into the brain, along with some neuroprotective properties (Elbatreek et al., 2023). CAPE also crosses the BBB, which means it can perform its antioxidant and anti-inflammatory functions in the brain.

Recently, more and more studies have been focusing on how to increase the absorption and distribution of propolis throughout the brain. One of these studies focuses on the use of nano-propolis as a brain nanocarrier system. These carriers are promising in the sense that they help with solubility, stability, and the ability to penetrate the BBB along with having controlled and targeted release of the active components in propolis (Babazadeh et al., 2020). In direct comparison of propolis nanoformulations to crude propolis extracts in both in vitro and in vivo models of neurodegeneration, nanoformulated propolis has shown improved neuroprotective outcomes and other neurophysiological parameters, which also included a decrease in protein aggregates (Wong et al., 2012).

Aside from the use of nanoformulations, other methods of delivery, such as protein conjugation with targeting ligands and the use of the intranasal route, are also being designed for a more efficient transport system to the brain. Simplistic approaches to increase the mitochondrial detoxification of propolis will have a huge impact, as they will provide synergy to counteract the problems of water insolubility and fast metabolism that propolis has to offer in order to maximize its use in treating neurodegenerative diseases.

8.3 MECHANISTIC INSIGHTS

8.3.1 Anti-Inflammatory Effects via NF-κB and Microglial Modulation

Bee propolis is a plant resin. It is especially created by stingless bees of the *Meliponini tribe* and *Apis mellifera* honeybees. It has bioactive components such as flavonoids like pinocembrin, galangin, chrysin, and the phenolic acid CAPE. It has broad anti-inflammatory activity linked with modulation of the NF-κB signaling pathway.

NF-κB is a primary regulator of pro-inflammatory cytokines (such as TNF-α and IL-1β) and other inflammation-related genes. Neurodegenerative disorders like Alzheimer's disease (AD) and Parkinson's disease (PD) are associated with persistent and abnormal NF-κB activation of inflammatory microglia. Neuroinflammation induced by activated microglia causes further damage and death of neurons. Seized inhibitory subunits of NF-κB (called IκB) are trapped antagonistically by flavonoids in propolis. NF-κB can be viewed as a major regulator of this gene.

In addition to NF-κB inhibition, bee propolis affects microglial activity by promoting the shift to the M2 anti-inflammatory state, as opposed to the M1 pro-inflammatory state. This transition restricts the production of damaging cytokines and oxidative stress, which helps to provide an environment conducive to the neurons and potentially slows the development of the disease.

Spagnuolo et al. (2017) established that flavonoids may significantly decrease the pro-inflammatory cues of IL-6, TNF-α, IL-1β, and COX-2. Mediating the modulatory effect is chiefly done by the changes in mitogen-activated protein kinase (MAPK) and NF-κB signaling pathways of microglial cells. This indicates that propolis and other flavonoid-rich compounds can be used as promising therapeutic agents against neuroinflammation in neurodegenerative conditions by blocking such pro-inflammatory cascades.

8.3.2 Antioxidant Defense through Nrf2-Activation and ROS-Neutralization

The lack of balance between the redox stress and antioxidant defense is also a contributing factor to mitochondrial dysfunction, protein misfolding, and chronic inflammation, which are all identified as hallmarks of NDDs. High levels of ROS are commonly found in Alzheimer's disease, Parkinson's disease, Huntington's disease, Friedreich ataxia, multiple sclerosis, and stroke.

Bee propolis is a resinous compound. It is high in flavonoids, phenolic acids, terpenoids, and strong antioxidants (Martinello & Mutinelli, 2021). Explaining its neuroprotective effects and the specific activation of nuclear factor erythroid 2-related factor 2 (Nrf2). Nrf2 is a transcription factor, and a cell's master regulator of the genes that maintain and redox balance. Under conditions of cellular oxidative stress, Nrf2 unbinds from a cytoplasmic tether and enters the nucleus. There, Nrf2 binds to the antioxidant response elements (AREs) of the DNA and activates transcription of cytoprotective genes. These include heme oxygenase-1 (HO-1), glutathione S-transferase (GST), and superoxide dismutase (SOD). Collectively, these cytoprotective enzymes and mitochondria superoxide dismutase (mtSOD) preserve cell intact and neutralize the ROS.

Propolis components trigger the activation of the Nrf2 factor and the direct scavenging of ROS species radicals such as hydroxyl radicals, superoxide anions, and extremely reactive nitrogen oxides. The concurrent execution of these functions helps minimize the oxidative neuronal damage associated with degenerative disorders. In clinical trials, the biological stress propolis supplements reduce within the body, including the stress markers malondialdehyde, carbonyls, and thiobarbituric acid reactive substances. Additionally, propolis supplements help increase the levels of antioxidant enzymes superoxide dismutase, glutathione peroxidase, and catalase in the body (Malekahmadi et al., 2023).

The use of bee propolis supplementation enhances the antioxidant defenses of the body and reduces the level of oxidative stress. Propolis has potential as a supportive natural agent by reversing this stress, which is central to the onset of most neurodegenerative diseases.

8.3.3 Neuroplasticity Support and Anti-Proteinopathy Actions (Ab, BDNF)

One of the most notable characteristics of neurodegenerative disorders is the unusual crystallization of some proteins and thus the concentration of toxic amyloid-beta and other misshapen proteins within the neurons. This accumulation is associated with neuronal damage, dysfunction, and eventual cell death. In the case of Alzheimer's disease (AD), the amyloid-beta (Aβ) peptides are deposited in the brain in insoluble aggregates that destroy the ability of the synapses to be active and cause the decline of the neurons. Prevention of disease progression must involve the use of strategies that prevent amyloid-beta (Aβ) buildup and promote neuronal survival.

Bee propolis has significant potential in this respect. Its bioactive compounds, especially flavonoids and phenolic acids, have been effective in treating protein-misfolding disorders, as well as maintaining the functionality of neurons. It has been experimentally shown that propolis extracts have the ability to inhibit Aβ aggregation and promote its clearance, highlighting its therapeutic benefits in the process of treating proteinopathies related to neurodegenerative diseases. Mechanistically, flavonoids such as pinocembrin and chrysin interfere with the formation of toxic Aβ oligomers by stabilizing monomeric forms or promoting disaggregation of existing fibrils. This reduces the neurotoxic burden and mitigates downstream inflammatory cascades triggered by Aβ deposits.

Propolis stimulates and strengthens cognitive functions, neuroplasticity, and anti-proteinopathy mechanisms. BDNF is key to synaptic neuroplasticity, memory consolidation, and neural survival, and is negatively correlated to cognitive decline in NDDs (Azman & Zakaria, 2022). Bioactives in propolis increase BDNF by activating CREB and the PI3K/Akt pathways, strengthening neurons and coupled synaptic, neural, and cellular defense mechanisms. This combined approach, countering protein misfolding and stimulating neuroplasticity, defines propolis as having broad-spectrum neuroprotective properties (Menezes da Silveira et al., 2020). The Aβ pathology in Alzheimer's disease means that targeting both mechanisms of action are essential, to the tune of propolis and synaptic dysfunction.

8.3.4 Propolis Synergy: A Systems Pharmacology View

As observed in network pharmacology, therapeutic actions of propolis components, including flavonoids (chrysin, pinocembrin), phenolic acids (CAPE), and terpenoids, work cumulatively and synergistically to multi-target various neuropharmacological mechanisms in Alzheimer's and Parkinson's, instead of isolated actions. This illustrates a systems biology approach.

Propolis interdependently affects four key proteostasis pathways at once: inflammation (NF-κB), oxidative stress (Nrf2), protein aggregation (involving Aβ, tau, α-synuclein), and neuroplasticity (BDNF). Propolis powerfully and simultaneously adjusted the flow through several of these systems (Figure 8.1). Complex bioactive propolis compounds in the cascades involve regulation of NF-κB, MAPK, and ROS pathways, advanced resolution and stability of mitochondrial processes, and control of apoptosis to reduce pro-inflammatory and oxidative damage (Bertolucci et al., 2025).

Beyond the interactions defining individual compounds, the components synergistically and significantly enhance and broaden an individual's bioavailability and pharmacodynamics. Systems pharmacology incorporates the experimental approach with computational modeling to define key therapeutic bioactive, rational pharmacology, standardized formulation, and design integration of calculations for dosage, drug combination predictions, and clinical rational translation.

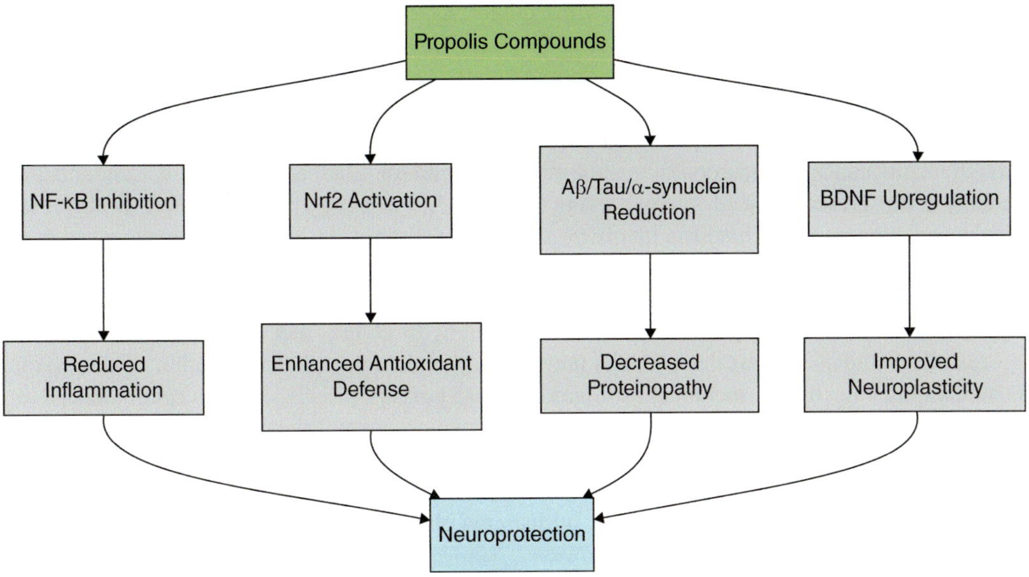

FIGURE 8.1 Propolis compounds inhibit four mechanisms of neurodegeneration: anti-inflammatory NF-κB inhibition, antioxidant Nrf2 activation, reduction of protein aggregates (anti-proteinopathy), and upregulation of BDNF (neuroplasticity). Pathways converge interdependently in the neuroprotection of propolis against neurodegenerative diseases.

8.4 PRECLINICAL EVIDENCE

Propolis has shown the capacity to counter adverse effects of neurodegenerative diseases through the action of its bioactive compounds employed through active compounds. A few of the most well-known of these constituents include certain flavonoids, phenolic acids, and terpenoids, all of which have been shown to posses antioxidant, pro-inflammation (especially terpenoids and flavonoids), and neuroprotective properties, which impact/make changes to the molecular processes of lesions and neurodegeneration. The therapeutic effects of propolis can also vary due to its source, as its geosource (botanical and geographical origin) also changes the chemical constituents within the propolis. Numerous studies have shown that in vitro and in vivo studies done with propolis have also shown some very positive effects on Alzheimer's disease and Parkinson's disease, as well as some initial studies with amyotrophic lateral sclerosis and multiple sclerosis. The findings sustain the use of propolis in the formulation of standardized therapeutic modalities for clinical use as multi-target approaches.

8.4.1 Alzheimer's Disease Models: Targeting Tau, Aβ, and Rescuing Memory

Alzheimers disease pathology is described as an accumulation of the pathological elements of the disease, amyloid-beta (Aβ) plaques, and the hyperphosphorylated tau protein (tau). Synaptic dysfunction, cognitive impairment, and neuronal loss are the consequences of clinical AD manifestation. Preclinical models of AD have shown the constitutive and multifaceted neuroprotective activities of propolis and the constituent elements.

For AD, the natural bioactive products research utilizes propolis based on various model systems for the disease. The propolis of the stingless bees *Plebeia catamarcensis* and *Tetragonisca fiebrigi* demonstrated antioxidant properties and an enhancement on the longevity of *C. elegans* models, while also diminishing the β-amyloid paralyzing models (Arteman et al., 2025). Research also demonstrates that pinocembrin, a flavonoid component of propolis, provides significant neuroprotective effects in AD models. In vitro studies show pinocembrin protects neurons against Aβ-induced toxicity through multiple mechanisms, including inhibition of RAGE-dependent signaling pathways, regulation of mitochondrion-mediated apoptosis, and activation of Nrf2/HO-1 antioxidant pathways. In APP/PS1 transgenic mice, a three-month pinocembrin (purity > 99%) prevented cognitive decline without altering Aβ burden, instead conferring neurovascular protection through maintenance of neuropil ultrastructure, reduction of glial activation and inflammatory mediators, and preservation of microvascular function (Liu et al., 2014). This effect is attributed to enhanced clearance mechanisms involving the upregulation of Aβ-degrading enzymes, such as neprilysin and insulin-degrading enzyme (IDE), as well as the inhibition of β-secretase activity, which reduces Aβ production. Additionally, propolis flavonoids exhibit direct anti-aggregation properties by interfering with Aβ oligomerization, thereby limiting synaptotoxic species formation.

Preclinical studies suggest that propolis may also target the hyperphosphorylation of tau protein by modulating key kinases, including glycogen synthase kinase-3β (GSK-3β) and cyclin-dependent kinase 5 (CDK5), leading to neurofibrillary tangle formation, a key driver of neuronal dysfunction in AD. Studies have established that protein phosphatase 2A (PP2A) and glycogen synthase kinase-3β (GSK-3β) are critical regulators of tau phosphorylation, with complex bidirectional interactions between these enzymes. PP2A serves as the primary tau phosphatase, while GSK-3β acts as a major tau kinase (Manoharan et al., 2024). However, Liu et al. (2018) showed that galangin and pinocembrin, two flavonoids found in propolis, ameliorate insulin resistance in HepG2 cells by upregulating phosphorylation of insulin receptor Akt and GSK3β while downregulating IRS phosphorylation via the Akt/mTOR signaling pathway. Importantly, propolis can increase PPP2R1A expression, a subunit of the PP2A enzyme, according to (Durmaz et al., 2025). This contributes to decreased tangle formation and improved neuronal integrity.

From a therapeutic perspective, propolis treatment has consistently demonstrated cognitive benefits in AD models. Behavioral assessments in AD models consistently indicate that propolis improves memory and learning deficits. Morris water maze and novel object recognition tests reveal that treated animals display enhanced spatial memory and cognitive flexibility. In APP-KI mice, propolis administration (300–1,000 mg/kg daily for 8 weeks) significantly prevented cognitive decline in Y-maze, novel object recognition, and passive avoidance tasks, particularly in younger (4-month) animals (Inagaki et al., 2024). These behavioral improvements correlated with restored hippocampal long-term potentiation (LTP) and enhanced synaptic plasticity. Mechanistically, propolis activated intracellular calcium signaling pathways, increasing phosphorylation of CaMKII, CaMKIV, CREB, and AMPAR subunit GluA1, while promoting BDNF production in the CA1 region (Inagaki et al., 2022).

These preclinical results provide strong evidence that bee propolis targets multiple AD pathologies, Aβ aggregation, tau phosphorylation, oxidative stress, and synaptic impairment, leading to functional recovery.

8.4.2 Parkinson's Disease Models: Protecting Dopaminergic Neurons and Targeting α-synuclein

Parkinson's disease (PD), the second-most common neurodegenerative disorder, is characterized by progressive loss of dopaminergic neurons in the substantia nigra pars compacta and the accumulation of α-synuclein protein aggregates forming Lewy bodies. Preclinical models, including neurotoxin-induced (e.g., MPTP, 6-OHDA) and genetic models, have been very important in assessing neuroprotective strategies. Propolis and its bioactive flavonoids exhibit significant neuroprotection in PD models by mitigating oxidative stress and neuroinflammation, two major contributors to dopaminergic cell death.

It has been demonstrated that stimulation of the Nrf2 antioxidant pathway is neuroprotective in Parkinsonism models. As an illustration, Jaźwa et al. (2011) showed that a Nrf2 activator, sulforaphane, rescued dopaminergic neurons against MPTP-induced toxicity in wild-type mice—but not in Nrf2-deficient mice—by inducing the phase II antioxidant enzyme heme oxygenase-1 (HO-1) and alleviating neuroinflammation. Correspondingly, Kaidery et al. (2013) indicated that Nrf2-dependent cytoprotective genes were induced by synthetic triterpenoids and that the toxicity of MPTP induced dopaminergic neurotoxicity in an Nrf2-dependent fashion.

Propolis also suppresses the activation of microglia and inhibits pro-inflammatory cytokine release, including TNF-α and IL-1β, and suppresses the neuroinflammatory mechanisms that lead to neuronal damage and disease progression. Following Wu et al. (2017), research on how Brazilian green propolis significantly inhibits the TNF-α, IL-1β, and IL-6 cytokines release from activated microglia clearly describes its anti-neuroinflammatory properties.

Recent preclinical evidence suggests that propolis may directly target α-synuclein pathology, a central feature of PD pathogenesis. Using in vitro fibrillation assays with recombinant α-synuclein, propolis extracts, and specifically chrysin and galangin, were shown to inhibit α-synuclein aggregation in a concentration-dependent manner (Limanaqi et al., 2019). The evidence supports that phytochemicals preserve neuronal function by stabilizing monomeric α-synuclein forms and enhancing cellular degradation mechanisms.

Additionally, Badban et al. (2015) showed that propolis treatment improved motor coordination and preserved tyrosine hydroxylase-positive neurons (107 $\pm$ 5.77 vs. 63 $\pm$ 6.58) compared to 6-OHDA alone, while enhancing antioxidant enzyme activities and reducing oxidative stress markers.

8.4.3 Emerging Evidence in Other Neurodegenerative Diseases (e.g., ALS, MS)

Outside of AD and PD, the initial preclinical studies indicate that propolis might provide some benefits in other neurodegenerative conditions, such as ALS and MS.

Amyotrophic lateral sclerosis (ALS) is a progressive neurodegenerative disorder that results in the degeneration of motor neurons in the spinal cord and brain stem, eventually leading to devastating consequences, such as wasting and paralysis of muscles, and death within two to three years (Barber & Shaw, 2010). One of the key factors that causes the development of ALS is oxidative stress, and damaged tissues, as well as the cerebrospinal fluid of ALS patients, display elevated oxidative damage markers. Although there is still a lack of detailed mechanistic studies, the antioxidant and anti-inflammatory action of propolis may explain the preservation of motor neurons.

Multiple sclerosis (MS) is an autoimmune inflammatory disease that is characterized by progressive central nervous system demyelination. The main animal model for assessing the pathophysiology of MS and its treatment is experimental autoimmune encephalomyelitis (EAE). Propolis and its active constituents have been documented to possess significant immunomodulatory properties in MS models. Inhibiting the main propolis constituent, caffeic acid phenethyl ester (CAPE), resulted in a significant reduction of the disease incidence and severity, as well as a decline in inflammatory cell infiltration, microglial activation, and injury associated with demyelination (Zhou et al., 2020).

Preclinical studies across different models of neurodegenerative diseases are yielding positive outcomes regarding the neuroprotective properties of propolis and its bioactive constituents. Propolis seems to be a natural therapeutic option, being able to mitigate some of the key pathogenic mechanisms, including addressing Aβ and tau pathologies in AD, loss of dopaminergic neurons and α-synuclein aggregation in PD, and oxidative stress and inflammation in ALS and MS. There is a need for rigorous preclinical and clinical studies to properly elucidate the therapeutic value of propolis in the treatment of neurodegenerative diseases.

8.5 CLINICAL EVIDENCE AND TRANSLATIONAL CHALLENGES

Interest in clinical assessment of propolis for NDDs, given the promising preclinical results, is well documented. Challenges remain, however, and these are primarily associated with the human therapeutic application of these results. The human studies are summarized in this section, with a focus on key gaps in translation and the potential contribution of biomarkers to the progression of clinical trials on propolis for NDDs.

8.5.1 Review of Human Trials and Observational Studies

There are only a few clinical studies on propolis and its bioactive components on NDDs, and for the most part, these are only in the preliminary stages. Most of the human studies to date have focused on AD and mild cognitive impairment (MCI), with a few exploratory studies on PD and other cognitive disorders. Table 8.1 summarizes key clinical trials and observational studies, highlighting population characteristics, dosing regimens, treatment durations, and main outcomes.

Asama et al. (2021) reported that in a placebo-controlled study of 79 elderly Japanese, daily supplementation with propolis extract for 24 weeks significantly improved cognitive function. Compared to the placebo group, the propolis group significantly showed better verbal memory and processing speed as measured by Cognitrax. Additionally, propolis improved cholesterol levels and kidney function markers, all within normal ranges. No side effects were reported. The data show that propolis improves memory, attention, and information processing in older adults.

A study over 24 months with a double-blind methodology involved older residents at high altitudes. This study showed improvements in cognitive function which varied with the amount of Brazilian green propolis used (0.33 g a day). Over 12 months, participants scored 1.64 points higher and 3.05 points higher over 24 months on the MMSE, and the ADAS-Cog scores dropped over the same periods by 1.67 and 2.20 points. These improvements in cognition were associated with a decrease in systemic inflammation: a decrease of 54–76% in serum IL-6 and an increase of 150–350% in TGFβ1 (Liu et al., 2018).

Overall, clinical data are encouraging but remain limited, as there is insufficient literature for review.

8.5.2 Key Translational Gaps: Standardization, Dosing, and Safety Profile

Clinical findings have shown great promise of bee propolis in neurodegenerative diseases (NDDs). However, despite its promise, several translational challenges are hindering its widespread use.

TABLE 8.1
Summary of Clinical Studies Evaluating Bee Propolis in Neurodegenerative Conditions

Study (Year)	Population	Dose and Formulation	Duration	Outcomes and Findings
Asama et al. (2021)	Japanese men and women aged 60–79 with MMSE scores of 24–29 (35)	Participants orally received a dietary supplement containing propolis extract for 24 weeks.	24 weeks	Propolis intake significantly improved verbal memory.
Liu et al. (2018)	Sample of 40 people with mean ages of 68.5 years (45% men) living in Qinghai province, China (at altitudes of 2,000–3,200 meters)	Participants received 0.33 g/day Brazilian green propolis extract capsules.	>12 months	Improves the cognitive functions in elderly people living at high altitude
Zhu et al. (2018)	Sample of 30 people with an average of 72.8 years living at an altitude of 2,260 meters	Participants received 0.83g/day of Brazilian green propolis.	24 months	Ingestion of propolis (>12 months) protected against cognitive decline after systemic inflammation was reduced.

8.5.2.1 Standardization of Propolis Preparations

One major limitation is due to the lack of standardized propolis products. Propolis composition varies widely depending on geographic origin, botanical sources, and extraction methods. This variability affects the concentration of active bioflavonoids and phenolic compounds (e.g., flavonoids, CAPE), leading to inconsistent pharmacological effects. Developing standardized, well-characterized propolis extracts with defined chemical profiles is essential to ensure reproducibility and regulatory approval in clinical settings.

8.5.2.2 Optimal Dosing and Treatment Duration

Another critical gap concerns the optimal dosing preparation. Human studies have employed a wide range of doses from 200 to 1,000 mg daily and durations ranging from a few weeks to several months. Meanwhile, preclinical studies have used doses ranging from 50 to 400 mg/kg, which is approximately equivalent to high doses of 4–32 g/day based on body surface area calculation in humans. The dose-response relationship remains poorly defined, and it is unclear whether higher doses or longer treatment durations would yield greater clinical benefits or pose safety concerns. Dose-finding studies and long-term trials are needed to establish effective and safe protocols.

8.5.2.3 Safety and Drug Interaction Profile

Propolis is generally considered safe with low toxicity based on its long history of traditional use; however, comprehensive safety evaluations in elderly or neurologically impaired populations are limited. Reported potential allergic reactions (particularly in individuals with bee product sensitivities) are mild, and gastrointestinal upset must be carefully monitored. Additionally, the possibility of interactions with conventional medications used by NDD patients (e.g., cholinesterase inhibitors, dopaminergic agents) requires systematic investigation. Pharmacokinetic and pharmacodynamic studies can clarify these aspects and support clinical guidelines for propolis use.

8.5.3 The Role of Biomarkers in Future Trial Design

The incorporation of developed sensitive and specialized biomarkers into clinical trials of bee propolis for NDDs offers a powerful approach to improve clinical trial design, patient stratification, and mechanistic understanding.

8.5.3.1 Neurofilament Light Chain (NfL)

NfL, a marker of axonal injury measurable in blood and cerebrospinal fluid (CSF), is emerging as a sensitive indicator of neurodegeneration across multiple disorders such as AD and PD. Monitoring NfL levels before and after propolis treatment could provide objective evidence of neuroprotective effects and disease-altering potential.

8.5.3.2 CSF Tau and Aβ42

In Alzheimer's disease, CSF biomarkers such as total tau, phosphorylated tau, and Aβ42 levels are still the standard for diagnosis and progression monitoring. Incorporating these biomarkers can help assess whether propolis interventions influence core AD pathologies, correlating biochemical changes (reduction of tau phosphorylation and amyloid accumulation) with cognitive outcomes.

8.5.3.3 Functional Magnetic Resonance Imaging (fMRI)

Advanced neuroimaging techniques, including functional MRI (fMRI), positron emission tomography (PET), diffusion tensor imaging (DTI), and other advanced MRI techniques, enable the assessment of brain activity, connectivity, and functional changes in response to treatment. Changes in hippocampal activation or default mode network connectivity could serve as early indicators of cognitive improvement or stabilization following propolis therapy.

8.5.3.4 Other Emerging Biomarkers

Additional biomarkers such as inflammatory cytokine panels, oxidative stress markers, and proteomic/metabolomic profiles may further elucidate propolis's multifaceted mechanisms. These could aid in identifying responsive patient subgroups and optimizing personalized treatment approaches.

8.6 CONCLUSION AND FUTURE DIRECTIONS

8.6.1 Summary: Propolis as a Mechanistically Validated Neuroprotective Agent

Bee propolis has appeared as a compelling natural product with multifaceted neuroprotective properties against NDDs. Supported by extensive preclinical evidence, propolis exerts antioxidant, anti-inflammatory, anti-proteinopathy, and neuroplasticity-promoting effects that address key pathological hallmarks of disorders such as AD and PD. Its bioactive compounds, including flavonoids and phenolic acids, modulate molecular targets ranging from amyloid-beta aggregation and tau phosphorylation to α-synuclein fibrillation and microglial activation. Animal and cell culture models consistently demonstrated bee propolis's ability to preserve neuronal viability, improve synaptic function, and rescue cognitive and motor deficits.

Early clinical trials and observational studies, though limited, support these findings by showing cognitive improvements, modulation of inflammatory biomarkers, and good tolerability in elderly populations. The merging of mechanistic insights and preliminary human data positions bee propolis as a mechanistically validated neuroprotective agent with promising translational potential.

8.6.2 Priorities for Translation: Formulation, Regulatory Strategy, and Biomarker-Based Clinical Trials

Despite the established encouraging data on the effectiveness of bee propolis, several critical priorities must be addressed to successfully incorporate propolis into clinical practice for NDDs.

8.6.2.1 Standardized, Bioavailable Formulations

The most immediate priority for clinical translation involves the development of standardized, pharmaceutical-grade propolis formulations suitable for clinical investigation. The compound and chemical heterogeneity of propolis, which differs based on geographic origin and extraction methods, necessitates the development of consistently characterized formulations with defined concentrations of active compounds. Enhancing bioavailability, stability, and blood-brain barrier penetration through advanced delivery systems (e.g., nanoformulations, liposomal encapsulation) will be essential to maximize therapeutic effects.

These advanced bee propolis formulations will require comprehensive preclinical evaluation, including pharmacokinetic studies and safety assessments, before advancing to human trials. Such formulations may achieve therapeutic levels within the desired range, allowing longer intervals between doses, which can improve adherence in the chronic treatment of NDDs.

8.6.2.2 Regulatory Pathway and Quality Control

One of the priorities to progress propolis-based therapeutics into clinical studies is a clearly defined regulatory pathway. The regulatory pathway for natural products, such as bee propolis, involves intensive control of quality and documented safety and efficacy.

In this context, early dialogue with regulatory bodies will be necessary to outline the likely pathways for development. Regulatory requirements will involve the establishment of standardized production processes and documented, reproducible pharmacokinetic profiles, as well as evidence from well-conducted clinical trials. This may be facilitated by coordinated activities between academia, industry, and regulatory bodies.

8.6.2.3 Biomarker-Guided, Well-Powered Clinical Trials

Future clinical trials should adopt biomarker-based approaches focusing on NfL, CSF tau, and amyloid-beta, and neuroimaging (fMRI) for participant selection and outcome measures. Efficacy determination and optimal dose adjustment warrant large-scale, randomized, placebo-controlled studies coupled with comprehensive cognitive tests.

8.6.3 Final Perspective on Propolis among Natural Therapeutics

Propolis stands out among natural therapeutics for its unique multi-target properties and safety profile. In contrast with most synthetic drugs, propolis acts on the key components of neurodegenerative disorders simultaneously (oxidative stress, inflammation, and aberrant protein misfolding). Propolis acts as a prevention adjunct to pharmacotherapy and, as clinical evidence and traditional usage, should be adopted integrally. Propolis can be used as a part of other therapies and lifestyle changes. In conclusion, with its natural origin and therapeutic versatility, propolis can be used as a part of personalized medicine for older patients suffering from cognitive decline.

REFERENCES

Arteman, K. S., Rocha, P. D. S., Leite, D. F., Oliveira, A. S., Silva, I. V., Jorge, N. G., Soutilha, A. S., Dos Santos, H. F., Baldivia, D., Balestieri, J., Cardoso, C., Souza, K. P., Dos Santos, E. L., & Campos, J. F. (2025). Plebeia catamarcensis and Tetragonisca fiebrigi propolis promotes longevity and anti-Alzheimer effects in *Caenorhabditis elegans*. *PLOS ONE*, *20*(1), e0321487. https://doi.org/10.1371/journal.pone.0321487

Asama, T., Hiraoka, T., Ohkuma, A., Okumura, N., Yamaki, A., & Urakami, K. (2021). Evidence-based complementary and alternative medicine. *Evidence-Based Complementary and Alternative Medicine*, *2021*, Article 6664217. https://doi.org/10.1155/2021/6664217

Azman, K. F., & Zakaria, R. (2022). Recent advances on the role of brain-derived neurotrophic factor (BDNF) in neurodegenerative diseases. *International Journal of Molecular Sciences*, *23*(12), 6827. https://doi.org/10.3390/ijms23126827

Babazadeh, A., Vahed, F. M., & Jafari, S. (2020). Nanocarrier-mediated brain delivery of bioactives for treatment/prevention of neurodegenerative diseases. *Journal of Controlled Release*, *321*, 211–221. https://doi.org/10.1016/j.jconrel.2020.02.015

Badban, L., Safari, M., Sameni, H., Bandegi, A., Vafaei, A. A., Rashidy-Pour, A., & Ghahari, L. (2015). Protective effects of water extract of propolis on dopaminergic neurons, brain-derived neurotrophic factor, and oxidative stress factors in the rat model of Parkinson's disease. *International Journal of Pharmacology*, *11*(3), 300–308. https://doi.org/10.3923/IJP.2015.300.308

Bankova, V. (2005). Chemical diversity of propolis and the problem of standardization. *Journal of Ethnopharmacology*, *100*(1–2), 114–117. https://doi.org/10.1016/J.JEP.2005.05.004

Barber, S. C., & Shaw, P. J. (2010). Oxidative stress in ALS: Key role in motor neuron injury and therapeutic target. *Free Radical Biology and Medicine*, *48*(5), 629–641. https://doi.org/10.1016/j.freeradbiomed.2009.11.018

Bertolucci, V., Ninomiya, A. F., Longato, G., Kaneko, L. O., Nonose, N., Scariot, P., & Messias, L. (2025). Bioactive compounds from propolis on bone homeostasis: A narrative review. *Antioxidants*, *14*(1), 81. https://doi.org/10.3390/antiox14010081

Del Fabbro, L., Bortolotto, V., Ferreira, L. M., Sari, M. H. M., & Furian, A. F. (2025). Chrysin's anti-inflammatory action in the central nervous system: A scoping review and evidence-gap mapping of its mechanisms. *European Journal of Pharmacology*, *963*, 177602. https://doi.org/10.1016/j.ejphar.2025.177602

Durmaz, B., Oktay Çelebi, L. M., Çekin, A., Ahadova, A., Günel, N., Kalkan Yıldırım, H., Özgönül, A., & Sözmen, E. Y. (2025). Effect of propolis on PPP2R1A and apoptosis in cancer cells. *Biochemistry Research International*, *2025*, 5538068. https://doi.org/10.1155/bri/5538068

Elbatreek, M. H., Mahdi, I., Ouchari, W., Mahmoud, M. F., & Sobeh, M. (2023). Current advances on the therapeutic potential of pinocembrin: An updated review. *Biomedicine & Pharmacotherapy*, *160*, 114032. https://doi.org/10.1016/j.biopha.2022.114032

Gardana, C., Scaglianti, M., Pietta, P., & Simonetti, P. (2007). Analysis of the polyphenolic fraction of propolis from different sources by liquid chromatography-tandem mass spectrometry. *Journal of Pharmaceutical and Biomedical Analysis*, *45*(3), 390–399. https://doi.org/10.1016/J.JPBA.2007.06.022

Ghallab, D. S., Shawky, E., Mohyeldin, M. M., Metwally, A. M., & Ibrahim, R. S. (2025). Propolis: An update on its chemical diversity, botanical origin, and biological activities. *Journal of Advanced Pharmaceutical Sciences*, *5*(1), 1–15. https://doi.org/10.21608/japs.2025.336494.1032

Inagaki, R., Yamakuni, T., & Moriguchi, S. (2022). Propolis ameliorates cognitive decline in Alzheimer's disease model mice. *Proceedings for Annual Meeting of the Japanese Pharmacological Society*, *96*, 3-b-o07-5. https://doi.org/10.1254/jpssuppl.96.0_3-b-o07-5

Inagaki, R., Yamakuni, T., Saito, T., Saido, T., & Moriguchi, S. (2024). Preventive effect of propolis on cognitive decline in Alzheimer's disease model mice. *Neurobiology of Aging*, *135*, 74–83. https://doi.org/10.1016/j.neurobiolaging.2024.03.002

Jaźwa, A., Rojo, A. I., Innamorato, N. G., Hesse, M., Fernández-Ruiz, J., & Cuadrado, A. (2011). Pharmacological targeting of the transcription factor Nrf2 at the basal ganglia provides disease-modifying therapy for experimental Parkinsonism. *Antioxidants and Redox Signaling*, *14*(12), 2347–2360. https://doi.org/10.1089/ars.2010.3731

Jorge, R. F., Furtado, N., Sousa, J., Fiho, A. A. D. S., Gregório, L. E., Martins, C., Soares, A., Bastos, J., Cunha, W., & Silva, M. (2008). Brazilian propolis: Seasonal variation of the prenylated p-coumaric acids and antimicrobial activity. *Pharmaceutical Biology*, *46*(7–8), 568–573. https://doi.org/10.1080/13880200802370373

Kaidery, N., Banerjee, R., Yang, L., Smirnova, N., Hushpulian, D., Liby, K., Williams, C. R., Yamamoto, M., Kensler, T., Ratan, R., Sporn, M., Beal, M., Gazaryan, I., & Thomas, B. (2013). Targeting Nrf2-mediated gene transcription by extremely potent synthetic triterpenoids attenuates dopaminergic neurotoxicity in the MPTP mouse model of Parkinson's disease. *Antioxidants and Redox Signaling*, *18*(2), 139–150. https://doi.org/10.1089/ars.2011.4491

Limanaqi, F., Biagioni, F., Busceti, C., Ryskalin, L., Polzella, M., Frati, A., & Fornai, F. (2019). Phytochemicals bridging autophagy induction and alpha-synuclein degradation in parkinsonism. *International Journal of Molecular Sciences*, *20*(13), 3274. https://doi.org/10.3390/ijms20133274

Liu, R., Li, J., Song, J., Zhou, D., Huang, C., Bai, X., Xie, T., Zhang, X., Li, Y., Wu, C., Zhang, L., Li, L., Zhang, T., & Du, G. (2014). Pinocembrin improves cognition and protects the neurovascular unit in Alzheimer related deficits. *Neurobiology of Aging*, *35*(6), 1275–1285. https://doi.org/10.1016/j.neurobiolaging.2013.12.031

Liu, Y., Liang, X., Zhang, G., Kong, L., Peng, W., & Zhang, H. (2018). Galangin and pinocembrin from propolis ameliorate insulin resistance in HepG2 cells via regulating Akt/mTOR signaling. *Evidence-Based Complementary and Alternative Medicine*, *2018*, 7971842. https://doi.org/10.1155/2018/7971842

Malekahmadi, M., Pahlavani, N., Heshmati, J., Clayton, Z., Beigmohammadi, M., Navashenaq, J. G., & Alavi-Naeini, A. (2023). Effect of propolis supplementation on oxidative stress markers: A systematic review of randomized controlled trials. *Journal of Herbal Medicine*, *36*, 100679. https://doi.org/10.1016/j.hermed.2023.100679

Manoharan, S. D., Hamid, H., Md Hashim, N. F., Cheema, M., Chiroma, S. M., Mustapha, M., & Mehat, M. Z. (2024). Could protein phosphatase 2A and glycogen synthase kinase-3 beta be targeted by natural compounds to ameliorate Alzheimer's pathologies? *Brain Research*, *1821*, 148793. https://doi.org/10.1016/j.brainres.2024.148793

Martinello, M., & Mutinelli, F. (2021). Antioxidant activity in bee products: A review. *Antioxidants*, *10*(1), 71. https://doi.org/10.3390/antiox10010071

Menezes da Silveira, C. C. S., Luz, D. A., da Silva, C. C. S., Prediger, R., Martins, M., Martins, M. A. T., Fontes-Júnior, E., & Maia, C. (2020). Propolis: A useful agent on psychiatric and neurological disorders? A focus on CAPE and pinocembrin components. *Medicinal Research Reviews*. Advance online publication. https://doi.org/10.1002/med.21757

Moise, A., & Bobiș, O. (2020). Baccharis dracunculifolia and Dalbergia ecastophyllum, main plant sources for bioactive properties in green and red Brazilian propolis. *Plants*, *9*(11), 1619. https://doi.org/10.3390/plants9111619

Shahinozzaman, M., Basak, B., Emran, R., Rozario, P., & Obanda, D. N. (2020). Artepillin C: A comprehensive review of its chemistry, bioavailability, and pharmacological properties. *Fitoterapia*, *147*, 104775. https://doi.org/10.1016/j.fitote.2020.104775

Shen, X., Liu, Y., Luo, X., & Yang, Z. (2019). Advances in biosynthesis, pharmacology, and pharmacokinetics of pinocembrin, a promising natural small-molecule drug. *Molecules*, *24*(12), 2323. https://doi.org/10.3390/molecules24122323

Spagnuolo, C., Moccia, S., & Russo, G. (2017). Anti-inflammatory effects of flavonoids in neurodegenerative disorders. *European Journal of Medicinal Chemistry*. Advance online publication. https://doi.org/10.1016/j.ejmech.2017.09.001

Trusheva, B., Popova, M., Koendhori, E., Tsvetkova, I., Naydenski, C., & Bankova, V. (2011). Indonesian propolis: Chemical composition, biological activity and botanical origin. *Natural Product Research*, *25*(6), 606–613. https://doi.org/10.1080/14786419.2010.488235

World Health Organization. (2017, December 7). *Dementia number of people affected to triple in next 30 years*. https://www.who.int/news/item/07-12-2017-dementia-number-of-people-affected-to-triple-in-next-30-years

Wong, H., Wu, X., & Bendayan, R. (2012). Nanotechnological advances for the delivery of CNS therapeutics. *Advanced Drug Delivery Reviews*, *64*(7), 686–700. https://doi.org/10.1016/j.addr.2011.10.007

Wu, Z., Zhu, A., Takayama, F., Okada, R., Liu, Y., Harada, Y., Wu, S., & Nakanishi, H. (2013). Brazilian green propolis suppresses the hypoxia-induced neuroinflammatory responses by inhibiting NF-κB activation in microglia. *Oxidative Medicine and Cellular Longevity*. Advance online publication. https://doi.org/10.1155/2013/906726

Wu, Z., Liu, Y., Zhu, A., & Wu, S. (2017). Brazilian green propolis suppresses microglia-mediated neuroinflammation by inhibiting NF-kB activation. *Journal of Neurological Sciences*, *381*, 678. https://doi.org/10.1016/j.jns.2017.08.1910

Yordanov, Y. (2019). Caffeic acid phenethyl ester (CAPE): Pharmacodynamics and potential for therapeutic application. *Pharmacia*, *66*(3), 107–114. https://doi.org/10.3897/pharmacia.66.e38573

Zhou, Y., Wang, J., Chang, Y., Li, R., Sun, X., Peng, L., Zheng, W., & Qiu, W. (2020). Caffeic acid phenethyl ester protects against experimental autoimmune encephalomyelitis by regulating T cell activities. *Oxidative Medicine and Cellular Longevity*. Advance online publication. https://doi.org/10.1155/2020/7274342

Zhu, A., Wu, Z., Zhong, X., Ni, J., Li, Y., Meng, J., Du, C., Zhao, X., Nakanishi, H., & Wu, S. (2018). Brazilian green propolis prevents cognitive decline into mild cognitive impairment in elderly people living at high altitude. *Journal of Alzheimer's Disease*, *63*(2), 551–560. https://doi.org/10.3233/JAD-170630

Zullkiflee, N., Taha, H. S., & Usman, A. (2022). Propolis: Its role and efficacy in human health and diseases. *Molecules*, *27*(18), 6120. https://doi.org/10.3390/molecules27186120

9 Bee Pollen Applications in Neurodegenerative Diseases

Joseph Igbo Enya, Naomi Joshua, Leviticus O. Arietarhire, Samson Ehindero Olorunnado, Toluwanimi Omolara Afolabi, Kelechi Roselyn Onyema, Sunday Okon Elijah, Babatunde Samuel Adeyemi, Michael Anozie Amadi, Ajirioghene E. Akpotu, Olusegun G. Adebayo, and Ijeoma Onyeleonu

9.1 INTRODUCTION

Neurodegenerative diseases (NDDs) which include Alzheimer's disease (AD), Parkinson's disease (PD), amyotrophic lateral sclerosis (ALS), and others, are characterized by the progressive dysfunction and loss of neurons, primarily caused by oxidative stress, neuroinflammation, mitochondrial dysfunction, and protein aggregation (Jan et al., 2022; Chaudhary et al., 2023). Despite the significant progress in symptomatic treatment for NDDs, there are no curative medicines available and therefore indicating the pressing necessity for innovative strategies toward a provision possible solution. Natural products that are rich in phytochemicals have emerged as potential neuroprotective agents (Asuku et al., 2024; Ajibare et al., 2025), with bee pollen attracting increased study due to its diverse bioactive contents (Chaudhary et al., 2023). Bee pollen is a composite of plant pollens gathered by bees (Čeksterytė et al., 2023; Lim et al., 2024), and it consists of 30–55% carbs, 15–25% protein, and 1–13% lipids (Table 9.1), in addition to lots of very important polyunsaturated fatty acids, minerals, flavonoids, amino acids, vitamins (particularly B-complex), and phenolic acids (De-Melo & de Almeida-Muradian, 2017; Pohl et al., 2023), as shown in Table 9.1. All of these substances hold significant antioxidant, anti-inflammatory, and also neuroprotective characteristics that are capable of enhancing cognitive resilience and neurotransmitter balance in the biological system (Lahoda Brodska et al., 2023; Lim et al., 2024).

Phytochemical compounds (flavonoids) such as quercetin, kaempferol, and rutin, neutralize reactive oxygen species (ROS), maintain mitochondrial function, and regulate inflammatory pathways associated with NDDs (Mahmud et al., 2023; Al Hoque et al., 2025). Also, micronutrients which include magnesium, zinc, selenium, and others, improve synaptic transmission and myelin sheet integrity (Ares et al., 2018; Ravikumar et al., 2022). Saral et al. previously reported in a preclinical study that bee pollen intake is capable of alleviating oxidative stress induced brain damage, diminishes neuroinflammatory effects, and also enhances learning and memory functions (Saral et al., 2022). Nonetheless, inadequate reports on compositional heterogeneity as well as the scarcity of human studies overs the years have limited translational progress. Taken together, this chapter consolidates accumulating preclinical and clinical information regarding the composition, phytochemistry, and bioactivities of bee pollen, with a focus on its mechanistic potential to mitigate NDD pathology.

9.1.1 Phytochemicals and Bioactivities of Bee Pollen

9.1.1.1 Flavonoids

This is the most studied component of bee pollen. They occur in high abundance and include rosmarinic acid, quercetin, myricetin, and naringenin (Kurtagić et al., 2015). Their antioxidant properties can scavenge ROS and regulate NF-κB signaling, thereby reducing oxidative stress and

 DOI: 10.1201/9781003685654-9

TABLE 9.1
The Macro- and Microelement Composition of Bee Pollen

Constituent Class	Representative Compounds	Percentage	Neuroprotective/Nutritional Application	References
Proteins and Amino Acids	Essential amino acids (e.g., leucine, phenylalanine, glutamic acid)	15–25%	Precursors for neurotransmitters, brain development, and repair mechanisms	Taha et al. (2019), Végh and Csóka (2023)
Carbohydrates	Glucose, fructose, polysaccharides	30–55%	Energy supply for neurons, supports glial function	Bertoncelj et al. (2023)
Lipids and Fatty Acids	Linoleic acid, α-linolenic acid, oleic acid	1–13%	Anti-inflammatory effects, supports neuronal membrane integrity	Čeksterytė et al. (2023); Kostić and Kilibarda (2023)
Vitamins	B-complex (B1, B2, B6, B9), C, D, E	Trace–Moderate	Cofactors in neurotransmitter synthesis and antioxidant enzymes	Kostić and Kilibarda (2023), El Ghouizi et al. (2023)
Minerals	Potassium, calcium, magnesium, zinc, selenium	Trace–Moderate	Neurotransmission, oxidative stress regulation, and enzyme activation	El Ghouizi et al. (2023)
Polyphenols and Flavonoids	Quercetin, kaempferol, rutin, apigenin	1–2% (total flavonoids)	Antioxidant, anti-inflammatory, mitochondrial protection	Kalaba et al. (2023).
Carotenoids	β-carotene, lutein, zeaxanthin	Trace	Protects against oxidative damage to retinal and neural tissues	Salazar-González et al. (2022)
Sterols	β-sitosterol, campesterol	Trace	Neuroendocrine balance, reduces neuroinflammation	Furse et al. (2023)
Enzymes and Coenzymes	SOD, catalase, glutathione peroxidase	Trace	Antioxidant defense, mitigation of ROS-related neurodegeneration	Urcan (2025).
Nucleic Acids	RNA, DNA fragments	Trace	Supports cellular repair and protein synthesis in neurons	Khalifa et al. (2021)

inflammation (Singh et al., 2022). Notably, quercetin has previously shown neuroprotective and possible therapeutic applications for NDDs (Rodríguez-Arce & Saldías, 2024).

9.1.1.2 Phenolic Acids

Ferulic, caffeic, p-coumaric, vanillic acids, and over 20 phenolic acids have been found in bee pollen (Kostić et al., 2024). All these compounds can inhibit pro-inflammation and modulate epigenetic factors linked with NDDs (Liberal et al., 2016). They also deplete ROS and show significant levels of apigenin derivatives, which strengthen their neuroprotective ability (Kacemi & Campos, 2025).

9.1.1.3 Tannin

Tannins, terpenoids, and flavonoids are common in bee pollen (Heleno et al., 2015). They regulate ROS and inflammatory activities and inhibit neurotoxic protein aggregation in NDDs (Heleno et al., 2015; Eshbah et al., 2021; Fadzil et al., 2023).

9.1.1.4 Terpenoid

Terpenoids have antioxidant, anti-inflammatory, and cognitive-enhancing potentials in abundance, capable of regulating synaptic activity through BDNF/TrkB pathways and TRPC3 modulation. This supports neuronal survival and plasticity (Rodríguez-Pólit et al., 2023; Morrison et al., 2025).

9.1.1.5 Carotenoid

Bee pollen carotenoids have antioxidant and anti-inflammatory properties (Salazar-González et al., 2022) capable of reducing anxiety and improve cognitive function, as previously reported in a drosophila model (Dos Santos et al., 2024).

9.1.1.6 Alkaloids

Bee pollen alkaloids can preserve neuronal structure (Table 9.2), balance excitatory/inhibitory neurotransmission, and improve brain redox homeostasis (Friedle et al., 2022; Kacemi & Campos, 2024).

TABLE 9.2
Bee Pollen Phytochemical Compounds

Phytochemical Class	Representative Compounds	Mechanisms of Action/ Bioactivity	Neuroprotective Roles in Neurodegenerative Diseases	References
Flavonoids	Quercetin, Kaempferol, Rutin, Apigenin	Antioxidant, anti-inflammatory, anti-apoptotic; inhibits NF-κB, MAPK	Reduces oxidative stress, suppresses neuroinflammation, preserves mitochondrial function	Pukalskas et al. (2016), Gülmez and Elmastaş (2025).
Phenolic Acids	Caffeic acid, Ferulic acid, Gallic acid	Scavenges ROS, chelates metals, upregulates SOD, CAT	Prevents oxidative neuronal damage, improves antioxidant defense	Nemś et al. (2025), Gercek et al. (2021).
Tannins	Ellagitannins, Catechins	Inhibits acetylcholinesterase, modulates redox balance	Enhances cholinergic transmission, prevents β-amyloid aggregation	Campos et al. (2021), Sobhani et al. (2021).
Terpenoids	β-Sitosterol, Lupeol, Triterpenes	Suppresses pro-inflammatory cytokines, modulates glial activity	Attenuates microglial activation, reduces neuro-inflammation	Kaplan et al. (2019), Salazar et al. (2020).
Carotenoids	β-Carotene, Lutein, Zeaxanthin	Prevents lipid peroxidation, anti-apoptotic	Reduces Aβ aggregation, protects neuronal membranes	Salazar-González et al. (2022)
Alkaloids (minor)	Harman derivatives	Inhibits MAO, neuromodulatory	Enhances dopaminergic activity, possible benefit in Parkinson's models	Friedle et al. (2022), Al-Salem et al. (2016).
Saponins	Not fully characterized	Anti-inflammatory, immunomodulatory	May suppress glial activation and oxidative damage	Ashwathi et al. (2024).
Coumarins	Esculetin, Umbelliferone	Inhibits ROS, modulates calcium influx	Protects against excitotoxicity and ischemia	Tel-Çayan et al. (2025)
Sterols and Phytosterols	Campesterol, Stigmasterol	Modulates cholesterol, stabilizes cell membranes	Preserves neuronal membrane integrity, reduces inflammatory signaling	Vanderplanck et al. (2020), Furse et al. (2023)
Polysaccharides	β-glucans, Arabinogalactans	Immunostimulatory, anti-apoptotic	Enhances neuron–glia communication, promotes neuroimmune regulation	Li et al. (2018).

(Continued)

Table 9.2 *(Continued)*

Phytochemical Class	Representative Compounds	Mechanisms of Action/ Bioactivity	Neuroprotective Roles in Neurodegenerative Diseases	References
Amino Acids	Glutamic acid, Phenylalanine, Leucine	Neurotransmitter precursors, synaptic plasticity	Supports memory, learning, and neuronal regeneration	Sommano et al. (2020), Al-Kahtani et al. (2020).
Fatty Acids	Linoleic acid, α-Linolenic acid, Oleic acid	Reduces pro-inflammatory cytokines, supports neuronal membranes	Maintains brain lipid homeostasis, modulates neuroinflammation	Thakur and Nanda (2018), Özcan et al. (2019).
Vitamins (B-complex)	B1, B6, B9, C, E	Cofactors in neurotransmitter synthesis and antioxidant function	Protects against neurodegeneration, improves cognitive function	Souza et al. (2018).
Trace Elements	Zinc, Selenium, Magnesium	Cofactors for antioxidant enzymes (e.g., GPx, SOD), modulate neurotransmission	Enhances antioxidant capacity, reduces neurotoxicity, supports neural repair	Altunatmaz et al. (2017), Habryka et al. (2020).
Enzymes	Superoxide dismutase (SOD), Catalase	Neutralizes ROS, supports redox homeostasis	Slows neuronal degeneration, improves endogenous antioxidant systems	Nikolić et al. (2016), Yazlovytska et al. (2023).

9.1.1.7 Saponins

Saponins found in bee pollen reduce oxidative stress (Table 9.2), decrease neuroinflammation, and have anxiolytic properties as reported in an NNDs animal study (Aabed et al., 2019).

9.1.1.8 Coumarins

Coumarins found in bee pollen, such as scopoletin and esculetin, inhibit pro-inflammatory mediators and acetylcholinesterase leading to NDD prevention (García-Flores et al., 2022). While coumarins's NDD effects have been reported, its direct evidence in bee pollen remains limited (Kostić et al., 2024).

9.1.1.9 Others

As shown in Table 9.2, other bee pollen phytochemicals, which are sterols, polysaccharides, amino acids, fatty acids, B-vitamins, trace elements, and enzymes, collectively protect against NDDs. Notably, phytosterols reduce neuroinflammation (Panigrahy et al., 2021; Kacemi & Campos, 2023), while bee pollen polysaccharides have antioxidant and immunomodulatory potentials (Wu et al., 2021; Dhahri et al., 2022). Furthermore, bee pollen amino acids promote neurotransmitter balance and enhance BDNF-mediated neuroplasticity (Al-Waili et al., 2021; Bryś & Strachecka, 2024). Polyunsaturated fatty acids (Aabed et al., 2019) strengthen neuronal membranes and reduce lipid peroxidation (Rupanar et al., 2024). Bee pollen B-complex vitamins (Mahdavifar et al., 2021; Zamri et al., 2023) potentially enhance energy metabolism and mood regulation. Notable, vital bee pollen minerals such as zinc, selenium, and iron (Ares et al., 2018; Anjum et al., 2024) enhance antioxidant defenses and promote cerebral neuronal differentiation and growth (Skalny et al., 2023). Also (Table 9.2), bee pollen enzymes improve nutrient bioavailability (Tutun et al., 2021), hence improving cognition, resilience, and neuroprotection (Ilie et al., 2025).

9.2 EXPERIMENTAL AND CLINICAL EVIDENCE OF BEE POLLEN APPLICATIONS IN NEURODEGENERATIVE DISEASES

Experimental investigations in animal models of AD and PD have shown that bee pollen can reduce oxidative stress, prevent neuroinflammation, and preserve neuronal function. While clinical research is rare, early-phase trials and observational reports suggest cognitive and mood-enhancing effects in the elderly; however, these findings are tentative and frequently influenced by the use of multi-herbal formulations. Furthermore, variations in pollen composition and a lack of standardization restrict translational potential. Despite promising preclinical findings, well-designed clinical trials are urgently required to demonstrate the efficacy, safety, and optimal dose of bee pollen in neurodegenerative disease treatment.

9.2.1 In Vitro Studies on Neuronal Protection

Various plant-derived bee pollen samples have been found to be rich in phenolic and flavonoid compounds (Mărgăoan et al., 2021; Boulfous et al., 2025). A study carried out in Turkey in 2021 profiled the various bee pollen samples using standard in vitro assays: CUPRAC, CERAC, and Trolox equivalent antioxidant capacity test. In this study, the botanical origin, total flavonoid and phenolic content, antioxidant activity, phenolic profile, and fatty acid composition of mixed bee pollen loads collected in Bayburt, Turkey, were determined (Gercek et al., 2021). The findings in this study indicate that bee pollen could be a potential source of antioxidants and antimicrobials, similar to a suggestion from a study that suggests that phytochemical-rich extracts of bee pollen, containing flavonoids and polyphenols, have demonstrated potent radical scavenging using in vitro DPPH, TPC, and FRAP assays (Lawag et al., 2021). In a study done by Dulger Altiner et al. (2021), it was demonstrated that commercial bee pollen products in Turkey possess high total phenolic content and antioxidant capacity, with bio-accessibility values reaching up to 95.76% in the CUPRAC assay (Table 9.1), suggesting that it could be efficiently available for cellular uptake in neural tissues. Similarly, comparative analysis by Mayda et al. (2020) revealed that bee pollen exhibited higher phenolic and flavonoid content and stronger radical scavenging activity through ABTS and DPPH assays than bee bread.

These in vitro studies are important because flavonoids such as Rutin, abundant in certain types of bee pollen, are known to modulate neuronal signaling pathways, reduce lipid peroxidation, and protect against neuroinflammatory damage (Kostić et al., 2024). From research, direct in vitro culture studies on bee pollen in neuronal cell lines are limited; however, some studies focus on the antioxidant and neuroprotective effects of other bee products. An example is the study on melittin from bee venom, which protects human neuroblastoma (SH-SY5Y) cells against oxidative insult via Nrf2/HO-1 and TrkB/CREB/BDNF signaling pathways (Nguyen & Lee, 2021). Although this is venom rather than pollen, it suggests that bee-derived bioactives engage neuroprotective molecular pathways.

9.2.2 In Vivo Studies on Neuronal Protection

Animal models have proved useful in understanding the neuroprotective properties of bee pollen in neurodegenerative disease research. Rodent and *Drosophila melanogaster* models, particularly those that mimic AD, PD, and chemically induced oxidative stress, have shown that bee pollen improves cognitive and motor skills (Liao et al., 2019; Saral et al., 2022; Dos Santos et al., 2024). In vivo studies have also demonstrated the ability of bee pollen to counteract oxidative and inflammatory insults in the central nervous system. In a mice model of scopolamine-induced cognitive impairment, bee pollen extract significantly improved performance in memory and learning tasks accompanied by increased hippocampal BDNF levels and activation of ERK/CREB signaling pathways (Liao et al., 2019). Oral administration of bee pollen has been reported to reverse scopolamine-induced memory deficits and improve hippocampal function (Campos et al., 2021; Khalifa et al., 2021). Furthermore, postnatal supplementation with bee pollen in offspring exposed to methylmercury in utero ameliorated oxidative stress, normalized energy metabolism enzymes, and corrected brain electrolyte imbalances linked to neuronal dysfunction (Al-Osaimi et al., 2018). In a *Drosophila* model of neurodegeneration in Alzheimer's disease, bee pollen extract enhanced survival rates, improved locomotor activity, and reduced histological signs of neuronal damage (Dos Santos et al., 2024).

9.2.2.1 Alzheimer's Disease (AD) Models

The study conducted on the antioxidant and anti-Alzheimer's potential of the stingless bees involved *Tetragonisca angustula* pollen extract in a *Drosophila melanogaster* AD model. For this purpose, the *D. melanogaster* AD-like model, which expresses genes related to the amyloidogenic pathway of AD, was used. Treatment improved survival and climbing ability and reduced neurodegeneration on histopathology, attributed to high flavonoid/polyphenol content and antioxidant action of bee pollen (Dos Santos et al., 2024). An extensive survey of ethanol extracts from 18 Korean bee-pollen samples found that the extract expressed inhibition of β-secretase (BACE1) and acetylcholinesterase (AChE), indicating that bee pollen extracts contain constituents capable of modulating the enzymatic processes involved in Aβ generation and cholinergic signaling in AD models (Zou et al., 2020). Another study in mice examined the oral administration of bee pollen extract (100 or 300 mg/kg/day, p.o.) ameliorated scopolamine-induced cognitive impairment. Bee pollen boosted hippocampal phosphorylation of ERK, Akt, CREB, and GSK-3β, and increased BDNF and tPA expression, through the ERK-CREB pathway or Akt-GSK-3β signaling pathway (Liao et al., 2019). Additionally, no clinical trials in human Alzheimer's patients using bee pollen have been reported. Thus, translational evidence remains limited to invertebrate and rodent models.

9.2.2.2 Parkinson's Disease (PD) Models

While there are currently no direct studies in a PD rodent model, some research on the hallmarks of the disease suggests that bee pollen may have an effect on PD because it is rich in polyphenolic compounds such as quercetin, kaempferol, and rutin, all of which exhibit strong ROS-scavenging properties (Tel-Çayan et al., 2025; Yang et al., 2020). Bee pollen is a potent source of quercetin, known to possess neuroprotective effects in Parkinson's models. In a study that used the 6-hydroxydopamine (6-OHDA) rat model, quercetin improved mitochondrial quality control, reduced oxidative stress and α-synuclein accumulation, and rescued motor impairments (Wang et al., 2021). Bee pollen is known to be rich in flavonoids, and several studies have explored the protective mechanisms of flavonoids against PD-related neuronal damage (de Andrade Teles et al., 2018; Jung & Kim, 2018). They exert antioxidant, anti-inflammatory properties, preserve mitochondrial function, and modulate gene expression relevant to dopaminergic neuroprotection (Magalingam et al., 2015).

9.2.2.3 Amyotrophic Lateral Sclerosis (ALS) Models

Despite ample evidence for other bee-derived products, such as bee venom and its components, providing neuroprotection and modulating inflammation in ALS animal models, bee pollen, although rich in neuroprotective flavonoids, has not been studied in ALS models (e.g., SOD1 mice or other genetic/chemical exposures). Several flavonoids and phenolic acids present in bee pollen have individually demonstrated neuroprotective effects in ALS studies (Novak et al., 2021). For instance, kaempferol has promoted autophagic clearance of misfolded SOD1 (Pilotto et al., 2025; Ueda et al., 2017), and gallic and p-coumaric acids have attenuated oxidative injury and improved locomotor performance in ALS-like in vivo and in vitro systems (Paul et al., 2022). Bee pollen, therefore, needs to be further studied in ALS models. In vivo studies, such as administering standardized bee pollen extracts to SOD1-G93A mice, or in vitro studies in motor neuron-like NSC-34 cells exposed to oxidative or excitotoxic injury, could clarify its role in ALS.

9.2.2.4 Evidence of Other Models

Studies (Al-Osaimi et al., 2018; Kocot et al., 2018; Kacemi & Campos, 2024) have shown that bee pollen possesses antioxidant, anti-inflammatory, and neuroprotective potentials. These effects on rodents subjected to chronic immobilization were examined in one study in which 200 mg/kg of bee pollen was administered daily for 10 days to mitigate anxiety-like behavior and restore hippocampal biochemical balance (Saral et al., 2022). These findings align with previous reporting in which bee pollen inhibited neuro-inflammatory cytokines, promoted antioxidant capacity, and reduced lipid peroxidation (Ueda et al., 2017; Pilotto et al., 2025). Moreover, bee pollen has been documented to exhibit neurotherapeutic potential in mitigating neurodevelopmental damage. Al-Osaimi reported in 2018 that bee pollen protective against prenatal exposure to methylmercury (MeHg) induced neurodevelopmental damage. However, 200 mg/kg of bee pollen normalized glutathione and catalase activity and restored metabolism markers and the regulation of magnesium and potassium ion levels (Al-Osaimi et al., 2018). Notable, the bee pollen anti-inflammatory qualities were evidenced in treating NNDs, basically, autism-related diseases. Bee pollen altered central and systemic cytokine in propionic acid (PPA)-induced neuroinflammation, a model used to simulate the pathophysiology of autism. Thus, bee pollen treatment reversed IL-6 and IL-10, TNF-α and vascular endothelial growth factor (VEGF) alteration (Alfawaz et al., 2022).

9.2.3 Human Clinical Studies

Despite some clinical studies that have been done using animal models, there is a significant gap in the clinical studies investigating the effects of bee pollen in humans with neurodegenerative disorders like AD, PD, ALS, or cognitive decline. Previously, a clinical report in Canada documented a case of rapid-onset anaphylaxis following oral ingestion of bee pollen in sensitized individuals, emphasizing the need for pre-screening and cautionary labeling and ongoing post-market safety surveillance (Jagdis & Sussman, 2012). While bee pollen has been studied in human populations for urological and metabolic disorders (Algethami et al., 2022), no trials to date have targeted neurodegenerative conditions. This indicates that the translation of bee pollen's potential against neurodegenerative diseases from animal models to clinical practice has not been explored. Therefore, translational studies adapting a dosing approach based on animal models and in vitro report are essential to transform bee pollen from potential laboratory findings into clinical applications for NDD treatment.

9.3 BEE POLLEN MECHANISMS IN NDDS

9.3.1 Biological Oxidative Stress

Inflammation is also important in both aerobic and arachidonic acid metabolism resulting in the production of ROS (Tutun et al., 2021). The interaction of oxidative stress in NDDs is complicated and relies on the delicate equilibrium of antioxidative agents and oxidative stressors. Accordingly, the accumulation of lipid peroxides caused by an increase in ROS can cause cell death by membrane lipid peroxidation, leading to cellular structure impairment and functional impairments (Asuku et al., 2023; Chaudhary et al., 2023). The pathophysiological peculiarity of NDDs and especially of Alzheimer's is the dysfunctional accumulation of iron deposits. Such an issue often accompanies GSH downfall, resulting in a reduction in the activity of Glutathione peroxidase 4 (GPX4). The subsequent disproportion causes the overactivation of NADPH oxidase, which intensifies the production of ROS. This cascade finally causes oxidative stress through mechanisms such as the Fenton reaction that make cells prone to ferroptosis, a regulated cell death that is characterized by iron-dependent lipid peroxidation (MDA). Inhibition of XC cystine/glutamate antiporter system reduces GSH levels, and therefore aggravates ferroptotic processes (Costa et al., 2023).

To establish an effective treatment approach to the management of NDDs, it is necessary to focus on the molecular pathways to relieve or impede such detrimental pathological mechanisms. Results of this review as published have highlighted the prospects of using bee pollen bioactive chemicals as potentials to treat the neurological ailments. The high carotenoids and vitamins beta-carotene, which also serve as a precursor to vitamin A in bee pollen, have the capability of scavenging free radicals through electron transfer, addition, and hydrogen elimination processes, neutralizing ROS. Vitamin C plays a crucial role as a water-soluble antioxidant and is capable of reacting with hydroxyl radicals to conserve nitric oxide (NO) levels in cases of oxidative stress, which plays a crucial role in a variety of cellular signaling pathways (Denisow & Denisow-Pietrzyk, 2016; Rzepecka-Anjum et al., 2024). Bee pollen as a source of quercetin has become of interest in the study of Parkinsonism because of its ability to cross the blood-brain barrier (Kostić et al., 2024) and enhance the structure and functioning of the mitochondria (Lin et al., 2022). Quercetin inhibits nuclear coactivator-1 (NCOA) (Jiang et al., 2023), which results in a decrease of ferroptosis, low levels of ROS (Nemś et al., 2025), and loss of neurons, therefore, providing protection. Quercetin and kaempferol action promote the induction of nuclear factor erythroid 2-related factor 2 (Nrf2) and GPX4, therefore, alleviating oxidative stress. In addition, myricetin has a high antioxidant activity with a significant decrease in MDA and ROS and restoration of the GSH levels, as reported by Gu et al. (2024).

Additionally, multifloral bee pollen, including *Fagopyrum esculentum* and *Schisandra chinensis*, have been reported to significantly reduce MDA levels. *Schisandra chinensis* notably elevates GSH levels, and it promotes downregulation of NADPH oxidase through Nrf2 activation (Kacemi & Campos, 2023). Moreover, bee pollen has demonstrated the ability to downregulate caspase-3, an enzyme integral to the mitochondrial apoptotic pathway (Huang et al., 2017; Li et al., 2023).

9.3.2 Biological Inflammation

Bee pollen functions as a natural supplement, potentially improving anti-inflammatory responses by increasing interleukin-10 (IL-10) and Heme Oxygenase 1 (HO-1) levels. The enzyme HO-1 is essential for macrophage polarization regulation (Kacemi & Campos, 2024; Kwak et al., 2024), playing an important role in the resolution of inflammation and the commencement of tissue-healing mechanisms. HO-1 modulation is closely associated with specialized lipid pro-resolving mediators (SPMs) present in bee pollen as well as docosapentaenoic acid (DPA) (Mohammad-Rafiei et al., 2024; Livshits & Kalinkovich, 2024). DPA, a product of omega-3 and omega-6 fatty acid metabolism, interacts with G-protein coupled receptors, thereby diminishing inflammatory cellular responses. Also, caffeic acid derived from bee pollen significantly modulates anti-inflammatory pathways, particularly JAK2/STAT3 and PI3K/AKT, via reducing interferon-gamma (IFN-γ) levels according to Farooqi et al., (2019). Also, caffeic acid phenethyl ester influences the expression of intracellular adhesion molecule 1 (ICAM-1), iNOS, and TNF-α, in neurovascular disorders and blocks NF-κB activity, thereby further mitigating inflammation by downregulation of proinflammatory miRNAs (miRNA-9, miRNA-125b, miRNA146a, and miRNA-155) from primary human neuronal–glial cells stressed with Alzheimer's disease (AD)-derived extracellular fluid (Khan et al., 2007; Alexandrov et al., 2012; Tolba et al., 2013).

Important flavonoids present in bee pollen, such as kaempferol and luteolin, facilitate the upregulation of HO-1, steering macrophage polarization toward the M2 phenotype, which is associated with anti-inflammatory functions, in contrast to the pro-inflammatory M1 phenotype (Boriero et al., 2021; Kacemi & Campos, 2024). In a similar vein, other polyphenolic compounds found in bee pollen, including phenolic acid, chlorogenic acid, resveratrol, and protocatechuic acid, contribute to the increase of HO-1 levels (Li et al., 2023; Yao et al., 2024; Han et al., 2026). Resveratrol, in particular, has been shown to counteract epigenetic alterations through the modulation of microRNA (miRNA) in neurodegenerative diseases, and upregulation of SIRT1, IGF-1, and mTOR pathways, which interact with the AMPK signaling cascade to regulate metabolic processes (Pyo et al., 2020; Griñán-Ferré et al., 2021). Secondly, linoleic acid is another essential fatty acid that has been obtained in bee pollen. It interacts with histamine H1 receptor to suppress the production of NO, which suppresses the inflammatory effect (Anjum et al., 2024). Anethole has also been found to suppress the action of NF-κB stimulated by TNF-α. This plays an important role in inflammatory signaling pathways (Han et al., 2026). The changes in the Wnt/2-catenin pathway can affect the activity of neuroimmune cells, hence facilitating protein aggregation and the development of abnormal amyloid-beta (Aβ) proteins. This may then further increase neuroinflammation (Ramakrishna et al., 2023). After that, bee pollen, which is a product of *Camellia sinensis*, includes abundant polyphenols. These polyphenols block the action of lipoxygenase and cyclooxygenase through interactions with amino acid residues. Such interaction suppresses arachidonic acid conversion, therefore significantly reducing inflammatory processes. Polyphenols have an effect on the biological processes through the inhibition of NF-κB signaling. This therefore downregulates the expression of the gene that governs inflammation, leading to a healthier and less inflammatory state (Li et al., 2019).

9.3.3 NDDS, Biological Serotonin and Dopamine

The crucial role in the immunological response is played by IFN-7 (Jorgovanovic et al., 2020): it stimulates indoleamine 2,3-dioxygenase (IDO). The resultant effect of this activation process is a reduction in serotonin levels, cognition, and many physiological processes. COMT is an important enzyme that breaks down dopamine and can be relevant in the pathophysiology of PD (Giri et al., 2022). COMT promotes transfer of a methyl group on S-adenosyl-methionine to catechol moiety of dopamine resulting in its inactivation and consequently, affecting the availability of L-DOPA (Nissinen & Männistö, 2009). Research has shown that bee pollen suppresses the production of IFN-7 (Al-Salem et al., 2016) and, thus, reduces its stimulating effect on IDO (Watanabe et al., 2024).

9.3.4 Biological BDNF Signaling Pathway of NNDs

The brain signaling pathways that involve MAPK/ERK and PI3K/AKT are also critical in learning and memory (Albert-Gascó et al., 2020; Razani et al., 2021). Long-term potentiation is promoted by ERK phosphorylation. CREB (cAMP response element-binding protein), on the other hand, is a downstream signaling mediator of ERK and the ERK-CREB pathway that is vital in the regulation of neural development and cognitive performance. As Liao et al. (2019) wrote, their study showed that bee pollen significantly enhances phosphorylation of ERK, CREB, and GSK3-b protein, which is an important AKT signaling substrate. Bee pollen has proven its ability to mediate miRNA increase in tissue plasminogen activator (tPA). The process of upregulation is mediated by the binding of cAMP response element to the tPA gene promoter region, which increases the conversion of pro-BDNF into m-BDNF. The stimulation of the ERK-CREB, MAPK/ERK, and PI3K/AKT pathways facilitates the promotion of the observed increase in the levels of BDNF expression, which leads to the enhancement of cognitive performance. Methanolic extract of the *T. angustifolia* pollen has demonstrated effectiveness in healing the brain tissue injury in a *Drosophila* Alzheimer's disease model. This effect is attributed to effects of flavonoids of bee pollen that improve hippocampus levels of monoamine and BDNF, and thus improves learning and memory and other behaviors of anxiety in such models (El-Seedi et al., 2020; Al Suhaibani et al., 2021; dos Santos et al., 2024).

The phenolic acid content in bee pollen inhibits monoamine oxidase (MAO). In the realm of NNDs, monoamine oxidase is associated with disease progression. The MAO-α and MAO-β isoforms specifically contribute to oxidative stress, amyloid beta accumulation, and neuroinflammation in AD; but in PD, MAO is linked to dopamine degradation, causing impaired motor function (Behl et al., 2021). The suppression of MAO promotes an elevation in monoamine levels (Nordin et al., 2021). Multifloral bee pollen extracts have been shown to exhibit superior efficacy compared to monofloral varieties in downregulating BACE-1 (beta-site APP-cleaving enzyme 1)

(Salazar-González et al., 2022). Quercetin promotes apoptosis in senescent cells, hence diminishing amyloid beta accumulation in animal models of Alzheimer's disease (Zhang et al., 2019; Özsoy Gökbilen et al., 2022). Padhi et al. (2024) have clarified the synergistic action of gallic acid and cyclic dipeptides, which implies their application in reducing amyloid-beta aggregation and tau pathology. Therefore, bee pollen has been identified as an anti-angiogenic promoter of NDDs by inhibiting VEGF, which provides a treatment option to NDDs (Roquetto et al., 2015; Aabed et al., 2019).

9.3.5 NNDs and Biological Genes and Epigenetic Factors

Change in DNA methylation is one of the contributors to NDDs, as it starts an epigenetic alteration process and especially a prominent cluster of CpG sites (Nigar & Shimosato, 2019; Jiang & Guo, 2020; Xie et al., 2023). The pathogenesis of a variety of NDDs is worsened by the deregulation of this immunological mechanism. Histone deacetylases (HDACs) overexpression makes significant contributions to the elevated levels of amyloid beta, tau, and alpha-synuclein, which are proteins closely related to AD and other NNDs (Kacemi & Campos, 2025). HDAC removes the acetyl group to the lysine residues on the histones, which allow for more condensed association of DNA with the proteins. In this way, gene accessibility and expression are reduced by this DNA compaction. On the contrary, when the histones are acetylated, the DNA structure becomes relaxed, which facilitates transcription and protein synthesis. The accumulation of misfolded proteins and stimulating NDDs is the result of the deacetylation process (Park & Kim, 2020; Shukla & Tekwani, 2020). The activity of HDACs is suppressed by bee pollen leading to the upregulation of histone acetylation stimulated by histone acetyltransferases (HATS) and a subsequent reduction in the levels of amyloid beta, tau, and alpha-synuclein (Shaikh et al., 2023), DNA methylation patterns are maintained by the action of DNA methyltransferases, especially DNMT1, DNMT3a, and DNMT3b, which are able to methylate hemimethylated CpG sites to maintain genomic stability. The inhibition of DNMT binding is one of the significant impacts of oxidative stress (Borsoi et al., 2023; Rubio et al., 2023). Kaempferol suppresses DNA methylation through reduction of the levels of oxidative stress and subsequently inhibits the activity of DNMTs (Rajendran et al., 2022). Increased oxidative stress and inflammation in NDDs may be a consequence of dysregulation of DNMTs that increases methylation of the Nrf2 gene. Substances such as apigenin, delphinidin, luteolin, and resveratrol, which are found in bee pollen, increase the expression of Nrf2 by suppressing its methylation condition through the inhibition of the DNMT activity (Divyajanani et al., 2023). This is an effective correction of dysregulated inflammation and oxidative reactions as a management of NDDs.

9.4 POTENTIALS OF BEE POLLEN TREATMENT

The results of other research reveal that bee pollen contains a valuable balanced composition, which makes it a suitable human food, as well as a source of nutritional supplement (Campos et al., 2021; Khalifa et al., 2021). This is because of the plethora of bioactive chemicals that facilitate a wide range of biological and pharmacological effects (Campos et al., 2021). The neuroprotective effects of these bioactive constituents have been demonstrated in the animal models by reducing neuroinflammation and oxidative stress. Oral administration of bee pollen has been reported to reduce memory deficits caused by scopolamine and to improve the performance of the hippocampus (Campos et al., 2021; Khalifa et al., 2021).

9.4.1 Standardizations and Formulations of Bee Pollen

9.4.1.1 Via Oral Route

Oral bee pollen is limited by poor gastrointestinal absorption caused by the tough outer pollen wall and stomach stability variation (Kacemi & Campos, 2024).

9.4.1.2 Via Encapsulation

Various technologies have been created in the processing of bee pollen. These methods include fermentation, drying, hydrolysis, and ultrasound treatment, which are aimed at breaking the pollen exine and improving the bioavailability of medicinal chemicals (Zuluaga-Domínguez et al., 2019). Bee pollen extract encapsulation has not been exploited.

9.4.1.3 Via Nano

Polymer-based nanoparticles and liposomes have proven to be effective in direct delivery of its content by intranasal route to the central nervous system. Mixed cationic liposomes are used to introduce acetylcholinesterase reactivator 2-PAM through the blood-brain barrier.

9.4.1.4 Bee Pollen Adjuvant Treatment Protocol

Clinical trial evidence on the application of bee pollen as an adjunctive treatment in neurodegenerative disorder is unavailable as of this writing. Pollen phytochemical properties in bees suggest that it can be used as a complementary intervention (Yang et al., 2020). Moreover, nano-preparations of bee pollen can be used to increase its medical potential in treating neurodegenerative diseases (Sanyal et al., 2023).

9.4.2 Bee Pollen Formulation and Standardization Problems

Pollen on bees can be classified as a superfood due to its high concentration of minerals and bioactive compounds. However, its bioavailability is limited by an outer layer that is resistant to degradation called exine. The latest developments in physical and biotechnological processes have been instituted to break down this coating and enhance the nutritional and functional quality of pollen (Alcalá-Orozco et al., 2024). Bee pollen also contains varying amounts depending on the geographic origin, which makes standardization and the development of reliable medicinal preparations difficult (Khalifa et al., 2021).

9.5 BEE POLLEN SAFETY PROFILE AND DOCUMENTED TOXICOLOGICAL EFFECTS

The anti-inflammatory, antioxidant, and immunomodulatory properties of bee pollen have led to increased popularity as a health supplement and in medical preparations (Kaškonienė et al., 2020; El Ghouizi et al., 2023; Rodríguez-Pólit et al., 2023). Preclinical studies show that there is minimal acute and sub-chronic toxicity. Nevertheless, side effects have also been reported, including allergic reactions ranging from moderate hypersensitivity to severe anaphylaxis. The animal-based research indicates that bee pollen has a considerable safety threshold in case of controlled administration (De Oliveira et al., 2013; Martinello & Mutinelli, 2021). Orally, acute toxicity tests have estimated LD 50 values above 10g/kg, which is considered low acute toxicity (Al-Salem et al., 2016; Khalifa et al., 2021; Saral et al., 2022). Subchronic toxicity studies (14–90 days) revealed no significant alteration in body mass, food intake, hematological, liver enzymes and histopathological changes in organs in oral intake of bee pollen in 250-2,000 mg/kg/day (Prakatur et al., 2019; Végh et al., 2021; Basavanna et al., 2023). Some studies have shown dose-related hepatic hypertrophy and some minor renal tubular changes with high dosage (>2 g/kg/day) (Matuszewska et al., 2022; Kaškonienė et al., 2020; Choi et al., 2015). The findings when applied to humans must be done cautiously because of the interspecies differences and different pollen formulations, as the long-term toxicity and cumulative effects have not been well established.

9.6 BEE POLLEN ALLERGY AND IMMUNO-TOXICITY

Bee pollen is known to have allergic reactions, which are one of the safety concerns. Bee pollen contains proteins and glycoproteins capable of triggering the occurrence of the type I reaction of hypersensitivity (Matuszewska et al., 2022; Basavanna et al., 2023). The recorded cases range from mild urticaria and rhinitis to severe anaphylaxis (Jagdis & Sussman, 2012). The reactions of IgE (immunoglobulin E) are often observed in individuals who are allergic to plant pollens, apian products, or allergenic fungi (Khalifa et al., 2021). Anti-allergic effects of bee pollen have also been shown by inhibiting mast cell degranulation in vitro when added to the cells during the process of IgE sensitization (Khalifa et al., 2021). Therefore, low dosages of bee pollen (0.11mg/mL) inhibited the production of TNF (tumor necrosis factor) by mast cells (El Ghouizi et al., 2023). However, consumption of bee pollen has been reported to pose a threat of allergy, particularly among people who are sensitive to pollens of weeds (Choi et al., 2015). Bee pollen contains fungi including *Aspergillus* and *Cladosporium*, which can potentially cause allergy symptoms (Beggs et al., 2023). The clinical evidence shows that bee pollen dietary supplements induce immediate anaphylaxis in susceptible individuals, which implies the need for pre-screening and precautious labeling (Jagdis & Sussman, 2012). The inconsistency of the allergic factors, especially in the multi-floral and wild-grown pollens, interferes with the safety assurances. This has led to the suggestion of precaution in the marketing of bee pollen without standardized allergen profiling by regulatory authorities (Végh et al., 2021; Matuszewska et al., 2022; Beggs et al., 2023).

9.7 RESEARCH GAPS AND FUTURE DIRECTIONS

The nutritional benefits and use in various medical settings have been extensively known about bee pollen (El Ghouizi et al., 2023). Consequently, it is increasingly attracting interest to be transformed into dietary supplements and survival meals. There is a lot of variance in its macro- and microelements structure, which serves as an obstacle to homogeneity of products (Pohl et al., 2024). Although things are seemingly promising with the application of bee pollen in the treatment of various NDDs, the standardization of products remains quite a problem. The lack of compositional consistency limits the reproducibility and reliability, and additional factors that increase the variability include processes of harvesting, storage, and contamination, thus changing the nutritional profile (Kacemi & Campos, 2024; Anjos et al., 2019). Negative effects due to the synergistic interactions of bee pollen with other compounds have been linked and could be increased in magnitude by contamination with bee venom (Sadek et al., 2024). Additionally, pesticide residues and the buildup of heavy metals in bee pollen pose additional health hazards. Recording allergic reactions includes urticaria, pruritus, dyspnoea, and glossal oedema (Baydaş et al., 2024). Lack of storage conditions may intensify microbiological contamination with molds and bacteria at the expense of food safety (Nuvoloni et al., 2021). The pharmacokinetic interactions of bee pollen bioactive flavonoids and phenolic acids with certain medicines can occur (Rzepecka-Stojko et al., 2015). There have been cases of liver poisoning manifested as jaundice, as well as pain in the abdomen in babies due to contaminated bee pollen (Nassar et al., 2020). Although bee pollen has the potential to be bioactive, the mechanisms that cause its neuroprotective effects are poorly determined. The outstanding questions that remain unanswered include its ability to cross the blood-brain barrier, the specific neuronal pathways that it activates, what effect it has on synaptic transmission, how it regulates gene expression, whether it interacts with certain receptors, and what effect it has on mitochondria (Kacemi & Campos, 2024). In addition, its safety, efficacy, and therapeutic relevance have not yet been scientifically planned and conducted on any human clinical trial as of this writing (Algethami et al., 2022). Lack of standardized dosing protocols, inter-batch composition heterogeneity, the possibility of allergic reactions, and poor understanding of mechanisms of action are significant barriers to clinical translation (Kacemi & Campos, 2024). These limitations make the use of bee pollen as a prophylaxis in neurodegenerative disease a wise endeavor.

These research challenges will be systematically addressed to transform bee pollen, which is a possible natural product, into an inseparable constituent of modern phytomedicine, which it has become in proving its neuroprotective effect. The main limitation of its therapeutic modification and standardization is the enormous heterogeneity of composition and, consequently, of the biological activity because of differences in botanical origin. Consistency, efficacy, and safety should be prioritized with pharmacokinetic profiling to make sure that all are consistent, effective, and safe. In addition, human clinical trials should be carefully designed to prove its therapeutic efficacy, assess both long-term effects, define optimal dosing regimens, and identify possible interactions with conventional therapies. One way to achieve standardization is the development of composite formulations to include a controlled mixture of multiple types of pollen, thereby ensuring the inclusion of vital bioactive ingredients in a prospective standardized bee pollen product. Better, strict quality control measures should be put in place to single out and define the respective bioactive chemicals, which are behind the neuroprotective and antioxidant effects of bee pollen. These could be then used to form phytopharmaceuticals or used as lead structures in drug discovery. The design of clinical trials should consider the safety and tolerability of the subjects, mostly those who are vulnerable and at-risk populations. It is absolutely necessary to evaluate the efficacy of bee pollen treatments with the need to focus on the dosage and decrease the adverse effects. The effectiveness of these interventions in enhancing memory and cognitive abilities as well as their effects on the levels of oxidative stress in the patients with neurodegenerative conditions should be tested in clinical trials. The assessment of the decreases of these markers will allow establishing a clear relationship between the treatment and biological outcomes in different demographic groups/populations. Lastly, ethical and regulatory authorities (National Agency for Food and Drug Administration and Control [NAFDAC] and the European Medicines Agency [EMA]) should ensure that they meet the requirements of the international safety standards and proper informed consent guidelines, and that they are complying with the global requirements.

9.8 CONCLUSION

Bee pollen is a nutritionally rich natural product that is abundant in many bioactive compounds that work together to overcome the most important neurodegenerative processes, including oxidative stress, mitochondrial damage, neuroinflammation, and protein aggregation. Preclinical studies show that bee pollen modulates the Nrf2, NF-κB, and BDNF/CREB pathways, restores neurotransmitter balance, and improves cognitive and motor abilities in animals with Alzheimer's, Parkinson's, and other diseases. Nevertheless, translational development is limited to compositional diversity, inaccessibility of standardized methods, and clinical research. The safety concerns, including the allergenicity and pyrrolizidine alkaloid levels, will require evaluation. Through meticulous standardization and comprehensive human research, bee pollen shows potential as a multi-target nutraceutical for delaying or alleviating the course of neurodegenerative diseases.

REFERENCES

Aabed, K., Bhat, R. S., Al-Dbass, A., Moubayed, N., Algahtani, N., Merghani, N. M., Alanazi, A., Zayed, N., & El-Ansary, A. (2019). Bee pollen and propolis improve neuroinflammation and dysbiosis induced by propionic acid, a short-chain fatty acid in a rodent model of autism. *Lipids in Health and Disease*, 18(1), 200. https://doi.org/10.1186/s12944-019-1150-0

Ajibare, A.J., Akintoye, O.O., Oriowo, O.A., Asuku, A.O., Oriyomi, I.A., Adedara, J.F., Ayoola, A.M., & Adejumo, O.M. (2025). Dietary virgin coconut exerted a neuroprotective effect via upregulation of nuclear erythroid factor and downregulation of nuclear factor KB expression in acrylamide-induced neurotoxicity: Role of BDNF and AchE pathways. *Tropical Journal of Natural Product Research*, 9(3), 1263–1269.

Al Hoque, A., Begum, S., & Dutta, D. (2025). Occurrence, chemical nature, and medicinal benefits of flavonoids—Rutin, kaempferol, quercetin, anthocyanidins, catechins, and flavones. In B. Mukherjee (Ed.), *Dietary supplements and nutraceuticals*. Springer. https://doi.org/10.1007/978-981-97-9936-7_5-1

Al Suhaibani, A., Ben Bacha, A., Alonazi, M., Bhat, R. S., & El-Ansary, A. (2021). Testing the combined effects of probiotics and prebiotics against neurotoxic effects of propionic acid orally administered to rat pups. *Food Science & Nutrition*, 9(8), 4440–4451. https://doi.org/10.1002/fsn3.2418

Albert-Gascó, H., Ros-Bernal, F., Castillo-Gómez, E., & Olucha-Bordonau, F. E. (2020). MAP/ERK signaling in developing cognitive and emotional function and its effect on pathological and neurodegenerative processes. *International Journal of Molecular Sciences*, 21(12), 4471. https://doi.org/10.3390/ijms21124471

Alcalá-Orozco, M., Lobo-Farfan, I., Tirado, D. F., & Mantilla-Escalante, D. C. (2024). Enhancing the nutritional and bioactive properties of bee pollen: A comprehensive review of processing techniques. *Foods*, 13(21), 3437. https://doi.org/10.3390/foods13213437

Alexandrov, P. N., Dua, P., Hill, J. M., Bhattacharjee, S., Zhao, Y., & Lukiw, W. J. (2012). microRNA (miRNA) speciation in Alzheimer's disease (AD) cerebrospinal fluid (CSF) and extracellular fluid (ECF). *International Journal of Biochemistry and Molecular Biology*, 3(4), 365–373.

Alfawaz, H. A., El-Ansary, A., Al-Ayadhi, L., Bhat, R. S., & Hassan, W. M. (2022). Protective effects of bee pollen on multiple propionic acid-induced biochemical autistic features in a rat model. *Metabolites*, 12(7), 571.

Algethami, J. S., El-Wahed, A. A. A., Elashal, M. H., Ahmed, H. R., Elshafiey, E. H., Omar, E. M., Naggar, Y. A., Algethami, A. F., Shou, Q., & Alsharif, S. M. (2022). Bee pollen: Clinical trials and patent applications. *Nutrients*, 14(14), 2858.

Al-Kahtani, S. N., Taha, E. K., Khan, K. A., Ansari, M. J., Farag, S. A., Shawer, D. M., & Elnabawy, E. S. M. (2020). Effect of harvest season on the nutritional value of bee pollen protein. *PLOS ONE*, 15(12), e0241393. https://doi.org/10.1371/journal.pone.0241393

Al-Osaimi, M., El-Ansary, A., Al-Daihan, S., Bhat, R. S., & Ben Bacha, A. (2018). Therapeutic and protective potency of bee pollen against neurotoxic effects induced by prenatal exposure of rats to methyl mercury. *Journal of Molecular Neuroscience*, 65(3), 327–335.

Al-Salem, H. S., Bhat, R. S., Al-Ayadhi, L., & El-Ansary, A. (2016). Therapeutic potency of bee pollen against biochemical autistic features induced through acute and sub-acute neurotoxicity of orally administered propionic acid. *BMC Complementary and Alternative Medicine*, 16, 120. https://doi.org/10.1186/s12906-016-1099-8

Altunatmaz, S. S., Tarhan, D., Aksu, F., Barutçu, U. B., & Or, M. E. (2017). Mineral element and heavy metal (cadmium, lead and arsenic) levels of bee pollen in Turkey. *Food Science and Technology*, 37, 136–141. https://doi.org/10.1590/1678-457x.31216

Al-Waili, N., Salom, K., & Al-Ghamdi, A. (2021). Bee pollen increases hippocampal brain-derived neurotrophic factor and suppresses neuroinflammation in adult rats with chronic immobilization stress. *Neuroscience Letters*, 763, 136159. https://doi.org/10.1016/j.neulet.2021.136159

Anjos, O., Paula, V., Delgado, T., & Estevinho, L. (2019). Influence of the storage conditions on the quality of bee pollen. *Žemdirbystė–Agriculture*, 106(1), 87–94. https://doi.org/10.13080/z-a.2019.106.011

Anjum, S. I., Ullah, A., Gohar, F., Raza, G., Khan, M. I., Hameed, M., Ali, A., Chen, C. C., & Tlak Gajger, I. (2024). Bee pollen as a food and feed supplement and a therapeutic remedy: Recent trends in nanotechnology. *Frontiers in Nutrition*, 11, 1371672. https://doi.org/10.3389/fnut.2024.1371672

Ares, A. M., Valverde, S., Bernal, J. L., Nozal, M. J., & Bernal, J. (2018). Extraction and determination of bioactive compounds from bee pollen. *Journal of Pharmaceutical and Biomedical Analysis*, 147, 110–124. https://doi.org/10.1016/j.jpba.2017.08.018

Ashwathi, K., Nandhini, V. M., Thirumalai, V., Ezhili, N., & Pushparaj, C. (2024). Phytochemical profiling of ethanolic extract of bee pollen using gas chromatography–mass spectroscopy and its in silico analysis of hypocholesterolemic activity. *Entomon*, 49(1), 55–64.

Asuku, A. O., Ayinla, M. T., Olajide, T. S., Yakub, Z., & Owoyele, B. V. (2023). Cola acuminata mitigates cognitive deficit and oxidative stress in mercury chloride -induced neurotoxicity in male Wistar rats. *Nigerian Journal of Neuroscience*, 14(2), 33–40. http://doi.org/10.47081/njn2023.14.2/001

Asuku, A. O., Ayinla, M. T., Ajibare, A. J., & Krishnamurthy. R. (2024). Modulatory and neuroprotective roles of marine green spirulina on the brain and neurodegeneration. In S. A. Aransiola, O. A. Oyewole, N. R. Maddela, U. J. J. Ijah, & B. Manjunatha (Eds.), *Marine greens: Environmental, agricultural, industrial, and biomedical applications*. CRC Press, Taylor and Francis Group, 17, 89. https://doi.org/10.1201/9781003369738

Basavanna, H., Kaur, J., Rasane, P., Singh, J., & Bhadariya, V. (2023). An overview of the health effects of bee pollen: A functional food ingredient. *The Pharma Innovation Journal*, 12(6), 4146–4150.

Baydaş, G., Aksoy, F. T., Dalar, Z. G., Aksakal, M. N., Çoruhlu, İ., Çıbık, S., … & Kul, Z. (2024). Bee products and the potential health issues arising from their uncontrolled use. *Türk Doğa ve Fen Dergisi*, 13(1), 176–188.

Beggs, P. J., Clot, B., Sofiev, M., & Johnston, F. H. (2023). Climate change, airborne allergens, and three translational mitigation approaches. *eBioMedicine*, 93, 104478. https://doi.org/10.1016/j.ebiom.2023.104478

Behl, T., Kaur, D., Sehgal, A., Singh, S., Sharma, N., Zengin, G., Andronie-Cioara, F. L., Toma, M. M., Bungau, S., & Bumbu, A. G. (2021). Role of monoamine oxidase activity in Alzheimer's disease: An insight into the therapeutic potential of inhibitors. *Molecules*, 26(12), 3724. https://doi.org/10.3390/molecules26123724

Bertoncelj, J., Lilek, N., & Korošec, M. (2023). Bee pollen carbohydrates composition and functionality. In N. E. Bayram, Ž. Kostic, & Y. Can Gercek (Eds.), *Pollen chemistry & biotechnology* (pp. 43–62). Springer. https://doi.org/10.1007/978-3-031-47563-4_3

Boriero, D., Carcereri de Prati, A., Antonini, L., Ragno, R., Sohji, K., Mariotto, S., & Butturini, E. (2021). The anti-STAT1 polyphenol myricetin inhibits M1 microglia activation and counteracts neuronal death. *The FEBS Journal*, 288(7), 2347–2359. https://doi.org/10.1111/febs.15577

Borsoi, F. T., Neri-Numa, I. A., De Oliveira, W. Q., De Araújo, F. F., & Pastore, G. M. (2023). Dietary polyphenols and their relationship to the modulation of non-communicable chronic diseases and epigenetic mechanisms: A mini-review. *Food Chemistry: Molecular Sciences*, 6, 100155. https://doi.org/10.1016/j.fochms.2022.100155

Boulfous, N., Belattar, H., Ambra, R., Pastore, G., & Ghorab, A. (2025). Botanical origin, phytochemical profile, and antioxidant activity of bee pollen from the Mila region, Algeria. *Antioxidants*, 14(3), 291. https://doi.org/10.3390/antiox14030291

Bryś, M. S., & Strachecka, A. (2024). The key role of amino acids in pollen quality and honey bee physiology—A review. *Molecules*, 29(11), 2605.

Campos, M. G., Frigerio, C., Bobiş, O., Urcan, A. C., & Gomes, N. G. (2021). Infrared irradiation drying impact on bee pollen: Case study on the phenolic composition of *Eucalyptus globulus labill* and *Salix atrocinerea Brot*. Pollens. *Processes*, 9(5), 890.

Čeksterytė, V., Bliznikas, S., & Jaškūnė, K. (2023). The composition of fatty acids in bee pollen, royal jelly, buckthorn oil and their mixtures with pollen preserved for storage. *Foods*, 12(17), 3164. https://doi.org/10.3390/foods12173164

Chaudhary, P., Janmeda, P., Docea, A. O., Yeskaliyeva, B., Abdull Razis, A. F., Modu, B., & Calina, D. (2023). Oxidative stress, free radicals and antioxidants: Potential crosstalk in the pathophysiology of human diseases. *Frontiers in Chemistry*, 11, 1158198. https://doi.org/10.3389/fchem.2023.1158198

Choi, J. H., Jang, Y. S., Oh, J. W., Kim, C. H., & Hyun, I. G. (2015). Bee pollen-induced anaphylaxis: A case report and literature review. *Allergy, Asthma & Immunology Research*, 7(5), 513–517. https://doi.org/10.4168/aair.2015.7.5.513

Costa, I., Barbosa, D. J., Benfeito, S., Silva, V., Chavarria, D., Borges, F., Remião, F., & Silva, R. (2023). Molecular mechanisms of ferroptosis and their involvement in brain diseases. *Pharmacology & Therapeutics*, 244, 108373. https://doi.org/10.1016/j.pharmthera.2023.108373

de Andrade Teles, R. B., Diniz, T. C., Costa Pinto, T. C., de Oliveira Junior, R. G., Gama e Silva, M., de Lavor, É. M., Fernandes, A. W. C., de Oliveira, A. P., de Almeida Ribeiro, F. P. R., & da Silva, A. A. M. (2018). Flavonoids as therapeutic agents in Alzheimer's and Parkinson's diseases: A systematic review of preclinical evidences. *Oxidative Medicine and Cellular Longevity*, 2018(1), 7043213.

De-Melo, A. A. M., & de Almeida-Muradian, L. B. (2017). Chemical composition of bee pollen. In J. Alvarez-Suarez (Ed.), *Bee products: Chemical and biological properties* (pp. 219–233). Springer. https://doi.org/10.1007/978-3-319-59689-1_11

Denisow, B., & Denisow-Pietrzyk, M. (2016). Biological and therapeutic properties of bee pollen: A review. *Journal of the Science of Food and Agriculture*, 96(13), 4303–4309. https://doi.org/10.1002/jsfa.7729

Dhahri, M., Alghrably, M., Mohammed, H. A., Badshah, S. L., Noreen, N., Mouffouk, F., Rayyan, S., Qureshi, K. A., Mahmood, D., Lachowicz, J. I., Jaremko, M., & Emwas, A.-H. (2022). Natural polysaccharides as preventive and therapeutic horizon for neurodegenerative diseases. *Pharmaceutics*, 14(1), 1. https://doi.org/10.3390/pharmaceutics14010001

Divyajanani, S., Harithpriya, K., Ganesan, K., & Ramkumar, K. M. (2023). Dietary polyphenols remodel DNA methylation patterns of NRF2 in chronic disease. *Nutrients*, 15(15), 3347. https://doi.org/10.3390/nu15153347

Dos Santos, N. C. L., Malta, S. M., Franco, R. R., Silva, H. C. G., Silva, M. H., Rodrigues, T. S., de Oliveira, R. M., Araújo, T. N., Augusto, S. C., & Espindola, F. S. (2024). Antioxidant and anti-Alzheimer's potential of *Tetragonisca angustula* (Jataí) stingless bee pollen. *Scientific Reports*, 14(1), 308.

Dulger Altiner, D., Sandikci Altunatmaz, S., Sabuncu, M., Aksu, F., & Sahan, Y. (2021). In-vitro bioaccessibility of antioxidant properties of bee pollen in Turkey. *Food Science and Technology*, 41.

El Ghouizi, A., Bakour, M., Laaroussi, H., Ousaaid, D., El Menyiy, N., Hano, C., & Lyoussi, B. (2023). Bee pollen as functional food: Insights into its composition and therapeutic properties. *Antioxidants*, 12(3), 557. https://doi.org/10.3390/antiox12030557

El-Seedi, H. R., Khalifa, S. A., El-Wahed, A. A., Gao, R., Guo, Z., Tahir, H. E., Zhao, C., Du, M., Farag, M. A., Musharraf, S. G., & Abbas, G. (2020). Honeybee products: An updated review of neurological actions. *Trends in Food Science & Technology*, 101, 17–27. https://doi.org/10.1016/j.tifs.2020.04.026

Eshbah, H., Mohamed, A., Zedan, O., & Ghanem, A. (2021). Phytochemical components in different types of pollen and their effects on honey bees. *Minia Journal of Agricultural Research and Development*, 41(2), 1–13.

Fadzil, M. A., Mustar, S., & Rashed, A. (2023). The potential use of honey as a neuroprotective agent for the management of neurodegenerative diseases. *Nutrients*, 15(7), 1558. https://doi.org/10.3390/nu15071558

Farooqi, A. A., Romero, M. A., Aras, A., Qureshi, M. Z., & Wakim, L. H. (2019). Honey and popolis-mediated regulation of protein networks in cancer cells. *Nutraceuticals and Natural Product Derivatives: Disease Prevention & Drug Discovery*, 137–144.

Friedle, C., Kapp, T., Wallner, K., Alkattea, R., & Vetter, W. (2022). High abundance of pyrrolizidine alkaloids in bee pollen collected in July 2019 from Southern Germany. *Environmental Monitoring and Assessment*, 194, 250. https://doi.org/10.1007/s10661-022-09907-8

Furse, S., Koch, H., Wright, G., & Stevenson, P. (2023). Sterol and lipid metabolism in bees. *Metabolomics*, 19, 39. https://doi.org/10.1007/s11306-023-02039-1

García-Flores, M., Juárez-Colunga, S., García-Casarrubias, A., Trachsel, S., Winkler, R., & Tiessen, A. (2022). Unveiling the techno-functional and bioactive properties of bee pollen as an added-value food ingredient. *Food Chemistry*, 396, 133651. https://doi.org/10.1016/j.foodchem.2022.133651

Gercek, Y. C., Celik, S., & Bayram, S. (2021). Screening of plant pollen sources, polyphenolic compounds, fatty acids and antioxidant/antimicrobial activity from bee pollen. *Molecules*, 27(1), 117.

Giri, B., Seamon, M., Banerjee, A., Chauhan, S., Purohit, S., Morgan, J., Baban, B., & Wakade, C. (2022). Emerging urinary alpha-synuclein and miRNA biomarkers in Parkinson's disease. *Metabolic Brain Disease*, 37(6), 1687–1696. https://doi.org/10.1007/s11011-021-00735-2

Griñán-Ferré, C., Bellver-Sanchis, A., Izquierdo, V., Corpas, R., Roig-Soriano, J., Chillón, M., Andres-Lacueva, C., Somogyvári, M., S⊠ti, C., Sanfeliu, C., & Pallàs, M. (2021). The pleiotropic neuroprotective effects of resveratrol in cognitive decline and Alzheimer's disease pathology: From antioxidant to epigenetic therapy. *Ageing Research Reviews*, 67, 101271. https://doi.org/10.1016/j.arr.2021.101271

Gu, S., Xie, Z., Gu, M., Wang, C., Xu, L., Gao, C., Yuan, X., Wu, Y., Hu, Y., Cao, Y., & Ye, Q. (2024). Myricetin mitigates motor disturbance and decreases neuronal ferroptosis in a rat model of Parkinson's disease. *Scientific Reports*, 14(1), 62910. https://doi.org/10.1038/s41598-024-62910-6

Gülmez, Y., & Elmastaş, M. (2025). Comparison of phenolic acid and flavonoid contents of sunflower (*Helianthus annuus* L.) bee pollen and bee bread taken from the same hives. *Bulletin of the Chemical Society of Ethiopia*, 39(9), 1675–1681. https://doi.org/10.4314/bcse.v39i9.6

Habryka, C., Socha, R., & Juszczak, L. (2020). The influence of honey enrichment with bee pollen or bee bread on the content of selected mineral components in multifloral honey. *Slovak Journal of Food Sciences*, 14(1), 624–632. https://doi.org/10.5219/1330

Han, H., Liu, X., Li, J., & Ma, C. (2026). Metabolomic and proteomic profiling of royal jelly from honeybee (Apis mellifera) colonies fed with camellia versus rapeseed bee pollen. *Journal of Agriculture and Food Research*, 26, 102669.

Heleno, S. A., Martins, A., Queiroz, M. J., & Ferreira, I. C. (2015). Bioactivity of phenolic acids: Metabolites versus parent compounds: A review. *Food Chemistry*, 173, 501–513. https://doi.org/10.1016/j.foodchem.2014.10.057

Huang, H., Shen, Z., Geng, Q., Wu, Z., Shi, P., & Miao, X. (2017). Protective effect of *Schisandra chinensis* bee pollen extract on liver and kidney injury induced by cisplatin in rats. *Biomedicine & Pharmacotherapy*, 95, 1765–1776. https://doi.org/10.1016/j.biopha.2017.09.083

Ilie, I., Păunescu, C., Păunescu, C., & Ciocoiu, M. (2025). Improved antibacterial properties of fermented and enzymatically hydrolyzed bee pollen and its combined effect with antibiotics. *Molecules*, 18(1), 15. https://doi.org/10.3390/molecules18010015

Jagdis, A., & Sussman, G. (2012). Anaphylaxis from bee pollen supplement. *Canadian Medical Association Journal*, 184(10), 1167. https://doi.org/10.1503/cmaj.112181

Jan, R., Khan, M., Asaf, S., Lubna, Asif, S., & Kim, K. M. (2022). Bioactivity and therapeutic potential of kaempferol and quercetin: New insights for plant and human health. *Plants*, 11(19), 2623. https://doi.org/10.3390/plants11192623

Jiang, S., & Guo, Y. (2020). Epigenetic clock: DNA methylation in aging. *Stem Cells International*, 2020, 1047896. https://doi.org/10.1155/2020/1047896

Jiang, Y., Xie, G., Alimujiang, A., Xie, H., Yang, W., Yin, F., & Huang, D. (2023). Protective effects of quercetin against MPP+-induced dopaminergic neuron injury via the Nrf2 signaling pathway. *Frontiers in Bioscience-Landmark*, 28(3), 42. https://doi.org/10.31083/j.fbl2803042

Jorgovanovic, D., Song, M., Wang, L., & Zhang, Y. (2020). Roles of IFN-γ in tumor progression and regression: A review. *Biomarker Research*, 8(1), 49.

Jung, U. J., & Kim, S. R. (2018). Beneficial effects of flavonoids against Parkinson's disease. *Journal of Medicinal Food*, 21(5), 421–432.

Kacemi, R., & Campos, M. G. (2023). Translational research on bee pollen as a source of nutrients: A scoping review from bench to real world. *Nutrients*, 15(10), 2413. https://doi.org/10.3390/nu15102413

Kacemi, R., & Campos, M. G. (2024). Bee pollen as a source of biopharmaceuticals for neurodegeneration and cancer research: a scoping review and translational prospects. *Molecules*, 29(24), 5893.

Kacemi, R., & Campos, M. G. (2025). Bee pollen phytochemicals and nutrients as unequaled pool of epigenetic regulators: Implications for age-related diseases. *Foods*, 14(3), 347. https://doi.org/10.3390/foods14030347

Kalaba, M., Tešić, Ž., & Blagojević, S. (2023). Flavonoids in pollen. In N. E. Bayram, Ž. Kostić, & Y. C. Gercek (Eds.), *Pollen chemistry and biotechnology* (pp. 115–142). Springer. https://doi.org/10.1007/978-3-031-47563-4_7

Kaplan, M., Demirtaş, İ., & Silici, S. (2019). Determination of phytosterols in beebread from different botanical origins. *Journal of Traditional Medical Complementary Therapies*, 2(3), 89–96.

Kaškonienė, V., Adaškevičiūtė, V., Kaškonas, P., Mickienė, R., & Maruška, A. (2020). Antimicrobial and antioxidant activities of natural and fermented bee pollen. *Food Bioscience*, 34, 100532. https://doi.org/10.1016/j.fbio.2019.100532

Khalifa, S. A., Elashal, M. H., Yosri, N., Du, M., Musharraf, S. G., Nahar, L., Sarker, S. D., Guo, Z., Cao, W., & Zou, X. (2021). Bee pollen: Current status and therapeutic potential. *Nutrients*, 13(6), 1876.

Khan, M., Elango, C., Ansari, M. A., Singh, I., & Singh, A. K. (2007). Caffeic acid phenethyl ester reduces neurovascular inflammation and protects rat brain following transient focal cerebral ischemia. *Journal of Neurochemistry*, 102(2), 365–377. https://doi.org/10.1111/j.1471-4159.2007.04526

Kocot, J., Kiełczykowska, M., Luchowska-Kocot, D., Kurzepa, J., & Musik, I. (2018). Antioxidant potential of propolis, bee pollen, and royal jelly: Possible medical application. *Oxidative Medicine and Cellular Longevity*, 2018, 7074209. https://doi.org/10.1155/2018/7074209

Kostić, A. Ž., Arserim-Uçar, D. K., Materska, M., Sawicka, B., Skiba, D., Milinčić, D. D., Pešić, M. B., Pszczółkowski, P., Moradi, D., & Ziarati, P. (2024). Unlocking quercetin's neuroprotective potential: A focus on bee-collected pollen. *Chemistry & Biodiversity*, 21(6), e202400114.

Kostić, A. Ž., Gercek, Y. C., & Bayram, N. E. (2023). Phenolic acids in pollen. In N. E. Bayram, Ž. Kostić, & Y. C. Gercek (Eds.), *Pollen chemistry and biotechnology* (pp. 103–125). Springer.

Kostić, A. Ž., & Kilibarda, S. (2023). Lipids in pollen. In N. E. Bayram, Ž. Kostić, & Y. C. Gercek (Eds.), *Pollen chemistry and biotechnology* (pp. 143–160). Springer. https://doi.org/10.1007/978-3-031-47563-4_4

Kurtagić, H., Barudanović, S., & Durmić, V. (2015). Determination of rutin, quercetin, naringenin and hesperetin in the honey from Bosnia and Herzegovina (B & H) in relation to the composition of pollen. *Journal of Environmental Engineering and Science*, 4(12), 615–622.

Kwak, J. E., Lee, J. Y., Baek, J. Y., Kim, S. W., & Ahn, M. R. (2024). The antioxidant and anti-inflammatory properties of bee pollen from acorn (*Quercus acutissima* Carr.) and Darae (*Actinidia arguta*). *Antioxidants*, 13(8), 981.

Lahoda Brodska, H., Klempir, J., Zavora, J., & Kohout, P. (2023). The role of micronutrients in neurological disorders. *Nutrients*, 15(19), 4129. https://doi.org/10.3390/nu15194129

Lawag, I. L., Yoo, O., Lim, L. Y., Hammer, K., & Locher, C. (2021). Optimisation of bee pollen extraction to maximise extractable antioxidant constituents. *Antioxidants*, 10(7). https://doi.org/10.3390/antiox10071113

Li, Q.-Q., Wang, K., Marcucci, M. C., Sawaya, A. C. H. F., Hu, L., Xue, X.-F., Wu, L.-M., & Hu, F.-L. (2018). Nutrient-rich bee pollen: A treasure trove of active natural metabolites. *Journal of Functional Foods*, 49, 472–484. https://doi.org/10.1016/j.jff.2018.08.020

Li, Q., Sun, M., Wan, Z., Liang, J., Betti, M., Hrynets, Y., Xue, X., Wu, L., & Wang, K. (2019). Bee pollen extracts modulate serum metabolism in lipopolysaccharide-induced acute lung injury mice with anti-inflammatory effects. *Journal of Agricultural and Food Chemistry*, 67(28), 7855–7868. https://doi.org/10.1021/acs.jafc.9b03082

Li, Q., Liu, X., Du, Y., Zhang, X., Xiang, P., Chen, G., Ling, W., & Wang, D. (2023). Protocatechuic acid boosts continual efferocytosis in macrophages by derepressing KLF4 to transcriptionally activate MerTK. *Science Signaling*, 16(786), eabn1372. https://doi.org/10.1126/scisignal.abn1372

Liao, Y., Bae, H. J., Zhang, J., Kwon, Y., Koo, B., Jung, I. H., Kim, H. M., Park, J. H., Lew, J. H., & Ryu, J. H. (2019). The ameliorating effects of bee pollen on scopolamine-induced cognitive impairment in mice. *Biological and Pharmaceutical Bulletin*, 42(3), 379–388.

Liberal, J., Ferreira, I. V., Cardoso, E. O., Silva, A., Bartolomeu, A. R., Martins, J., Santiago, K. B., Conti, B. J., Neves, B. M., & Batista, M. T. (2016). Anti-Inflammatory Activity of the honeybee plant derived products honey, Pollen and Propolis. In *Chemistry, Biology and Potential Applications of Honeybee Plant-Derived Products* (pp. 313–346). Bentham Science Publishers.

Lim, D. W., Lee, J. E., Lee, C., & Kim, Y. T. (2024). Natural products and their neuroprotective effects in degenerative brain diseases: A comprehensive review. *International Journal of Molecular Sciences*, 25(20), 11223. https://doi.org/10.3390/ijms252011223

Lin, Z. H., Liu, Y., Xue, N. J., Zheng, R., Yan, Y. Q., Wang, Z. X., Li, Y. L., Ying, C. Z., Song, Z., Tian, J., Pu, J. L., & Zhang, B. R. (2022). Quercetin protects against MPP+/MPTP-induced dopaminergic neuron death in Parkinson's disease by inhibiting ferroptosis. *Oxidative Medicine and Cellular Longevity*, 2022, 7769355. https://doi.org/10.1155/2022/7769355

Livshits, G., & Kalinkovich, A. (2024). Restoration of epigenetic impairment in the skeletal muscle and chronic inflammation resolution as a therapeutic approach in sarcopenia. *Ageing Research Reviews*, 96, 102267. https://doi.org/10.1016/j.arr.2024.102267

Magalingam, K. B., Radhakrishnan, A. K., & Haleagrahara, N. (2015). Protective mechanisms of flavonoids in Parkinson's disease. *Oxidative Medicine and Cellular Longevity*, 2015(1), 314560.

Mahdavifar, B., Hosseinzadeh, M., Salehi-Abargouei, A., Mirzaei, M., & Vafa, M. (2021). Dietary intake of B vitamins and their association with depression, anxiety, and stress symptoms: A cross-sectional, population-based survey. *Journal of Affective Disorders*, 288, 92–98. https://doi.org/10.1016/j.jad.2021.04.034

Mahmud, A. R., Ema, T. I., Siddiquee, M. F., et al. (2023). Natural flavonols: Actions, mechanisms, and potential therapeutic utility for various diseases. *Beni-Suef University Journal of Basic and Applied Sciences*, 12, 47. https://doi.org/10.1186/s43088-023-00387-4

Mărgăoan, R., Özkök, A., Keskin, Ş., Mayda, N., Urcan, A. C., & Cornea-Cipcigan, M. (2021). Bee collected pollen as a value-added product rich in bioactive compounds and unsaturated fatty acids: A comparative study from Turkey and Romania. *LWT*, 149, 111925. https://doi.org/10.1016/j.lwt.2021.111925

Martinello, M., & Mutinelli, F. (2021). Antioxidant activity in bee products: A review. *Antioxidants*, 10(1), 71. https://doi.org/10.3390/antiox10010071

Matuszewska, E., Plewa, S., Pietkiewicz, D., Kossakowski, K., Matysiak, J., Rosiński, G., & Matysiak, J. (2022). Mass spectrometry-based identification of bioactive bee pollen proteins: Evaluation of allergy risk after bee pollen supplementation. *Molecules*, 27(22), 7733. https://doi.org/10.3390/molecules27227733

Mayda, N., Özkök, A., Ecem Bayram, N., Gerçek, Y. C., & Sorkun, K. (2020). Bee bread and bee pollen of different plant sources: Determination of phenolic content, antioxidant activity, fatty acid and element profiles. *Journal of Food Measurement and Characterization*, 14(4), 1795–1809.

Mohammad-Rafiei, F., Negahdari, S., Tahershamsi, Z., & Gheibihayat, S. M. (2024). Interface between resolvins and efferocytosis in health and disease. *Cell Biochemistry and Biophysics*, 82(1), 53–65. https://doi.org/10.1007/s12013-023-01187-4

Morrison, B., Newburn, L. R., & Fitch, G. (2025). Food as medicine: A review of plant secondary metabolites from pollen, nectar, and resin with health benefits for bees. *Insects*, 16(4), 414. https://doi.org/10.3390/insects16040

Nassar, A. M., Salim, Y. M., Eid, K. S., Shaheen, H. M., Saati, A. A., Hetta, H. F., . . . & Batiha, G. E. S. (2020). Ameliorative effects of honey, propolis, pollen, and royal jelly mixture against chronic toxicity of sumithion insecticide in white albino rats. *Molecules*, 25(11), 2633. https://doi.org/10.3390/molecules25112633

Nemś, A., Lachowicz-Wiśniewska, S., Kapusta, I. T., Miedzianka, J., Kita, A., & Carbonell-Barrachina, Á. A. (2025). Bee pollen as a source of phenolic compounds in potato snacks. *Scientific Reports*, 15(1), 24203. https://doi.org/10.1038/s41598-025-09776-4

Nguyen, C. D., & Lee, G. (2021). Neuroprotective activity of melittin—the main component of bee venom—against oxidative stress induced by Aβ25–35 in in vitro and in vivo models. *Antioxidants*, 10(11), 1654. https://doi.org/10.3390/antiox10111654

Nigar, S., & Shimosato, T. (2019). Cooperation of oligodeoxynucleotides and synthetic molecules as enhanced immune modulators. *Frontiers in Nutrition*, 6, 140. https://doi.org/10.3389/fnut.2019.00140

Nikolić, T. V., Kojić, D., Orčić, S., Batinić, D., Vukašinović, E., Blagojević, D. P., & Purać, J. (2016). The impact of sublethal concentrations of Cu, Pb and Cd on honey bee redox status, superoxide dismutase and catalase in laboratory conditions. *Chemosphere*, 164, 98–105. https://doi.org/10.1016/j.chemosphere.2016.08.084

Nissinen, E., & Männistö, P. T. (2009). Biochemistry and pharmacology of catechol-O-methyltransferase inhibitors. *International Review of Neurobiology*, 95, 73–118. https://doi.org/10.1016/B978-0-12-381326-8.00005-3

Nordin, A., Saim, A. B., & Idrus, R. B. H. (2021). Honey ameliorates negative effects in neurodegenerative diseases: An evidence-based review. *Sains Malaysiana*, 50(3), 791–801. https://doi.org/10.17576/jsm-2021-5003-22

Novak, V., Rogelj, B., & Župunski, V. (2021). Therapeutic potential of polyphenols in amyotrophic lateral sclerosis and frontotemporal dementia. *Antioxidants*, 10(8), 1328.

Nuvoloni, R., Meucci, V., Turchi, B., Sagona, S., Fratini, F., Felicioli, A., … & Pedonese, F. (2021). Bee-pollen retailed in Tuscany (Italy): Labelling, palynological, microbiological, and mycotoxicological profile. *LWT*, 140, 110712. https://doi.org/10.1016/j.lwt.2020.110712

de Oliveira, M. C., da Silva, D. M., Loch, F. C., Martins, P. C., Dias, D. M. B., & Simon, G. A. (2013). Effect of bee pollen on the immunity and tibia characteristics in broilers. Brazilian *Journal of Poultry Science*, 15(4), 323–327. https://doi.org/10.1590/S1516-635X2013000400005

Özcan, M. M., Aljuhaimi, F., Babiker, E. E., Uslu, N., Ceylan, D. A., Ghafoor, K., … & Alsawmahi, O. N. (2019). Determination of antioxidant activity, phenolic compound, mineral contents and fatty acid compositions of bee pollen grains collected from different locations. *Journal of Apicultural Science*, 63(1), 69–79.

Özsoy Gökbilen, S., Becer, E., & Vatansever, H. S. (2022). Senescence-mediated anticancer effects of quercetin. *Nutrition Research*, 104, 82–90. https://doi.org/10.1016/j.nutres.2022.04.007

Padhi, D., Baruah, P., Ramesh, M., Moorthy, H., & Govindaraju, T. (2024). Hybrid molecules synergistically mitigate ferroptosis and amyloid-associated toxicities in Alzheimer's disease. *Redox Biology*, 71, 103119. https://doi.org/10.1016/j.redox.2024.103119

Panigrahy, D., Gilligan, M. M., Serhan, C. N., & Kashfi, K. (2021). Resolution of inflammation: An organizing principle in biology and medicine. *Pharmacology & Therapeutics*, 227, 107879.

Park, Y., & Kim, S. (2020). A short guide to histone deacetylases including recent progress on class II enzymes. *Experimental & Molecular Medicine*, 52(2), 204. https://doi.org/10.1038/s12276-020-0382-4

Paul, A., Guha, N., & Zaman, M. K. (2022). Nanotherapeutics of phytoantioxidants for CNS disorders. *Phytoantioxidants and Nanotherapeutics*, 449–476.

Pilotto, F., Smeele, P. H., Scheidegger, O., Diab, R., Schobesberger, M., Sierra-Delgado, J. A., & Saxena, S. (2025). Kaempferol enhances ER-mitochondria coupling and protects motor neurons from mitochondrial dysfunction and ER stress in C9ORF72-ALS. *Acta Neuropathologica Communications*, 13(1), 21.

Pohl, P., et al. (2023). Macro-, micro-, trace, and toxic elements of pollen. In N. E. Bayram, Ž. Kostić, & Y. C. Gercek (Eds.), *Pollen chemistry and biotechnology* (pp. 85–102). Springer. https://doi.org/10.1007/978-3-031-47563-4_5

Pohl, P., Dzimitrowicz, A., Jamroz, P., Lesniewicz, A., Szymczycha-Madeja, A., Welna, M., & Greda, K. (2024). Macro-, micro-, trace, and toxic elements of pollen. *Pollen Chemistry & Biotechnology*, 85–102.

Prakatur, I., Miskulin, M., Pavic, M., Marjanovic, K., Blazicevic, V., Miskulin, I. & Domacinovic, M. (2019). Intestinal morphology in broiler chickens. *Animals*, 9, 301.

Pukalskas, A., Kazernavičiuté, R., & Balžekas, J. (2016). Evaluation of antioxidant activity and flavonoid composition in differently preserved bee products. *Czech Journal of Food Sciences*, 34(2), 133–142.

Pyo, I. S., Yun, S., Yoon, Y. E., Choi, J. W., & Lee, S. J. (2020). Mechanisms of aging and the preventive effects of resveratrol on age-related diseases. *Molecules (Basel, Switzerland)*, 25(20), 4649. https://doi.org/10.3390/molecules25204649

Rajendran, P., Abdelsalam, S. A., Renu, K., Veeraraghavan, V., Ben Ammar, R., & Ahmed, E. A. (2022). Polyphenols as potent epigenetics agents for cancer. *International Journal of Molecular Sciences*, 23(19), 11712. https://doi.org/10.3390/ijms231911712

Ramakrishna, K., Nalla, L. V., Naresh, D., Venkateswarlu, K., Viswanadh, M. K., Nalluri, B. N., Chakravarthy, G., Duguluri, S., Singh, P., Rai, S. N., Kumar, A., Singh, V., & Singh, S. K. (2023). WNT-β catenin signaling as a potential therapeutic target for neurodegenerative diseases: Current status and future perspective. *Diseases (Basel, Switzerland)*, 11(3), 89. https://doi.org/10.3390/diseases11030089

Ravikumar, N., Chegukrishnamurthi, M., Gadde Venkata, S. (2022). Role of micronutrients in neurological development. In: Rajagopal, S., Ramachandran, S., Sundararaman, G., Gadde Venkata, S. (Eds.), *Role of nutrients in neurological disorders: Nutritional neurosciences*. Springer, Singapore. https://doi.org/10.1007/978-981-16-8158-5_9

Razani, E., Pourbagheri-Sigaroodi, A., Safaroghli-Azar, A., Zoghi, A., Shanaki-Bavarsad, M., & Bashash, D. (2021). The PI3K/Akt signaling axis in Alzheimer's disease: A valuable target to stimulate or suppress? *Cell Stress and Chaperones*, 26(6), 871–887. https://doi.org/10.1007/s12192-021-01231-3

Rodríguez-Arce, E., & Saldías, M. (2024). Unlocking quercetin's neuroprotective potential: a focus on bee-collected pollen. *Current Pharmaceutical Design*, 30(12), 923–935.

Rodríguez-Pólit, C., Gonzalez-Pastor, R., Heredia-Moya, J., Carrera-Pacheco, S. E., Castillo-Solis, F., Vallejo-Imbaquingo, R., Carlos Barba-Ostria, C., & Guamán, L. P. (2023). Chemical properties and biological activity of bee pollen. *Molecules*, 28(23), 7768. https://doi.org/10.3390/molecules28237768

Roquetto, A. R., Monteiro, N. E. S., Moura, C. S., Toreti, V. C., de Pace, F., Santos, A. D., Park, Y. K., & Amaya-Farfan, J. (2015). Green propolis modulates gut microbiota, reduces endotoxemia and expression of TLR4 pathway in mice fed a high-fat diet. *Food Research International*, 76(Pt 3), 796–803. https://doi.org/10.1016/j.foodres.2015.07.026

Rubio, K., Hernández-Cruz, E. Y., Rogel-Ayala, D. G., Sarvari, P., Isidoro, C., Barreto, G., & Pedraza-Chaverri, J. (2023). Nutriepigenomics in environmental-associated oxidative stress. *Antioxidants (Basel, Switzerland)*, 12(3), 771. https://doi.org/10.3390/antiox12030771

Rupanar, S. V., Satpute, R., Gadhave, P., & Kakade, N. (2024). Bee pollen beyond nutrition: An integrated review of its chemical, pharmacological, and biological properties. *Journal of Drug Research in Ayurvedic Sciences*, 9(4), 211–221.

Rzepecka-Stojko, A., Stojko, J., Kurek-Górecka, A., Górecki, M., Kabała-Dzik, A., Kubina, R., Moździerz, A., & Buszman, E. (2015). Polyphenols from Bee Pollen: Structure, Absorption, Metabolism and Biological Activity. *Molecules (Basel, Switzerland)*, 20(12), 21732–21749. https://doi.org/10.3390/molecules201219800

Sadek, K. M., Shib, N. A., Taher, E. S., Rashed, F., Shukry, M., Atia, G. A., … & Abdeen, A. (2024). Harnessing the power of bee venom for therapeutic and regenerative medical applications: An updated review. *Frontiers in Pharmacology*, 15, 1412245. https://doi.org/10.3389/fphar.2024.1412245

Salazar, J. R., Loza-Mejía, M. A., & Soto-Cabrera, D. (2020). Chemistry, biological activities and in silico bioprospection of sterols and triterpenes from Mexican columnar cactaceae. *Molecules*, 25(7), 1649. https://doi.org/10.3390/molecules25071649

Salazar-González, C. Y., Stinco, C. M., Rodríguez-Pulido, F. J., Díaz-Moreno, C., Fuenmayor, C., Heredia, F. J., & González-Miret, M. L. (2022). Characterization of carotenoid profile and α-tocopherol content in Andean bee pollen influenced by harvest time and particle size. *LWT*, 170, 114065. https://doi.org/10.1016/j.lwt.2022.114065

Sanyal, A., Ghosh, A., Roy, C., Mazumder, I., & Marrazzo, P. (2023). Revolutionizing the use of honeybee products in healthcare: a focused review on using bee pollen as a potential adjunct material for biomaterial functionalization. *Journal of Functional Biomaterials*, 14(7), 352.

Saral, Ö., Şahin, H., Saral, S., Alkanat, M., Akyıldız, K., Topçu, A., & Yılmaz, A. (2022). Bee pollen increases hippocampal brain-derived neurotrophic factor and suppresses neuroinflammation in adult rats with chronic immobilization stress. *Neuroscience Letters*, 766, 136342.

Shaikh, A., Ahmad, F., Teoh, S. L., Kumar, J., & Yahaya, M. F. (2023). Honey and Alzheimer's disease—Current understanding and future prospects. *Antioxidants*, 12(2), 427. https://doi.org/10.3390/antiox12020427

Shukla, S., & Tekwani, B. L. (2020). Histone deacetylases inhibitors in neurodegenerative diseases, neuroprotection and neuronal differentiation. *Frontiers in Pharmacology*, 11, 537. https://doi.org/10.3389/fphar.2020.00537

Singh, A., Singh, D. K., Kharwar, R. N., White, J. F., & Gond, S. K. (2022). Bioactivity and therapeutic potential of kaempferol and quercetin: New insights for plant and human health. *Plants*, 11(19), 2623. https://doi.org/10.3390/plants11192623

Skalny, A., Chernova, L., & Skalnaya, A. (2023). Essential trace elements in neurodevelopment: An updated narrative. In *Vitamins and minerals in neurological disorders* (pp. 673–690). Academic Press. https://doi.org/10.1016/B978-0-323-89835-5.00033-8

Sobhani, M., Farzaei, M. H., Kiani, S., & Khodarahmi, R. (2021). Immunomodulatory, anti-inflammatory, and antioxidant effects of polyphenols: A comparative review on the parental compounds and their metabolites. *Food Reviews International*, 37(8), 759–811. https://doi.org/10.1080/87559129.2020.1725961

Sommano, S. R., Bhat, F. M., Wongkeaw, M., Sriwichai, T., Sunanta, P., Chuttong, B., & Burgett, M. (2020). Amino acid profiling and chemometric relations of black dwarf honey and bee pollen. *Frontiers in Nutrition*, 7, 558579. https://doi.org/10.3389/fnut.2020.558579

Souza, B. R. D., Sattler, J. A. G., Mello, A. A. M. D., Granato, D., Araújo, E. D. S., Freitas, A. D. S. D., . . . & Muradian, L. B. D. A. (2018). Physicochemical parameters and content of B-complex vitamins: An exploratory study of bee pollen from southern Brazilian states. *Revista Chilena de Nutrición*, 45(3), 232–242. https://doi.org/10.4067/S0717-75182018000300232

Taha, E. A., Al-Kahtani, S., & Taha, R. (2019). Protein content and amino acids composition of bee pollens from major floral sources in Al-Ahsa, eastern Saudi Arabia. *Saudi Journal of Biological Sciences*, 26(2), 232–237. https://doi.org/10.1016/j.sjbs.2017.06.003

Tel-Çayan, G., Çayan, F., Deveci, E., Karataş, Ş., & Duru, M. E. (2025). Characterization of chemical compounds of fractions from four different bee pollens with chemometric classification. *European Food Research and Technology*, 251(5), 667–683. https://doi.org/10.1007/s00217-024-04645-2

Thakur, M., & Nanda, V. (2018). Assessment of physico-chemical properties, fatty acid, amino acid and mineral profile of bee pollen from India with a multivariate perspective. *Journal of Food & Nutrition Research*, 57(4), 328–340.

Tolba, M. F., Azab, S. S., Khalifa, A. E., Abdel-Rahman, S. Z., & Abdel-Naim, A. B. (2013). Caffeic acid phenethyl ester, a promising component of propolis with a plethora of biological activities: A review on its anti-inflammatory, neuroprotective, hepatoprotective, and cardioprotective effects. *IUBMB Life*, 65(8), 699–709. https://doi.org/10.1002/iub.1189

Tutun, H., Kaya, M. M., Usluer, M. S., & Kahraman, H. A. (2021). Bee pollen: Its antioxidant activity. *Uludağ Arıcılık Dergisi*, 21(1), 119–131.

Ueda, T., Inden, M., Shirai, K., Sekine, S., Masaki, Y., Kurita, H., Ichihara, K., Inuzuka, T., & Hozumi, I. (2017). The effects of Brazilian green propolis that contains flavonols against mutant copper-zinc superoxide dismutase-mediated toxicity. *Scientific Reports*, 7(1), 2882.

Urcan, A. C. (2025). Chemical composition of bee bread. In *Bee products – Chemical and biological properties* (pp. 419–472). Springer. https://doi.org/10.1007/978-3-031-89049-9_13

Vanderplanck, M., Zerck, P. L., Lognay, G., & Michez, D. (2020). Sterol addition during pollen collection by bees: Another possible strategy to balance nutrient deficiencies? *Apidologie*, 51(5), 826–843. https://doi.org/10.1007/s13592-020-00765-1

Végh, R., & Csóka, M. (2023). Amino acids, peptides, and proteins of pollen. In N. E. Bayram, A. Ž. Kostic, & Y. Can Gercek (Eds.), *Pollen chemistry and biotechnology* (pp. 23–49). Springer. https://doi.org/10.1007/978-3-031-47563-4_2

Végh, R., Csóka, M., Sörös, C., & Sipos, L. (2021). Food safety hazards of bee pollen—A review. *Trends in Food Science & Technology*, 114, 490–509. https://doi.org/10.1016/j.tifs.2021.06.002

Wang, W.-W., Han, R., He, H.-J., Li, J., Chen, S.-Y., Gu, Y., & Xie, C. (2021). Administration of quercetin improves mitochondria quality control and protects the neurons in 6-OHDA-lesioned Parkinson's disease models. *Aging (Albany NY)*, 13(8), 11738.

Watanabe, C., Yanagihara, A., Miyata, R., Mitsui, T., Honda, C., Fujinami, D., & Kumazawa, S. (2024). Catechol-O-methyltransferase and monoamine oxidase B inhibitory activities of Australian bee pollen. *Bioscience, Biotechnology, and Biochemistry*, 88(6), 665–670. https://doi.org/10.1093/bbb/zbae041

Wu, X., Cai, X., Ai, J., Zhang, C., Liu, N., & Gao, W. (2021). Extraction, structures, bioactivities and structure-function analysis of the polysaccharides from safflower (*Carthamus tinctorius* L.). *Frontiers in Pharmacology*, 12, 767947. https://doi.org/10.3389/fphar.2021.767947

Xie, J., Xie, L., Wei, H., Li, X. J., & Lin, L. (2023). Dynamic regulation of DNA methylation and brain functions. *Biology*, 12(2), 152. https://doi.org/10.3390/biology12020152

Yang, Y., Liu, M., Wang, K., Yang, Y., Su, N., Huang, W., & Wu, Y. (2020). Chemical and cytological evaluation of honeybee pollen antioxidant ability. *Journal of Food Science*, 85(3), 824–833.

Yao, X., Liu, Y., Mao, M., Yang, L., Zhan, Q., & Xiao, J. (2024). Calorie restriction mimetic, resveratrol, attenuates hepatic ischemia and reperfusion injury through enhancing efferocytosis of macrophages via AMPK/STAT3/S1PR1 pathway. *The Journal of Nutritional Biochemistry*, 126, 109587. https://doi.org/10.1016/j.jnutbio.2024.109587

Yazlovytska, L. S., Karavan, V. V., Domaciuk, M., Panchuk, I. I., Borsuk, G., & Volkov, R. A. (2023). Increased survival of honey bees consuming pollen and beebread is associated with elevated biomarkers of oxidative stress. *Frontiers in Ecology and Evolution*, 11, 1098350. https://doi.org/10.3389/fevo.2023.1098350

Zamri, N. A., Ghani, N., Ismail, C. A. N., Zakaria, R., & Shafin, N. (2023). Honey on brain health: A promising brain booster. *Frontiers in Aging Neuroscience*, 14, 1092596. https://doi.org/10.3389/fnagi.2022.1092596

Zhang, P., Kishimoto, Y., Grammatikakis, I., Gottimukkala, K., Cutler, R. G., Zhang, S., Abdelmohsen, K., Bohr, V. A., Misra Sen, J., Gorospe, M., & Mattson, M. P. (2019). Senolytic therapy alleviates Aβ-associated oligodendrocyte progenitor cell senescence and cognitive deficits in an Alzheimer's disease model. *Nature Neuroscience*, 22(5), 719–728. https://doi.org/10.1038/s41593-019-0372-9

Zou, Y., Hu, J., Huang, W., Zhu, L., Shao, M., Dordoe, C., Ahn, Y.-J., Wang, D., Zhao, Y., Xiong, Y., & Wang, X. (2020). The botanical origin and antioxidant, anti-BACE1 and antiproliferative properties of bee pollen from different regions of South Korea. *BMC Complementary Medicine and Therapies*, 20(1), 236. https://doi.org/10.1186/s12906-020-03023-1

Zuluaga-Domínguez, C., Castro-Mercado, L., & Quicazán, M. C. (2019). Effect of enzymatic hydrolysis on structural characteristics and bioactive composition of bee pollen. *Journal of Food Processing and Preservation*, 43(7), e13983. https://doi.org/10.1111/jfpp.13983

10 Bee Royal Jelly Applications in Neurodegenerative Diseases

Nyejirime Young Wike

10.1 INTRODUCTION

Bee royal jelly (BRJ) is a nutritious natural product that is widely utilized in the healthcare sector, both as personal products and as supplements (Oršolić & Jazvinšćak, 2024). BRJ is a yellowish-white, creamy protein produced by hypopharyngeal and mandibular glands (Guo et al., 2021). Apart from BRJ, honey, propolis, pollen, and bee bread are some foods from the beehive that have potential health benefits (Collazo et al., 2021; Oršolić & Jazvinšćak, 2024). BRJ is a mixture of important compounds, such as proteins, vitamins, lipids, minerals, hormones, neurotransmitters, flavonoids, and polyphenols, which underlie its remarkable biological and therapeutic activities. Some of the biologically active components of BRJ include antibacterial protein, jelleins, apisin, royalactin, and defensin-1 (Kunugi & Mohammed Ali, 2019; Oršolić & Jazvinšćak, 2024). Bee royal jelly has an exceptional value, and its consumption offers numerous health benefits for humans (Collazo et al., 2021). It has a wide range of immunomodulatory, antioxidant, neurotropic, anticancer, and anti-inflammatory properties, which makes its application in neurodegenerative diseases an interesting topic (Abu-Serie & Habashy, 2019; Petelin et al., 2019; Fan et al., 2020).

Recently, consumers and food manufacturers have become increasingly aware of functional foods, which can be natural or produced by some biological processes, and how they can contribute to sustaining human health. The important role of a diet in the prevention and treatment of various illnesses is, nowadays, widely accepted. Furthermore, some ingredients, such as omega-3 fatty acids, vitamins, probiotics, fibers, bioactive peptides, and fitosterols, can form supplements to help improve the functionality and benefit of the functional foods.

A progressive loss of neurons is a hallmark of neurodegenerative diseases (NDs), which typically result in mortality (Lamptey et al., 2022; Gadhave et al., 2024). Among these are Alzheimer's disease (AD), multiple sclerosis (MS), Parkinson's disease (PD), amyotrophic lateral sclerosis (ALS), Huntington's disease (HD), and several other neuropsychiatric disorders (Hui et al., 2023). Frequently, NDs are associated with the gradual deterioration of synaptic links and neurons, most of which occur later in life (Pant et al., 2022). The diagnosis of NDs is determined by the clinical prevalence of the patient and supporting data of magnetic resonance imaging (MRI) (Huang & Zhang, 2023).

The development and progression of clinical symptoms are strongly associated with the extent of neuronal loss. People with NDs are said to live shorter lives, according to studies by Ding et al. (2022) and Huang et al. (2023). A 2019 study found that severe neurological illnesses impacted 349.2 million people worldwide and caused 10 million deaths (Ding et al., 2022; Huang et al., 2023).

Although there are numerous approved treatments for NDs at the moment, the majority merely target the symptoms that are related to the condition. Due to various obstacles related to ND therapies, including the blood-brain barrier and numerous undesirable side effects of available treatments, the available therapies have failed to control the progression of NDs, resulting in a lower survival rate (Gadhave et al., 2019; Gadhave et al., 2023; Niazi, 2023). Many researchers have focused on the use of natural products to treat the central nervous system (CNS) conditions such as NDs. According to several recent articles, natural products have the potential to offer neuroprotective benefit against NDs (Ayinla et al., 2022; Asuku et al., 2024, 2025a).

 DOI: 10.1201/9781003685654-10

10.2 BEE ROYAL JELLY AND ITS CHEMICAL COMPONENTS

BRJ contains the following: water (60–70%), proteins (9–18%), amino acids, sugars (7–18%), lipids (3–8%), minerals (0.7–1.5%), and vitamins (336–351 mg/100 g) (Ahmad et al., 2020; Collazo et al., 2021; Botezan et al., 2023). These constituents do not maintain a constant value as they are dependent on plant/bee species, geographical location, and harvesting method as well as season of the year. BRJ also contains a variety of phenolic compounds, flavonoids, organic acids, enzymes, acetylcholine and its precursor choline, as well as sex hormones (estradiol, testosterone, and progesterone).

10.3 BIOLOGICAL ACTIVITIES OF BRJ

Biological activities of BRJ as a functional food denote not only the biologically active components but also their antioxidant, anti-inflammatory, and antitumor effects and other protection against the toxic side effects of chemotherapy and radiation (Table 10.1).

10.3.1 Antioxidant Activities

Phenols and flavonoids are the most crucial antioxidants in BRJ (Kocot et al., 2018). Fatty acids and fatty acid esters are also sources of the antioxidant effect of BRJ (Bagameri et al., 2023; Botezan et al., 2023; Li et al., 2023). Besides, the antioxidant property of BRJ may be linked to free amino acids (Kocot et al., 2018). It is also observed that antioxidant capacity and overall sum of polyphenols of BRJ is contingent to the collection time. Moreover, it was noted that the antioxidant activities were maximized after 24 hours (Kocot et al., 2018). Its antioxidative effect has been demonstrated in liver and kidney tissues that were exposed to various harmful stimuli with pro-oxidant effect. This observation indicated that BRJ had a desirable impact on the incidence of lipid peroxidation and the activity of the principal antioxidative enzymes glutathione reductase, glutathione peroxidase, superoxide dismutase, and catalase (Chi et al., 2021).

10.3.2 Anti-Inflammatory Activity

It has been noted that BRJ has a potent impact on the life expectancy of patients with rheumatoid arthritis and inflammatory bowel diseases (Nattagh-Eshtivani et al., 2021; Ahmad et al., 2020). It also increases the plasma glutathione peroxidase activity, TNF damage, and major inflammatory mediators' production during the plasma glutathione peroxidase, COX-2, and NF-κB (Guo et al., 2021; Guo et al., 2022).

TABLE 10.1
Some Biological Activities of Bee Royal Jelly

Biological activities	Effects	References
Antioxidant	Reduces lipid peroxidation; improves activities of the antioxidative enzymes, e.g., glutathione reductase, glutathione peroxidase, superoxide dismutase, and catalase	Chi et al. (2021)
Antibacterial	Alleviates fatal diseases	Bagameri et al. (2022)
Anti-inflammatory	Improves the activity of plasma glutathione peroxidase	Guo et al. (2021); Guo et al. (2022).

10.3.3 Antibacterial Activity

Bee royal jelly has also been reported to alleviate the development of certain terminal diseases that are caused by different pathogenic agents in both human and animal studies (Bagameri et al., 2022). It has been put to the test against other bacterial strains such as *E.coli*, *S. aureus*, and *P. gingivalis*, among others (Khosla et al., 2020; Bagameri et al., 2022). Its activity can be explained by the presence of its biologically active substance such as proteins, peptide, and 10-HAD (Khosla et al., 2020; Botezan et al., 2023).

10.4 CLASSIFICATIONS OF NEURODEGENERATIVE DISORDERS

10.4.1 Alzheimer's Disease (AD)

Alzheimer's disease (AD) is characterized by buildup of protein deposits in the brain. It is a progressive neurological disease, which is mainly based in the brain, resulting in the deterioration of memory, intellectual functions, and behavioral patterns (Asuku et al., 2025b). Lamptey et al. (2022) assert that it is the main cause of dementia among seniors in the population. The deposits of protein aggregates consist of beta-amyloid plaques and tau tangles that block the smooth flow of impulses between nerve cells, causing their death and progressive loss of brain tissue (Breijyeh & Karaman, 2020). AD is characterized by memory loss, cognitive deterioration, behavior change, and functional limitation. A detailed analysis of the patient's medical history, cognitive tests, and their elimination of any possible reasons that may possibly lead to a decline in their cognitive abilities helps in arriving at the diagnosis. Few cases are subjected to advanced imaging methods and cerebrospinal fluid examination. AD can be divided into three phases—mild (early stage), moderate (middle stage), and severe (late stage). All stages are further defined by specific symptoms and various degrees of dysfunction (Breijyeh & Karaman, 2020). At present there is no known treatment for AD. Nonetheless, drugs like cholinesterase inhibitors and memantine are effective in alleviating the symptoms and reducing the rate of cognitive decline. Additional therapeutic interventions, including occupational therapy and cognitive training, can also offer an advantage.

10.4.2 Parkinson's Disease (PD)

Parkinson's disease (PD) is a progressive loss of motor function. PD is often brought on by the progressive death of dopamine-producing neurons in the brain, particularly in the substantia nigra (Lamptey et al., 2022). Environmental and genetic variables have been connected to PD. Tremors, bradykinesia, stiffness, and postural instability are among the symptoms of PD, according to Khatri et al. (2020). Due to the lack of a conclusive diagnostic test, the diagnosis is primarily based on evaluating clinical symptoms and medical history. To rule out other disorders, neurological testing and occasionally imaging techniques may be utilized. A medication called levodopa, which replaces dopamine, is frequently used to treat motor problems. Supplementary medications, like dopamine agonists and MAO-B inhibitors, can also be prescribed by medical practitioners (Kulisevsky, 2022).

10.4.3 Amyotrophic Lateral Sclerosis (ALS)

Another name for this is Lou Gehrig's disease. Amyotrophic lateral sclerosis (ALS) is a degenerative neurological disorder characterized by a slow loss of brain and spinal cord nerve cells.

Motor neurons, which are nerve cells that regulate voluntary muscular contraction, are lost as a result of ALS (Lamptey et al., 2022). The degeneration of the motor neurons leads to the inability of the brain to begin and control muscular movement. ALS etiology is not well understood, and both genetic and non-genetic environmental factors may play a role in the pathogenesis of the disease. The common symptoms of this condition include weakness of the muscles, their wasting, speech and swallowing problems, and complications related to breathing. There are two subtypes of ALS, known as sporadic ALS and familial ALS (Masrori & Van Damme, 2020). Diagnosis of ALS is largely based on clinical symptoms and systematic ruling out of other diseases, as a diagnostic test has not yet been developed to detect this condition. Electromyography (EMG) and nerve conduction studies are commonly employed in assessing the efficacy of motor neurons and muscles. Riluzole is an FDA-approved pharmaceutical that could slow the progression of ALS by reducing the damage caused to the motor neurons. Also, an adjuvant drug approved by the FDA is edaravone, which is prescribed to treat ALS and can slow down the progressive loss of physical functions (Neupane et al., 2023).

10.4.4 Huntington's Disease (HD)

Huntington's disease (HD) is an inheriting and progressive neurological disorder that involves deficiencies in cognitive and physical functioning of a person. The disease is caused by genetic mutation of huntingtin gene (*HTT*), which produces a defective form of huntingtin protein. This mutation causes the breakdown of certain areas of the brain, leading to diverse physical, cognitive, as well as mental dysfunction. As much as HD is incurable, the symptoms could be managed with the help of symptomatic medication and supportive treatment. Prescriptions can be provided to minimize motor symptoms and mental issues and improve the overall quality of life. A multidisciplinary approach involving neurologists, psychiatrists, physical and occupational therapists, and genetic counselors may be necessary to treat the complex needs of HD patients. Genetic testing for HD may reveal a person's propensity to have or transmit the disease; those who are considering genetic testing are advised to undergo genetic counseling to understand the potential consequences (McGarry et al., 2020).

10.4.5 Multiple Sclerosis (MS)

Multiple sclerosis (MS) is a neurological chronic, immune-mediated disease impacting the brain and the spinal cord. It is a disorder that occurs when the immune system assaults the protective layer of nerve tissues (Lamptey et al., 2022). This is normally the cause of inflammation, the loss of the underlying neurons and interference with the normal conduction of electrical impulses along the nerves. The symptoms of MS are cognitive impairment, fatigue, lack of coordination, sensory dysfunctions, and visual defects.

The three primary forms of MS are relapsing-remitting MS, primary progressive MS, and secondary progressive MS. Relapsing-remitting multiple sclerosis is affected by frequent outbursts or recurrence of the disease followed by partial or total recovery. Primary progressive multiple sclerosis is characterized by a gradual deterioration of the condition at its onset without apparent flare-ups or remissions. Finally, secondary progressive MS starts as relapsing-remitting MS and proceeds to the transition stage with increasing symptoms and disability (Eva et al., 2023). Although the exact etiology of MS remains ambiguous, it is believed to be a combination of an interaction between genes and the environment. Smokers and people with low concentrations of vitamin D are prone to the onset of MS (Gombash et al., 2022).

Diagnosing MS appears to be challenging as a result of lack of a definitive diagnostic test. Neurosurgeons usually apply a combination of medical resonance imaging (MRI) scans, physical examination, and clinical history. Despite the fact that there is no known cure for MS, therapeutic strategies and lifestyle modifications are usually applied to manage the symptoms.

10.5 APPLICATIONS OF BRJ IN SOME NEURODEGENERATIVE DISEASES

10.5.1 Alzheimer's Disease (AD)

AD is a neurodegenerative disorder characterized by progressive decline in cognitive function. BRJ has been documented to promote beneficial effects on learning impairments and also exhibit neuroprotective action in STZ experimental model of sporadic AD by attenuating oxidative stress (Ali & Kunugi, 2020; Guardia de Souza e Silva et al., 2020; Raoufi et al., 2023).

10.5.2 Multiple Sclerosis (MS)

The central nervous system is usually attacked by inflammatory cells like lymphocytes, macrophages, and microglia in MS. This invasion results in the destruction of myelin, damage to nerve fibers, and impairment of neurological function (Faissner et al., 2019). The progression of MS is usually marked by inflammation through a series of pathogenic activities that often result in destruction and injury in the mitochondria, especially in the early phase of the disease. Oxidative stress has also been linked to the development of demyelination in MS (Pegoretti et al., 2020; Dive et al., 2022). BRJ supplementation was reported to have beneficial effects on inflammatory and oxidative stress markers and also increase the lifespan of MS patients (Eslamifar et al., 2025).

10.5.3 Parkinson's Disease (PD)

PD is one of the most important neurodegenerative diseases in developed countries due to their prevalence among aged people and their specific symptoms (Balestrino & Schapira, 2020). BRJ and its bioactive components reduce oxidative damage directly by scavenging free radicals through the release of an electron from their phenolic group and indirectly through activation of extracellular signal-regulated kinase/mitogen-activated protein kinase (ERK/MAPK) signaling, which deactivates the Keap1 molecule. Keap1 is responsible for the degradation of NRF2, resulting in NRF2 translocation into the nucleus to activate antioxidant response elements (AREs). Activation of AREs will stimulate the expression of antioxidant genes such as heme oxygenase 1 (HO-1). Nuclear factor erythroid 2 (NRF2) and HO-1 prevent the transcription of inflammatory pathways such as nuclear factor kappa-B (NF-κB), resulting in decreased production of inflammatory cytokines. Mitigation of oxidative stress and neuroinflammation is associated with less mitochondrial respiration and less production of apoptotic molecules such as caspase-3 and bax, eventually leading to decreased neurodegeneration (Ali & Kunugi, 2020).

10.6 CONCLUSION

BRJ contains bioactive components which have been identified to possess functions such as anti-inflammatory, bacterial, and oxidative effects. It exhibits neuroprotective effects in neurodegenerative diseases such as Parkinson's disease, Alzheimer's disease, and multiple sclerosis, among others. BRJ has the potential to be used as a valuable natural remedy for neurodegenerative diseases and warrants further exploration into its clinical applications.

REFERENCES

Abu-Serie, M. M., & Habashy, N. H. (2019). Two purified proteins from royal jelly with in vitro dual anti-hepatic damage potency: Major royal jelly protein 2 and its novel isoform X1. *International Journal of Biological Macromolecules*, 128, 782–795.

Ahmad, S., Campos, M. G., Fratini, F., Altaye, S. Z., & Li, J. (2020). New insights into the biological and pharmaceutical properties of royal jelly. *International Journal of Molecular Sciences*, 21(2), 382.

Ali, A. M., & Kunugi, H. (2020). Apitherapy for Parkinson's disease: A focus on the effects of propolis and royal jelly. *Oxidative Medicine and Cellular Longevity*, 2020, 1727142.

Asuku, A. O., Ayinla, M. T., Ajibare, A. J., & Olajide, T. S. (2024). Mercury chloride causes cognitive impairment, oxidative stress and neuroinflammation in male Wistar rats: The potential protective effect of 6-gingerol-rich fraction of Zingiber officinale via regulation of antioxidant defense system and reversal of pro-inflammatory markers increase. *Brain Research*, 148741, 1826, ISSN 0006-8993, https://doi.org/10.1016/j.brainres.2023.148741

Asuku, A. O., Alege, D. O., Ogungbangbe, G. O., Ajibare, A. J., Adeyemo, O. R., & Sabinari, I. W. (2025a). Neuroprotective properties of marine bioresources. In S. A. Aransiola, S. Bhatnagar, & N. R. Maddela (Eds.), *Marine Bioresources: Prospects and Obstacles*. Elsevier, 2026, 369–378, https://doi.org/10.1016/B978-0-443-36320-7.00019-3

Asuku, A. O., Ogungbangbe, G. O., & Olajide, T. S. (2025b). Effect of happiness and positive emotions on dementia. In *Neuroscience of Feelings*. Progress in Brain Research, Elsevier, https://doi.org/10.1016/bs.pbr.2025.03.001

Ayinla, M. T., Asuku, A. O., Bangwai, T. B., & Owoyele, B. V. (2022). Aqueous leaf extract of Cola acuminata (Malvaceae) alleviate Scopolamine-induced memory impairment in Wistar rats by modulating Cholinergic Neurotransmission. *Biokemistri*, 34(3), 34248–34259.

Bagameri, L., Baci, G. M., & Dezmirean, D. S. (2022). Royal jelly as a nutraceutical natural product with a focus on its antibacterial activity. *Pharmaceutics*, 14(6), 1142.

Bagameri, L., Botezan, S., Bobis, O., Bonta, V., & Dezmirean, D. S. (2023). Molecular insights into royal jelly anti-inflammatory properties and related diseases. *Life*, 13(7), 1573.

Balestrino, R., & Schapira, A. H. (2020). Parkinson disease. *European Journal of Neurology*, 27(1), 27–42.

Botezan, S., Baci, G. M., Bagameri, L., Paşca, C., & Dezmirean, D. S. (2023). Current status of the bioactive properties of royal jelly: A comprehensive review with a focus on its anticancer, anti-inflammatory, and antioxidant effects. *Molecules*, 28(3), 1510.

Breijyeh, Z., & Karaman, R. (2020). Comprehensive review on Alzheimer's disease: Causes and treatment. *Molecules*, 25(24), 5789.

Chi, X., Liu, Z., Wang, H., Wang, Y., Wei, W., & Xu, B. (2021). Royal jelly enhanced the antioxidant activities and modulated the gut microbiota in healthy mice. *Journal of Food Biochemistry*, 45(5), e13701.

Collazo, N., Carpena, M., Nuñez-Estevez, B., Otero, P., Simal-Gandara, J., & Prieto, M. A. (2021). Health promoting properties of bee royal jelly: Food of the queens. *Nutrients*, 13(2), 543.

Ding, C., Wu, Y., Chen, X., Chen, Y., Wu, Z., Lin, Z., Kang, D., Fang, W., & Chen, F. (2022). Global, regional, and national burden and attributable risk factors of neurological disorders: The Global Burden of Disease study 1990–2019. *Frontiers in Public Health*, 10, 952161.

Dive, D., Dauby, S., Lommers, E., Hansen, I., Groenendyck, D., Ernon, C., & Maquet, P. (2022). Multiple sclerosis: A neurological dysimmune inflammatory disease. *Revue Medicale de Liege*, 77(5–6), 384–391.

Eslamifar, Z., Manucheri, F., Ghanbari, S., Zandi, M., & Sabbagh, S. (2025). Impact of royal jelly consumption on oxidative stress, anti-oxidant markers and physical activities of patients with multiple sclerosis: a randomized double-blind placebo-controlled study. *Irish Journal of Medical Science*, 194(6), 2307–2318. https://doi.org/10.1007/s11845-025-04009-z.

Eva, L., Pleş, H., Covache-Busuioc, R. A., Glavan, L. A., Bratu, B. G., Bordeianu, A., Dumitrascu, D. I., Corlatescu, A. D. and Ciurea, A. V. (2023). A comprehensive review on neuroimmunology: Insights from multiple sclerosis to future therapeutic developments. *Biomedicines*, 11(9), 2489.

Faissner, S., Plemel, J. R., Gold, R., & Yong, V. W. (2019). Progressive multiple sclerosis: from pathophysiology to therapeutic strategies. *Nature Reviews Drug Discovery*, 18(12), 905–922.

Fan, P., Han, B., Hu, H., Wei, Q., Zhang, X., Meng, L., Nie, J., Tang, X., Tian, X., Zhang, L., & Wang, L. (2020). Proteome of thymus and spleen reveals that 10-hydroxydec-2-enoic acid could enhance immunity in mice. *Expert Opinion on Therapeutic Targets*, 24(3), 267–279.

Gadhave, D. G., Tagalpallewar, A. A., & Kokare, C. R. (2019). Agranulocytosis-protective olanzapine-loaded nanostructured lipid carriers engineered for CNS delivery: optimization and hematological toxicity studies. *AAPS PharmSciTech*, 20(1), 22.

Gadhave, D., Gupta, A., Khot, S., Tagalpallewar, A., & Kokare, C. (2023). Nose-to-brain delivery of paliperidone palmitate poloxamer-guar gum nanogel: Formulation, optimization and pharmacological studies in rats. In *Annales Pharmaceutiques Françaises* (Vol. 81, No. 2, pp. 315–333). Elsevier Masson.

Gadhave, D. G., Sugandhi, V. V., & Kokare, C. R. (2024). Potential biomaterials and experimental animal models for inventing new drug delivery approaches in the neurodegenerative disorder: Multiple sclerosis. *Brain Research*, 1822, 148674.

Gombash, S. E., Lee, P. W., Sawdai, E., & Lovett-Racke, A. E. (2022). Vitamin D as a risk factor for multiple sclerosis: Immunoregulatory or neuroprotective? *Frontiers in Neurology*, 16(13), 796933. https://doi.org/10.3389/fneur.2022.796933

Guardia de Souza e Silva, T., do Val de Paulo, M. E. F., da Silva, J. R. M., da Silva Alves, A., Britto, L. R. G., Xavier, G. F., & Lopes Sandoval, M. R. (2020). Oral treatment with royal jelly improves memory and presents neuroprotective effects on icv-STZ rat model of sporadic Alzheimer's disease. *Heliyon*, 6(2), e03281.

Guo, J., Wang, Z., Chen, Y., Cao, J., Tian, W., Ma, B., & Dong, Y. (2021). Active components and biological functions of royal jelly. *Journal of Functional Foods*, 82, 104514.

Guo, J., Ma, B., Wang, Z., Chen, Y., Tian, W., & Dong, Y. (2022). Royal jelly protected against dextran-sulfate-sodium-induced colitis by improving the colonic mucosal barrier and gut microbiota. *Nutrients*, 14(10), 2069.

Huang, P., & Zhang, M. (2023). Magnetic resonance imaging studies of neurodegenerative disease: From methods to translational research. *Neuroscience Bulletin*, 39(1), 99–112.

Huang, Y., Li, Y., Pan, H., & Han, L. (2023). Global, regional, and national burden of neurological disorders in 204 countries and territories worldwide. *Journal of Global Health*, 13, 04160.

Hui, B. S. M., Zhi, L. R., Retinasamy, T., Arulsamy, A., Law, C. S. W., Shaikh, M. F., & Yeong, K. Y. (2023). The role of interferon-α in neurodegenerative diseases: A systematic review. *Journal of Alzheimer's Disease*, 94(s1), S45–S66.

Khatri, D. K., Choudhary, M., Sood, A., & Singh, S. B. (2020). Anxiety: An ignored aspect of Parkinson's disease lacking attention. *Biomedicine & Pharmacotherapy*, 131, 110776.

Khosla, A., Gupta, S. J., Jain, A., Shetty, D. C., & Sharma, N. (2020). Evaluation and comparison of the antimicrobial activity of royal jelly–A holistic healer against periodontopathic bacteria: An in vitro study. *Journal of Indian Society of Periodontology*, 24(3), 221–226.

Kocot, J., Kiełczykowska, M., Luchowska-Kocot, D., Kurzepa, J., & Musik, I. (2018). Antioxidant potential of propolis, bee pollen, and royal jelly: Possible medical application. *Oxidative Medicine and Cellular Longevity*, 2018(1), 7074209.

Kulisevsky, J. (2022). Pharmacological management of Parkinson's disease motor symptoms: Update and recommendations from an expert. *Revista de neurologia*, 75(Supl 4), S1.

Kunugi, H., & Mohammed Ali, A. (2019). Royal jelly and its components promote healthy aging and longevity: From animal models to humans. *International Journal of Molecular Sciences*, 20(19), 4662.

Lamptey, R. N., Chaulagain, B., Trivedi, R., Gothwal, A., Layek, B., & Singh, J. (2022). A review of the common neurodegenerative disorders: Current therapeutic approaches and the potential role of nanotherapeutics. *International Journal of Molecular Sciences*, 23(3), 1851.

Li, Q., Zhang, W., Zhou, E., Tao, Y., Wang, M., Qi, S., Zhao, L., Tan, Y. & Wu, L. (2023). Integrated microbiomic and metabolomic analyses reveal the mechanisms by which bee pollen and royal jelly lipid extracts ameliorate colitis in mice. *Food Research International*, 171, 113069.

Masrori, P., & Van Damme, P. (2020). Amyotrophic lateral sclerosis: A clinical review. *European Journal of Neurology*, 27(10), 1918–1929.

McGarry, A., Auinger, P., Kieburtz, K., Geva, M., Mehra, M., Abler, V., Grachev, I. D., Gordon, M. F., Savola, J. M., Gandhi, S. & Papapetropoulos, S. (2020). Additional safety and exploratory efficacy data at 48 and 60 months from open-HART, an open-label extension study of pridopidine in Huntington disease. *Journal of Huntington's Disease*, 9(2), 173–184.

Nattagh-Eshtivani, E., Jokar, M., Tabesh, H., Nematy, M., Safarian, M., Pahlavani, N., Maddahi, M., & Khosravi, M. (2021). The effect of propolis supplementation on inflammatory factors and oxidative status in women with rheumatoid arthritis: Design and research protocol of a double-blind, randomized controlled. *Contemporary clinical trials communications*, 23, 100807.

Neupane, P., Thada, P. K., Singh, P., Faisal, A. R., Rai, N., Poudel, P., Waleed, M. S., Quinonez, J., Ruxmohan, S., & Jain, E. (2023). Investigating edaravone use for management of amyotrophic lateral sclerosis (ALS): A narrative review. *Cureus*, 15(1), e33746.

Niazi, S. K. (2023). Non-invasive drug delivery across the blood–brain barrier: A prospective analysis. *Pharmaceutics*, 15(11), 2599.

Oršolić, N., & Jazvinšćak, J. M. (2024). Royal jelly: Biological action and health benefits. *International Journal of Molecular Sciences*, 25(11), 6023.

Pant, S., Kapri, A., & Nain, S. (2022). Pyrimidine analogues for the management of neurodegenerative diseases. European *Journal of Medicinal Chemistry Reports*, 6, 100095.

Pegoretti, V., Swanson, K. A., Bethea, J. R., Probert, L., Eisel, U. L., & Fischer, R. (2020). Inflammation and oxidative stress in multiple sclerosis: Consequences for therapy development. *Oxidative Medicine and Cellular Longevity*, 2020(1), 7191080.

Petelin, A., Kenig, S., Kopinč, R., Deželak, M., Černelič Bizjak, M., & Jenko Pražnikar, Z. (2019). Effects of royal jelly administration on lipid profile, satiety, inflammation, and antioxidant capacity in asymptomatic overweight adults. *Evidence-Based Complementary and Alternative Medicine*, 2019(1), 4969720.

Raoufi, S., Salavati, Z., Komaki, A., Shahidi, S., & Zarei, M. (2023). Royal jelly improves learning and memory deficits in an amyloid β-induced model of Alzheimer's disease in male rats: Involvement of oxidative stress. *Metabolic Brain Disease*, 38(4), 1239–1248.

11 Bee Venom Applications in Brain Cancer

Solihu Olashile Ibrahim, Suleiman Muhammed Eze, Fatimat Omowumi Hamzat, Aminat Temitayo Atoyebi, Ismail Adetayo Lawal, and Babatunde Joseph Dare

11.1 INTRODUCTION

Bee venom has been used as a medicine for millennia as a tool of apitherapy along with honey, propolis, and royal jelly (Duffy et al., 2020). It has a complex biomolecular composition mainly comprising bioactive peptides and enzymes, the majority of which are melittin, which is about 40–50% of dry matter, apamin, and phospholipase A_2, which have a wide range of physiological effects such as anti-inflammatory, antimicrobial, and neuroactive activities (Lim et al., 2019; Yu et al., 2020; Tiwari et al., 2022). Recent research also revealed that bee venom and melittin possess an anticancer effect that is active against a wide range of tumors (Małek et al., 2023). It has given rise to interest in the use of bee venom on central nervous system (CNS) tumors. Some of the brain cancers listed, including gliomas and medulloblastoma, are some of the deadliest malignancies in part because of the blood-brain barrier (BBB), heterogeneity of the tumor, and resistance to conventional treatments (Małek et al., 2022; Chahla et al., 2024). It is worth noting that the elements of bee venom can enter or induce the BBB; for example, melittin and apamin can cross the BBB and even temporarily open endothelial junctions (Oller Salvia, 2015; Linville et al., 2021), indicating that they may target intracranial tumors. The other borderline is with respect to apitherapy of neurological diseases, which involves the use of bee venom to attack brain cancer. Evidence relating to the mechanisms of action of bee venom in brain tumors, preclinical and any clinical studies, and challenges and future directions are discussed in this chapter.

11.2 MECHANISMS OF ACTIONS

Melittin is the major anticancer substance of bee venom. It is a highly amphipathic linear peptide that is composed of 26 amino acids: the N-terminal portion is constituted mainly of hydrophobic amino acids, and the C-terminal portion of the peptide is highly composed of positively charged amino acids (Wehbe et al., 2019; Duffy et al., 2020). This dual character permits melittin to penetrate lipid bilayers and take an α-helical structure when attached to the membrane, and it perturbs the usual membrane structure (Hong et al., 2019; Tissera et al., 2019). In aqueous solution, melittin is largely unstructured, whereas on interacting with lipid membranes, the protein adopts a highly ordered helix that penetrates the interior of a bilayer. In sufficiently high concentrations, Melittin creates temporary cell membrane pores, exposing cells to increased permeability to ions and small molecules (Moretta et al., 2021). Melittin interferes with the cell membranes by inserting and creating approximately 4.4 nM size pores that cause cells to lyse at high concentrations (Duffy et al., 2020; Li et al., 2022).

Free melittin is cell-toxic because of its general cytolytic effect. However, its application in drug delivery by the membrane activity has also been a stimulus, and a carefully prepared dosage of carrier formulation has the ability to utilize the pore-forming propensity of melittin without causing

DOI: 10.1201/9781003685654-11

permanent damage. Melittin causes programmed cell death pathways at sub-lytic doses. It causes both intrinsic (mitochondrial) and extrinsic (death-receptor) induction of apoptosis (Gajski et al., 2016; Sisakht et al., 2017). Melittin increases pro-apoptotic factors and cleaves PARP in cancer cells (Li et al., 2022; Chahla et al., 2024). Melittin treatment resulted in the fragmentation of DNA and Annexin-V positivity indicative of apoptosis in glioma cells, but acute permeabilization of cell membranes indicative of necrosis occurs with increased doses (Chahla et al., 2024).

Besides direct cytolysis, melittin controls the discharge of key indicative results in cancerous cells. It is able to inhibit receptor tyrosine kinase and resultant signaling: an instance of this is that melittin is able to inhibit HER2/EGFR activation and PI3K/Akt/mTOR in various cancers (Rady et al., 2017; Chen et al., 2019). Melittin suppresses pro-survival factors like HIF-1α, STAT3, and NF-κB and inhibits cell-cycle proteins in glioma models, halting cell proliferation (Choi et al., 2014). Notably, melittin has a potent anti-angiogenic and anti-invasive effect. It suppresses vascular endothelial growth factor (VEGF) and matrix metalloproteins (MMPs), which are enzymes responsible for tumor invasion. Indicatively, MMP-2 and MMP-9 activity serves as a factor in downregulation of glioblastoma cells (Gu et al., 2020; Huang et al., 2021; Małek et al., 2022). VEGF and FLT-1 were also inhibited by melittin in glioma cells of rats (Li et al., 2022). Bee venom inhibits tumor vascularization and metastasis in brain tumors by inhibiting MMPs and VEGF.

The cationic property of melittin is also applied to ion channels: recent studies indicate that melittin is able to activate the Ca^{2+} permeable TRPM2 channel in glioblastoma cells, causing calcium overload, zinc imbalance, and a burst of reactive oxygen species (ROS), all of which can induce tumor cell death (Ertilav & Nazıroğlu, 2023). This can be improved by upregulating the cytotoxicity of chemotherapeutic agents such as cisplatin in glioma cells mediated by this TRPM2 (Ertilav & Nazıroğlu, 2023). Other peptides of bee venom can play an indirect role: phospholipase A2 is pro-apoptotic with immunogenic effects, whereas apamin, an SK potassium channel blocker, can have an anticancer effect, but its effect on glioma excitability is less clear (Lv et al., 2021; Małek et al., 2022; Shi et al., 2022). Collectively, the bee venom attack on cancer causes lysis of cancer membranes, apoptosis, and growth-inhibitory signaling, anti-angiogenesis, and modulation of the tumor microenvironment. Figure 11.1 shows the applications of bee venom in brain cancer management.

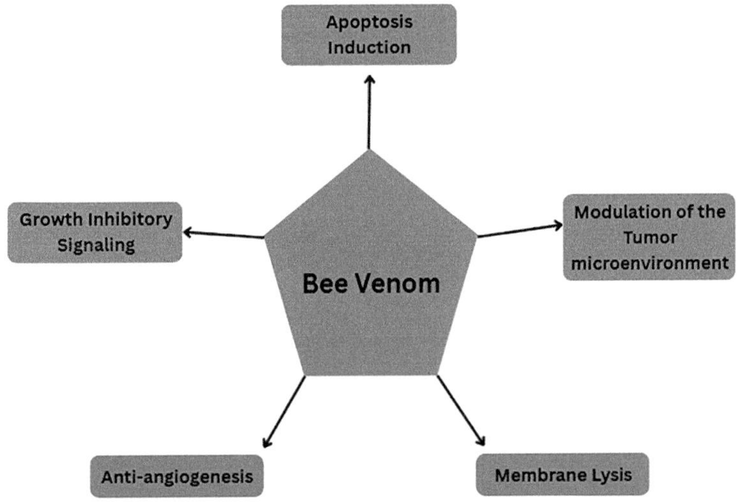

FIGURE 11.1 Schematic diagram illustrating bee venom applications in brain cancer.

11.3 THE BLOOD-BRAIN BARRIER

The blood-brain barrier is an extremely selective permeable border between the central nervous system and the blood circulatory system (Wu et al., 2023; Asuku et al., 2024). It is not a single layer but a porous and multicellular membrane, devoid of pores and of minimal pinacolysis, which filters foreign agents or the vast majority of blood-borne substances in the blood of the central nervous system (Upton et al., 2022). This obstacle has advantages and disadvantages. Among its advantages is the stabilization of the internal environment of the brain by the neurovascular system, hence avoiding its damage (Cai et al., 2018); whereas its disadvantages are that therapeutic intervention in brain ailments would be a challenge due to the constraint of drug entry into the CNS (Saraiva et al., 2016; Pandit et al., 2020).

Structurally, the blood-brain barrier is differentiated from the peripheral endothelial cells due to the presence of special protein complexes such as claudins, occludins, and junctional adhesion molecules, commonly known as tight junctions and adhesive junctions, respectively, which prevent the permeability of largely hydrophilic substances through paracellular pathways (Wu et al., 2023). Other contributory structures to BBB are the basal lamina that envelops the endothelial cells, and within which are astrocytic end feet and pericytes. Collectively, the structures form a neurovascular unit (NVU) that maintains the brain homeostasis (Cai et al., 2018; Kadry et al., 2020) (Figure 11.2).

Only the lipid-soluble drugs that are below the molecular weights of 400Da will be allowed to diffuse through the tight junctions. In this manner, BBB protects the neural tissue from toxins and pathogens and enhances homeostasis (Kadry et al., 2020; Upton et al., 2022). Laboratory studies have demonstrated that the Crizotinib drugs have been modified in laboratories to augment their lipophilic properties, which has increased their permeability into the brain (Alauddin et al., 2019). Thus, it provides information on developments in treatment plans against disease of the brain.

The same feature of BBB that provides the brain defense makes the penetration of the drugs extremely hard. In practice, only about 98% of the small-molecule drugs have the ability to pass through the BBB at therapeutic concentrations. Also, active efflux pumps, especially in endothelial cells, P-glycoprotein, and breast cancer resistance protein (BCRP), rapidly empty most of the chemotherapeutics that do enter the CNS, and further reduce CNS access (Kadry et al., 2020).

CNS access is complicated by the blood-tumor barrier (BTB). Nonhomogeneous and leaky capillaries in some regions are common outcomes of tumor-induced angiogenesis; nevertheless, the

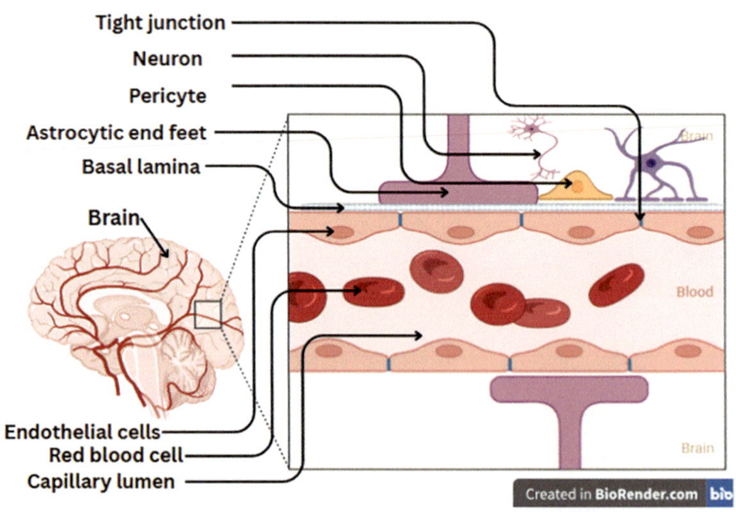

FIGURE 11.2 Diagram illustrating the blood-brain barrier, together with the components of the neurovascular unit (NVU).

BTB is restraining in others (Upton et al., 2022). Thus, uneven distribution and insufficiency of drugs even in glioma are present. To the point, the BBB (and BTB) inflicts immense obstacles on the management of brain cancer—its resistance to most drugs and the necessity to devise new delivery mechanisms.

11.4 MELITTIN-INDUCED OPENING OF THE BLOOD-BRAIN BARRIER

Researchers have applied the membrane activity of melittin to permeabilize the BBB temporally and reversibly. It has been demonstrated in vitro and in animal experiments that melittin applied at the appropriate dose can interact with the endothelial tight junctions without irreversible damage. As a pointer, Linville et al. established that the intravascular injection of melittin in mice led to transient and focal breakage of blood vessels in the brain (Linville et al., 2021). Melittin joins the brain endothelial membranes mechanically and interferes with the cell-cell junctions; temporarily increasing the permeability of the paracellular membranes (Figure 11.3). Melittin exposure in human BBB tissue-engineered models resulted in an area of local junctional breakdown, and endothelial viability was essentially preserved at the effective dose. Notably, this opening could be reversed after hours to a day with a minimum effective dose, 3μM, as revealed in the study and reported to be neurologically safe (Linville et al., 2021; Galpayage Dona et al., 2023). In conclusion, melittin is a transmembrane-active peptide that transiently disrupts tight junction (TJ) integrity, and this offers an access point for the drug.

11.5 MELITTIN IN TARGETED BRAIN CANCER DRUG DELIVERY

Based on the aforementioned mechanism, there are a number of strategies currently using melittin to improve the delivery of drugs to brain tumors. In one of them, melittin is loaded into nanoscale carriers that also contain anticancer drugs. An example is the engineering of Angiopep-2-modified liposomes, which were loaded with melittin, as an α-helical conjugate, and the chemotherapeutic resveratrol (Zhang et al., 2025). Such dual-loaded liposomes were found to be more penetrative in the BBB and had targeting to glioblastoma (GB) cells: mice that were fed the melittin/resveratrol liposomes had improved tumor uptake and longer survival compared to controls (Zhang et al., 2025). It is important to note that systemic toxicity (hemolysis) of melittin was also reduced by

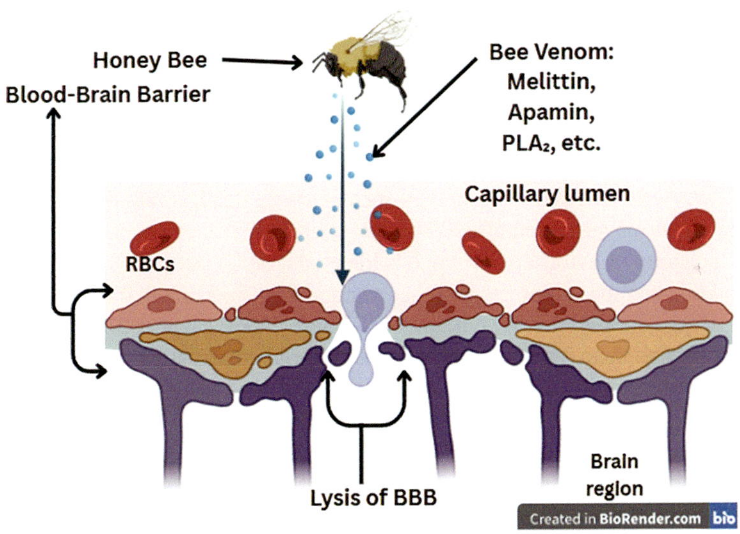

FIGURE 11.3 Diagram demonstrating how bee venom creates pores in the blood-brain barrier.

co-delivery, which, in combination with resveratrol, caused GB cell apoptosis. In another study, Wang et al. (2019) developed glycopeptide-targeted lipodisks that contain paclitaxel and melittin (Wang et al., 2019). This glioma-targeted nano-formulation enabled stable co-loading of agents, in vitro paclitaxel, and melittin synergistically acted on U87 glioblastoma cells, and in vivo, the targeted lipodisks elicited a much higher level of tumor suppressive effect and increased survival of glioma-bearing mice.

They have also used nanoparticles as well as transferrin-based conjugates to deliver melittin and drug cargos across the BBB. As an example, the nanoparticles of iron oxide (melittin-capped with transferrin) that have been designed in the framework of Alzheimer's disease models could be demonstrated to enter the brain lesions with melittin with considerable ease (Liu et al., 2025). These studies, though neurodegenerative designed, highlight the fact that melittin can be taken into the CNS by receptor-mediated transport. In totality, these and the other reports provide support of the idea that melittin is capable of aiding in the target transduction of anticancer agents to the brain. The presence of melittin-bearing carriers permit pharmacological agents to be made at increased concentrations in brain tumor, by a temporary increase in BBB permeability, and by cytolytic activity of its own (Wang et al., 2019; Zhang et al., 2025).

11.6 PRECLINICAL STUDIES

Vast laboratory tests indicate that bee venom is selective in killing brain tumor cells. As an example, *Apis mellifera syriaca* venom was discovered by Chahla et al. (2024) to cause severe reduction in viability of human U87 glioblastoma cells, with IC_{50} values of the low-microgram range (about 7–14 μg/mL by LDH and MTT assays). In vitro membrane permeabilization and necrotic cleavage patterns of PARP were observed in vitro in treated glioma cells, which indicated direct lytic damage (Chahla et al., 2024). In a similar way, Małek et al. (2022) exposed tumor cells in human glioblastoma GAMG and 8-MG-BA cells to crude bee venom: lower doses of venom, i.e. less than 1 ug/mL, induced more than 50–80% cell death in tumor cells, whereas normal neuronal (HT-22) cells were far less susceptible ($IC_{50} > 1.3$ μg/mL). The culture of glioma in the presence of venom also produced much lower concentrations of MMP-2 and MMP-9 than controls, but the hippocampal cells of the normal brain did not (Małek et al., 2022). The implication of these results is the selective anticancer effect against glioma as opposed to the normal brain cells.

Glioblastoma lines, such as U87, A1235, and GAMG, or even metastatic CNS tumor models are regularly sensitive to bee venom or melittin. As an example, Małek et al. found that 20 μg/mL of bee venom had a strong negative impact on the viability of A1235 glioblastoma cells, but not on normal colon epithelial cells (Małek et al., 2022; Małek et al., 2023). The reason behind conflicting results on apoptosis versus necrosis could be due to the dose and exposure time differences: in a number of studies, apoptotic signs can be found at higher doses or longer exposure, and pure acute lytic death can be found at extremely high doses (Chahla et al., 2024; Pandey et al., 2023). Bee venom has likewise been demonstrated to cause autophagy-mediated cell death in various cancers, and this can be used to complement its apoptotic effects (Małek et al., 2023; Pandey et al., 2023).

There is still a paucity of evidence in vivo. Chahla et al. (2024) put EGFP-expressing U87 cells in mouse brains and injected the tumor site with bee venom (25 μg). Mice that had been treated showed considerably less tumor cells ($EGFP^{+}$) in the brain and more caspase-3-positive tumor cells as compared to the control mice. The presence of smaller tumor masses was determined by the use of hematoxylin/eosin staining following the application of venom (Chahla et al., 2024). The rationale is that local bee venom can reduce intracranial tumors in vivo according to this preliminary model of glioblastoma. Elsewhere in the literature of cancer in animals, there are numerous reports of slowed tumor growth, angiogenesis retardation, and enhanced survival following the administration of melittin or bee venom (Liu et al., 2016; El Bakary et al., 2020; Chahla et al., 2024). It is worth noting that concentrations of 0.5–1.0 mg/kg of the venom have been safely administered to mice (Lee et al., 2015), which is sufficient to get therapeutic levels in tumors in local or through nanoparticles.

11.7 CLINICAL STUDIES

No reported clinical trial has so far been done to test bee venom therapy in brain cancer. A current review of the database of ClinicalTrials.gov confirms that as of the middle of 2023, it has no completed or ongoing trial expressly testing the application of bee venom in brain cancer (Małek et al., 2023). Thus, patient testimony is at best anecdotal. In practice, some patients may prefer to receive complementary apitherapy (needle acupuncture or injections) in order to heal cancer, which is not supported by clinical evidence in brain tumors. The nearest human testing that has been conducted is to use bee venom on other diseases other than Parkinson's disease and multiple sclerosis. The absence of clinical data is an indication of enormous gap: all the available information is preclinical. This would then involve Phase I trials that are very strict before the translation to ascertain safety, dosing, and delivery in the cancer patients.

11.8 CHALLENGES

There are various challenges that make it difficult to use bee venom on brain cancer. In the first place, there is toxicity and safety. Melittin breaks down cell membranes indiscriminately; hence, a high level in the systemic circulation leads to hemolysis and destruction of normal tissues (Małek et al., 2023). It has been estimated that to realize approximately 1 μg/mL of venom in human blood, which is around the typical, effective in vitro level, injection of on the order of 75 mg of venom would be required, which is much higher than amounts safely injected into the body (El Bakary et al., 2020). This was not used in animal studies by injecting locally, intratumorally, or encapsulating the venom into carriers, but this should be carefully transferred to patients. Bee venom contains allergic substances as well; actually, bee stings allergy is common and may even be life-threatening (Elieh Ali Komi et al., 2018; Burzyńska & Piasecka-Kwiatkowska, 2021). A survey of patients receiving apitherapy to treat cancer shows that the patients lack awareness of the allergy risks of apitherapy products. To be able to use apitherapy products appropriately, patients should be educated about the mechanism, potential risks, and the age groups of the products. In this regard, a great role belongs to healthcare professionals (Kavurmacı & Yıldız, 2024). Any use of the drug in medicine would involve strict allergy tests and most likely desensitization therapy (Małek et al., 2022).

Third, the heterogeneity of bee venom is a feasible issue. The composition of the venom differs depending on the bee subspecies, location, and season. Apitoxin preparations have not been standardized easily, and concentrations of melittin, PLA_2, etc., can vary (Duffy et al., 2020). Purified or recombinant melittin can be a good choice of crude venom in order to receive consistent therapy. Also, the majority of the available research involves honeybee venom of the wild type; the therapeutic value of minor constituents (e.g., apamin, MCD peptide) in brain cancer treatment is not well known.

Finally, regulatory and ethical issues restrict the speed of translation. As a natural product, bee venom therapy falls between the drug and alternative medicine systems. Good clinical trials are not available, and therefore low insurance coverage and medical acceptance. Patients are more likely to consider apitherapy a complementary and alternative medicine (CAM), and most of them would administer apitherapy without a doctor's recommendation (Kavurmacı & Yıldız, 2024). This raises the problems of unregulated use that are ethical. Bee venom is not yet a proven and safe treatment when it comes to trials, and is therefore an experimental treatment rather than a general treatment.

11.9 FUTURE DIRECTION

In view of its possibilities and challenges, researchers are conducting research to locate sophisticated ways of delivery and modification to make the application of bee venom in the treatment of cancer a reality. One of the strategies, as mentioned earlier, is targeted nanoparticle encapsulation.

Based on the example of Lv et al. (2021), D-amino-acid melittin peptides are designed to be incorporated into pH-sensitive nanoparticles (DMM), which strongly reduce nonspecific immune reactions and are cytotoxic. α-melittin-lipid nanoparticles were also able to target lymph nodes and promote antigen-specific T-cell immunity in tumors in murine models (Yu et al., 2020). In addition, cell growth and hemolysis studies revealed that the modified MpG@LPN (polyethylene glycol [PEG]-modified melittin-loaded lipid-coated polymeric nanoparticle) was not specific to cytotoxicity and did not have hemolytic activity at high concentration. The change in the targeting molecules on the MpG@LPNs enabled generation of apoptosis and targeting of single tumor cells. In vivo experiments showed that MpG@LPNs were capable of inhibiting tumor growth without causing tissue damage and hemolysis.

The results showed that the proposed MpG@LPN with a core–shell structure has great potential as a cancer treatment and could effectively overcome the main issues melittin faces in clinical applications (Ye et al., 2021). Other groups have conjugated melittin to polymers or liposomes modified with tumor-targeting ligands, for example, RGD peptides and aptamers. They consequently cause a reduction in hemolytic toxicity and localize the venom in tumor tissue. As indicated by Bayat et al. (2022), the AS1411 aptamer-melittin-gold nanoparticle complex is one such example, wherein killing was selective to breast cancer cells but not normal fibroblasts; similar designs would be developed with glioma-specific markers.

The other treatment method is the use of bee venom alongside other treatments. Preclinical studies demonstrate that there is synergy between melittin and chemotherapeutics: melittin increases the ROS-based killing of cisplatin in glioblastoma and similarly boosts lung cancer chemo (Oršolić et al., 2005). Bee venom could be combined with temozolomide (TMZ) to treat glioma, or it could be used with radiation to enhance the tumor-targeting ability of oncolytic viruses containing melittin (Lv et al., 2021; Bahreyni et al., 2023). Notably, the immunogenic cell death capability of bee venom is also used: D-melittin nanoparticles have been demonstrated to cause robust dendritic cell activation and tumor cell T-cell response (Lv et al., 2021; Jafari et al., 2024). This implies that bee venom might stimulate the immune system so that hidden or resistant cancers become visible to the immune system, thereby making the immune system fight them more effectively, and enhancing the success of modern immunotherapies.

The blood-brain barrier (BBB) will also be a companion in future strategies. Johns Hopkins researchers had been able to prove that the tight junctions in the BBB can be opened temporarily with the assistance of melittin when a limited dosage is utilized (Johns Hopkins University, 2021). Low-dose melittin could be used in the future not to kill the cells, but to deliver bigger drugs or nanoparticles into the brain tumor safely. Continued efforts will focus on finding the best dosing that is nontoxic to neurons, but one that will increase CNS drug delivery (Ma et al., 2018; Johns Hopkins University, 2021; Małek et al., 2022).

Finally, it will be necessary to undergo rigorous trials in clinical translation. Phase I dose-escalation trials of tumor-targeted melittin could be their first step, with keen attention to the toxicity. With the mortality of brain cancer approaching high, the risk-benefit calculus can perhaps prove the point by making an earlier rather than a later exploratory human experiment. Simultaneously, an extended period of basic research is recommended to maximize the peptide (e.g., D-enantiomers, pro-drug form activated by tumor enzymes), and test their effects in high-level models of glioma, like patient-derived xenografts and organoids, among others. Provided that the initial clinical outcomes reveal some indication of activity, bigger studies might proceed.

11.10 CONCLUSION

Bee venom and its active ingredients, and especially melittin, hold a bright prospect in brain cancer. Bee venom has high anticancer potential as a standalone cytotoxic agent and as an adjuvant to standard chemotherapy, as it has multiple effects on hallmarks of tumor progression, such as proliferation, angiogenesis, and immune evasion. It is important to note that its potential to temporarily permeabilize the BBB presents some special chances to achieve chemotherapeutic delivery to previously inaccessible tumor locations or immunomodulatory applications. Early research has indicated promising results of effectiveness, but the next step of transferring the results to clinical practice is limited by the issues of toxicity, dosage-effect optimization, and targeted delivery.

The focus of future directions should be the creation of nanoformulations, conjugates, and delivery systems that would help in improving therapeutic precision with fewer adverse effects. Furthermore, a combination of bee venom with immunotherapies and cancer vaccines would turn cold tumors into hot ones, improving the level of immune recognition and response. Although ethical, safety, and regulatory factors are to be considered, the accumulating evidence makes the bee venom an excellent candidate concerning new multimodal treatment approaches in brain cancer. Through further interdisciplinary studies, it has the potential to close the divide between laboratory discoveries and clinical application, which will eventually provide hope to patients with one of the most aggressive and treatment-resistant forms of malignancy.

REFERENCES

Alauddin, M. M., Radaram, B., Rao, Y., Yang, P., & Piwnica-Worms, D. (2019). Novel derivatives of anaplastic lymphoma kinase inhibitors: Synthesis, radiolabeling and preliminary biological studies on crizotinib and alectinib analogues. *Cancer Research*, *79*(13_Supplement), 10–10.

Asuku, A. O., Ayinla, M. T., Aransiola, S. A., Ajibare, A. J., Adeyemo, M. B., & Adeyemo, R. O. (2024). Chitosan-based nanoparticles to bypass the blood-brain barrier for the treatment of neurological diseases: A review. *Pacific Journal of Medical Sciences*, 25(1), 2072–2625

Bahreyni, A., Liu, H., Mohamud, Y., Xue, Y. C., Fan, Y. M., Zhang, Y. L., & Luo, H. (2023). A combination of genetically engineered oncolytic virus and melittin-CpG for cancer viro-chemo-immunotherapy. *BMC Medicine*, *21*(1), 193.

Bayat, P., Abnous, K., Balarastaghi, S., Taghdisi, S. M., Saeedi, M., Yazdian-Robati, R., & Mahmoudi, M. (2022). Aptamer AS1411-functionalized gold nanoparticle-melittin complex for targeting MCF-7 breast cancer cell line. *Nanomedicine Journal*, *9*(2), 164–169.

Burzyńska, M., & Piasecka-Kwiatkowska, D. (2021). A review of honeybee venom allergens and allergenicity. *International Journal of Molecular Sciences*, *22*(16), 8371.

Cai, Z., Qiao, P. F., Wan, C. Q., Cai, M., Zhou, N. K., & Li, Q. (2018). Role of blood–brain barrier in Alzheimer's disease. *Journal of Alzheimer's Disease*, *63*(4), 1223–1234.

Chahla, C., Rima, M., Mouawad, C., Roufayel, R., Kovacic, H., El Obeid, D., Sabatier, J. M., Luis, J., Fajloun, Z., & El-Waly, B. (2024). Effect of *Apis mellifera syriaca* bee venom on glioblastoma cancer: In vitro and in vivo studies. *Molecules*, *29*(16), 3950. https://doi.org/10.3390/molecules29163950

Chen, Q., Lin, W., Yin, Z., Zou, Y., Liang, S., Ruan, S., . . . Ling, C. (2019). Melittin inhibits hypoxia-induced vasculogenic mimicry formation and epithelial-mesenchymal transition through suppression of HIF-1α/Akt pathway in liver cancer. *Evidence-Based Complementary and Alternative Medicine*, *2019*, 9602935.

Choi, K. E., Hwang, C. J., Gu, S. M., Park, M. H., Kim, J. H., Park, J. H., & Hong, J. T. (2014). Cancer cell growth inhibitory effect of bee venom via increase of death receptor 3 expression and inactivation of NF-κB in NSCLC cells. *Toxins*, *6*(8), 2210–2228.

Duffy, C., Sorolla, A., Wang, E., Golden, E., Woodward, E., Davern, K., . . . Blancafort, P. (2020). Honeybee venom and melittin suppress growth factor receptor activation in HER2-enriched and triple-negative breast cancer. *NPJ Precision Oncology*, *4*(1), 24.

El Bakary, N. M., Alsharkawy, A. Z., Shouaib, Z. A., & Barakat, E. M. (2020). Role of bee venom and melittin on restraining angiogenesis and metastasis in γ-irradiated solid Ehrlich carcinoma-bearing mice. *Integrative Cancer Therapies*, *19*, 1534735420944476.

Elieh Ali Komi, D., Shafaghat, F., & Zwiener, R. D. (2018). Immunology of bee venom. *Clinical Reviews in Allergy & Immunology*, *54*(3), 386–396.

Ertilav, K., & Nazıroğlu, M. (2023). Honey bee venom melittin increases the oxidant activity of cisplatin and kills human glioblastoma cells by stimulating the TRPM2 channel. *Toxicon*, *222*, 106993. https://doi.org/10.1016/j.toxicon.2022.106993

Gajski, G., Čimbora-Zovko, T., Rak, S., Osmak, M., & Garaj-Vrhovac, V. (2016). Antitumour action on human glioblastoma A1235 cells through cooperation of bee venom and cisplatin. *Cytotechnology*, *68* (4), 1197–1205.

Galpayage Dona, K. N. U., Ramirez, S. H., & Andrews, A. M. (2023). A next-generation 3D tissue-engineered model of the human brain microvasculature to study the blood–brain barrier. *Bioengineering*, *10*(7), 817.

Gu, H., Han, S. M., & Park, K. K. (2020). Therapeutic effects of apamin as a bee venom component for non-neoplastic disease. *Toxins*, *12*(3), 195.

Hong, J., Lu, X., Deng, Z., Xiao, S., Yuan, B., & Yang, K. (2019). How melittin inserts into cell membrane: Conformational changes, inter-peptide cooperation, and disturbance on the membrane. *Molecules*, *24*(9), 1775.

Huang, J. Y., Peng, S. F., Chueh, F. S., Chen, P. Y., Huang, Y. P., Huang, W. W., & Chung, J. G. (2021). Melittin suppresses epithelial–mesenchymal transition and metastasis in human gastric cancer AGS cells via regulating Wnt/BMP-associated pathway. *Bioscience, Biotechnology, and Biochemistry*, *85*(11), 2250–2262.

Jafari, Z., Sadeghi, S., Dehaghi, M. M., Bigham, A., Honarmand, S., Tavasoli, A., . . . & Varma, R. S. (2024). Immunomodulatory activities and biomedical applications of melittin and its recent advances. *Archiv der Pharmazie*, *357*(4), 2300569.

Johns Hopkins University. (2021, December 3). *Bee venom crosses blood–brain barrier*. Johns Hopkins University Hub. https://hub.jhu.edu/2021/12/03/bee-venom-crosses-blood-brain-barrier

Kadry, H., Noorani, B., & Cucullo, L. (2020). A blood-brain barrier overview on structure, function, impairment, and biomarkers of integrity. *Fluids and Barriers of the CNS*, *17*(1), 69.

Kavurmacı, M., & Yıldız, İ. (2024). Cancer patients' use of apitherapy as supportive care: An exploratory study. *The Eurasian Journal of Medicine*, *56*(1), 15.

Lee, H. L., Park, S. H., Kim, T. M., Jung, Y. Y., Park, M. H., Oh, S. H., . . . & Hong, J. T. (2015). Bee venom inhibits growth of human cervical tumors in mice. *Oncotarget*, *6*(9), 7280.

Li, R., Tao, T., Ren, Q., Xie, S., Gao, X., Wu, J., Chen, D., & Xu, C. (2022). Key genes are associated with the prognosis of glioma, and melittin can regulate the expression of these genes in glioma U87 cells. *BioMed Research International*, *2022*, 7033478. https://doi.org/10.1155/2022/7033478

Lim, H. N., Baek, S. B., & Jung, H. J. (2019). Bee venom and its peptide component melittin suppress growth and migration of melanoma cells via inhibition of PI3K/AKT/mTOR and MAPK pathways. *Molecules*, *24*(5), 929.

Linville, R. M., Komin, A., Lan, X., DeStefano, J. G., Chu, C., Liu, G., Walczak, P., Hristova, K., & Searson, P. C. (2021). Reversible blood–brain barrier opening utilizing the membrane active peptide melittin in vitro and in vivo. *Biomaterials*, *275*, 120942. https://doi.org/10.1016/j.biomaterials.2021.120942

Liu, C. C., Hao, D. J., Zhang, Q., An, J., Zhao, J. J., Chen, B., . . . & Yang, H. (2016). Application of bee venom and its main constituent melittin for cancer treatment. *Cancer Chemotherapy and Pharmacology*, *78*(6), 1113–1130.

Liu, S., Jin, X., Ge, Y., Dong, J., Liu, X., Pei, X., . . . & Yu, X. A. (2025). Advances in brain-targeted delivery strategies and natural product-mediated enhancement of blood–brain barrier permeability. *Journal of Nanobiotechnology*, *23*(1), 382.

Lv, S., Sylvestre, M., Song, K., & Pun, S. H. (2021). Development of D-melittin polymeric nanoparticles for anti-cancer treatment. *Biomaterials*, *277*, 121076.

Ma, X., Song, Q., & Gao, X. (2018). Reconstituted high-density lipoproteins: Novel biomimetic nanocarriers for drug delivery. *Acta Pharmaceutica Sinica B*, *8*(1), 51–63.

Małek, A., Kocot, J., Mitrowska, K., Posyniak, A., & Kurzepa, J. (2022). Bee venom effect on glioblastoma cells viability and gelatinase secretion. *Frontiers in Neuroscience*, *16*, 792970. https://doi.org/10.3389/fnins.2022.792970

Małek, A., Strzemski, M., Kurzepa, J., & Kurzepa, J. (2023). Can bee venom be used as anticancer agent in modern medicine? *Cancers*, *15*(14), 3714.

Moretta, A., Scieuzo, C., Petrone, A. M., Salvia, R., Manniello, M. D., Franco, A., . . . & Falabella, P. (2021). Antimicrobial peptides: A new hope in biomedical and pharmaceutical fields. *Frontiers in Cellular and Infection Microbiology*, *11*, 668632.

Oller Salvia, B. (2015). From bee venom to blood–brain barrier shuttles: Development of minimized apamin derivatives for brain delivery of antibodies and other cargoes. [Doctoral dissertation, University of Barcelona].

Oršolić, N., Terzić, S., Šver, L., & Bašić, I. (2005). Honey-bee products in prevention and/or therapy of murine transplantable tumours. *Journal of the Science of Food and Agriculture*, *85*(3), 363–370.

Pandey, P., Khan, F., Khan, M. A., Kumar, R., & Upadhyay, T. K. (2023). An updated review summarizing the anticancer efficacy of melittin from bee venom in several models of human cancers. *Nutrients*, *15*(14), 3111.

Pandit, R., Chen, L., & Götz, J. (2020). The blood–brain barrier: Physiology and strategies for drug delivery. *Advanced Drug Delivery Reviews*, *165*, 1–14.

Rady, I., Siddiqui, I. A., Rady, M., & Mukhtar, H. (2017). Melittin, a major peptide component of bee venom, and its conjugates in cancer therapy. *Cancer Letters*, *402*, 16–31.

Saraiva, C., Praça, C., Ferreira, R., Santos, T., Ferreira, L., & Bernardino, L. (2016). Nanoparticle-mediated brain drug delivery: Overcoming blood–brain barrier to treat neurodegenerative diseases. *Journal of Controlled Release*, *235*, 34–47.

Shi, P., Xie, S., Yang, J., Zhang, Y., Han, S., Su, S., & Yao, H. (2022). Pharmacological effects and mechanisms of bee venom and its main components: Recent progress and perspective. *Frontiers in Pharmacology*, *13*, 1001553.

Sisakht, M., Mashkani, B., Bazi, A., Ostadi, H., Zare, M., Avval, F. Z., ... & Soukhtanloo, M. (2017). Bee venom induces apoptosis and suppresses matrix metalloprotease-2 expression in human glioblastoma cells. *Revista Brasileira de Farmacognosia*, *27*(3), 324–328.

Tissera, M. J. E., Disalvo, E. A., Martini, M. F., & Cutró, A. C. (2019). Filling gaps in the knowledge of melittin on lipid membranes. *Colloids and Surfaces A: Physicochemical and Engineering Aspects*, *561*, 136–146.

Tiwari, R., Tiwari, G., Lahiri, A., Ramachandran, V., & Rai, A. (2022). Melittin: A natural peptide with expanded therapeutic applications. *The Natural Products Journal*, *12*(2), 13–29.

Upton, D. H., Ung, C., George, S. M., Tsoli, M., Kavallaris, M., & Ziegler, D. S. (2022). Challenges and opportunities to penetrate the blood–brain barrier for brain cancer therapy. *Theranostics*, *12*(10), 4734.

Wang, H., Wang, S., Wang, R., Wang, X., Jiang, K., Xie, C., Zhan, C., Wang, H., & Lu, W. (2019). Co-delivery of paclitaxel and melittin by glycopeptide-modified lipodisks for synergistic anti-glioma therapy. *Nanoscale*, *11*(27), 13069–13077. https://doi.org/10.1039/c9nr01820d

Wehbe, R., Frangieh, J., Rima, M., El Obeid, D., Sabatier, J. M., & Fajloun, Z. (2019). Bee venom: Overview of main compounds and bioactivities for therapeutic interests. *Molecules*, *24*(16), 2997.

Wu, D., Chen, Q., Chen, X., Han, F., Chen, Z., & Wang, Y. (2023). The blood–brain barrier: Structure, regulation and drug delivery. *Signal Transduction and Targeted Therapy*, *8*(1), 217.

Ye, R., Zheng, Y., Chen, Y., Wei, X., Shi, S., Chen, Y., ... Peng, J. (2021). Stable loading and delivery of melittin with lipid-coated polymeric nanoparticles for effective tumor therapy with negligible systemic toxicity. *ACS Applied Materials & Interfaces*, *13*(47), 55902–55912.

Yu, R., Wang, M., Wang, M., & Han, L. (2020). Melittin suppresses growth and induces apoptosis of non-small-cell lung cancer cells via down-regulation of TGF-β-mediated ERK signal pathway. *Brazilian Journal of Medical and Biological Research*, *54*(2), e9017.

Zhang, H. Q., Wang, Y., Geng, X., Dong, M., Liu, Z., Sun, C., Yu, K., Xin, W., Xu, Y., Xu, N., & Liu, W. (2025). ANG-modified liposomes coloaded with α-melittin and resveratrol induce apoptosis and pyroptosis in glioblastoma cells by impeding Wnt/β-catenin signaling. *CNS Neuroscience & Therapeutics*, *31*(5), e70437. https://doi.org/10.1111/cns.70437

12 Bee Bread Applications in Neurodegenerative Diseases

Abdulmajeed Opeyemi Agboola, Jamiu Sulaiman Jamiu, Abdulwarith Ayomide Bamidele, and Uthman Okikiola Adebayo

12.1 INTRODUCTION

The loss of neuron structure and functional status (progressive) resulting in irreversible impairment of cognitive, motor, or behavioral functions is called neurodegenerative diseases (NDs). Examples are Alzheimer's and Parkinson's diseases, but the list also includes amyotrophic lateral sclerosis, frontotemporal dementia, and other related diseases (Asuku et al., 2022).

Early intervention is vital, and research and public health should be a priority because many disorders share a lengthy preclinical phase during which genetic and cellular abnormalities build up before symptoms appear (WHO, 2025). According to Ilie et al. (2024) and Khalifa et al. (2020), there is growing interest in complementary interventions, especially apitherapy (the therapeutic use of bee products), because many bee products, such as bee bread (BB), are rich in vitamins, polyphenols, and other bioactive compounds with anti-inflammatory and antioxidant properties that may help counteract early pathogenic processes in NDs.

The prevalence of neurodegenerative diseases is rapidly increasing globally because of demographic shifts and population aging. Most dementia cases are caused by Alzheimer's disease, which also contributes to societal expenses in care and loss of productivity (Asuku et al., 2024). In 2021, dementia impacted around 57 million people globally, with almost 10 million new cases reported each year (Alzheimer's Association, 2023; WHO, 2025). According to recent study, the rate of Parkinson's disease increased worldwide between 1990 and 2021. This increase is expected to continue in the upcoming decades, primarily as a result of aging populations (Li et al., 2025). Safe, cost-effective therapy is needed that can hinder prognosis, reduce symptoms, or increase resilience to neurodegeneration in a variety of settings, as evidenced by the diseases' rising prevalence and their financial and career implications.

In neurodegenerative diseases, oxidative stress, mitochondrial dysfunction, neuroinflammation, and protein misfolding are some of the causes that lead to neuronal death (Asuku et al., 2022). Apoptosis is encouraged and bioenergetics are hindered by mitochondrial malfunction, whereas oxidative stress destroys proteins, lipids, and DNA. Microglia and astrocytes mediate neuroinflammation, which intensifies neuronal damage. Despite the lack of thorough research, bee bread's anti-inflammatory and antioxidant qualities may be able to prevent these harmful pathways (Ilie et al., 2024).

12.2 APITHERAPY AND BEE PRODUCTS

Apitherapy bee-derived products have been used as a traditional and alternative medicine for millennia. Some of the most studied bee products include honey, propolis, royal jelly, bee venom, and bee bread due to their unique bioactive chemicals and their potential to be used as medicines. Honey possesses antibacterial, wound healing, and anti-inflammatory properties because it is rich in sugars, enzymes, and polyphenols. Propolis contains flavonoids and phenolic acids, which are also

DOI: 10.1201/9781003685654-12

resinous substances collected by the bees, who extract them by beetle exudates; it is an effective antioxidant and neuroprotectant (Diniz et al., 2020). The neurogenesis and memory, along with overall brain health, are supported by peptides, fatty acids, and vitamins secreted in royal jelly by nurse bees. Apamin and melittin, bee venom proteins that are commonly associated with allergy responses, have been investigated to have anti-inflammatory and neuro-modulatory activities in neurodegenerative diseases. Bee bread is high in concentration of polyphenols, amino acids, and vitamins; it is a fermented mixture of pollen, nectar, and bee saliva, and this is unique in its neuroprotective and antioxidant properties. One of the proposed solutions for neurodegenerative disorders is bee bread, which would help address the most common neurodegenerative disorders (Popova et al., 2021).

There is curiosity associated with the idea of using natural substances with anti-inflammatory and antioxidant properties to counter the rise in neurodegenerative disorders. They are perceived as products that would facilitate the health of the brain and prevent aging-related cognitive impairment. Nutraceutical studies are being incorporated in products derived by the bees, particularly bee bread. Bee-derived products contain polyphenols, carotenoids, and bioactive peptides that eliminate free radicals and modify oxidative processes (Silva et al., 2020).

12.3 BEE BREAD APPLICATIONS

Bees gather pollen in the hive, combine it with honey and salivary enzymes, and ferment it in the lactic acid to create bee bread (Figure 12.1). This fermentation process is affected by the microbial activity and converts the raw pollen to a more bioavailable form. This preserves nutrients, enhances digestion, and provides extra metabolites to the final product. Bee bread made nutrients more accessible and was thus a more efficient functional diet in comparison with raw pollen, which is not absorbed because of the hard exine (Kieliszek et al., 2018). Its probiotic qualities are enhanced by its microbial nature and ability to avoid spoilage; thus it may have some neuroprotective effects.

Some of the nutritional and rich ingredients that make up bee bread include polyphenolic chemicals, vitamins, minerals, vital amino acids, and polyunsaturated fatty acids, which offer strong anti-inflammatory, immunomodulatory, and antioxidant effects as well as bioactive effects. They are able to reduce oxidative stress, inhibit protein buildup, and prevent apoptosis in neuronal cells.

FIGURE 12.1 Stages in the formation of bee bread in the hive.

Polyphenols include flavonoids and phenolic acids, which are especially topical to neurodegenerative diseases. B vitamins, magnesium, and zinc enhance the production of neurotransmitters and synaptic plasticity; the two processes are always impaired in Parkinson's disease (PD) and Alzheimer's disease (AD). Being a nutraceutical, having the potential to positively impact neurodegenerative and cognitive functioning, bee bread is more distinctive compared to other bee products because of the high concentration of these chemicals following the fermentation process (Uţoiu et al., 2018).

The purpose of the research is to clarify the interaction between bee bread complex composition and neurological pathways through which the prognosis of the disease can be slowed down. Some of the processes that have been reported include the regulation of mitochondrial activity, neuroinflammation inhibition, oxidative stress control, and neuron loss (Popova et al., 2021). Trial models of AD and PD are undergoing therapeutic investigation and have demonstrated promising results of neuro protection. Bee bread is not easily standardized because the bioactive properties of the product can be changed according to the situation in the hives, processing methods, and even the botanical source. Moreover, it requires a lot of human testing to establish its safety and efficacy, and no clear clinical data can be offered (Diniz et al., 2020).

12.4 BIOACTIVE COMPOSITION OF BEE BREAD AND NEUROPROTECTIVE MECHANISMS

12.4.1 Nutritional Constituents

Bee bread is a good food supplement because of its numerous nutrients, including proteins, amino acids, fatty acids, carbohydrates, mineral salts, polyphenols, and flavonoids. Its nutritional and functional characteristics are globally reported and also depend on the source of the bee pollen. Bee bread consists of water, proteins, free amino acids, bioactive compounds, fatty acids, and carbohydrates (URCAN et al., 2017). Bee bread has a protein content of 14.1 to 37.3 g/100 g (dry basis). It contains amino acids such as glutamic acid, aspartic acid, and proline in larger quantities. The carbohydrate content of bee breed is around 34.80%; fructose, glucose, and maltose are the main sugars. Bee bread has phosphorus in the highest concentrations and also contains minerals such as calcium, magnesium, iron, aluminum, and copper. It is also a source of polyunsaturated fatty acids (PUFAs), all of which are essential for human nutrition (URCAN et al., 2017).

12.4.2 Polyphenolic Compounds and Antioxidants

Bee bread contains polyphenols, which are part of its chemical composition, and they are differentiated according to year, botanical origin, and collection place. The antioxidant properties of bee bread samples have been discussed in different studies. However, results show differences in the amount and content of phenolic compounds, which is due to the variation of the floral (Zhang et al., 2019). Solvents like methanol, ethanol, and water are used to prepare extracts and find their influence on the concentration of bioactive compounds. Methanol shows the best extraction efficiency. Bee bread also contains enzymatic and non-enzymatic antioxidants, including glucose oxidase, catalase, ascorbic acid, flavonoids, and phenolic and other organic acids, that produce nutritional and biological effects such as antimicrobial, antioxidant, and anti-inflammatory qualities. HPLC and GS-MS methods are used to differentiate bioactive compounds such as p-coumaric acid and isorhamnetin. The antioxidant power of bee pollen in different botanical origins is related to specific flavone and phenolic acid profiles rather than the total content of flavones (URCAN et al., 2017).

12.4.3 Fermentation and Probiotic Activity

Bee bread is formed through an anaerobic fermentation process carried out by bees, where bee pollen is transformed with the help of enzymes and microorganisms like *Pseudomonas* and

Saccharomyces. This is the process by which the pollen wall is disrupted, and this thereby improves nutrient bioavailability. During this process, lactic acid is produced, and it lowers the PH; the PH acts as a natural preservative. Carbohydrates are used as a source of oxygen by *Lactobacillus* species and produce lactic acid up to a concentration of 3.2%, which inactivates the *Lactobacillus* sp. The breakdown of bee bread is promoted by the fermentation process, with in vitro studies showing up to 94.7% protein breakdown compared to 38.7% to 85.3% for bee pollen. This bioavailability increase and the production of bioactive metabolites make bee bread a valuable food supplement (URCAN et al., 2017).

12.4.4 Neuroprotective Mechanisms

Bee breeds have antioxidant properties (Figure 12.2). Meliponini honey of stingless bees is high on flavonoids, phenolic acids, and ascorbic acid; these characteristics have the effect of cleaning bad molecules in the body known as reactive oxygen species (ROS), and unless they are cleansed, they may cause destruction of the cells. They also prevent breakdown of the fats in the cell membranes, which is known as lipid peroxidation. For example, an apoptotic signal-reducing effect and nuclear morphological protection of neuronal cells were claimed as a result of ursolic and gallic acid in stingless bee honey (SBH). SBH can also be used as a method of reducing brain inflammation as it lessens the concentration of various cytokines such as TNF-6 and IL-6 which are harmful to the body. It is also rich in luteolin, which prevents activation of the NF-κB. This aids in safeguarding the brain against the damages associated with the diseases like Alzheimer's. Luteolin has been demonstrated to have a calming effect on overreactive animal brain immune cells (Zulkifli et al., 2023). Another essential role of phenylalanine is enhanced performance of the brain as it activates the factor that is critical to learning and memory of the brain known as the brain-derived neurotrophic factor (BDNF) pathway. This helps the brain to be more flexible and increases brain functions such as concentration, memory retention, and clarity (Zulkifli et al., 2023).

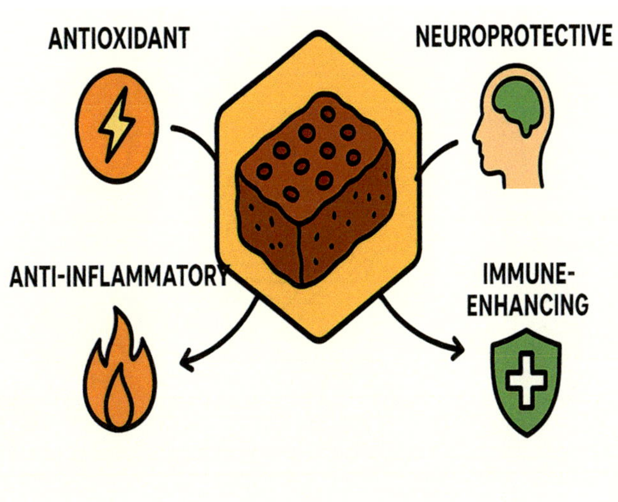

FIGURE 12.2 Properties and neuroprotective mechanisms of bee bread.

12.5 APPLICATIONS OF BEE BREAD IN NEURODEGENERATIVE DISEASES

Considering the antioxidant, anti-inflammatory, immune-enhancing, and neuroprotective properties of bee bread, it has potential to be used in the treatment and management of neurodegenerative diseases like AD, PD, and Huntington's disease (Figure 12.3). Bee bread can serve as a therapeutic agent to AD, especially in treating neuroinflammation and products of the pathological processes. Research shows that bee bread may enhance memory and decrease disorders of sensory systems, which can be interpreted as evidence that this food supplement may have protective qualities. Bee bread regulates the kynurenine (KYN) pathway, which plays a key role in AD neuroinflammation. The administration of bee bread also contributes to the reduction of hippocampal IL-6 and caspase-3 in STZB rats in relation to STZ rats. Moreover, the level of A2 was reduced in the brain of STZB rats with the use of bee bread treatment as well as the level of NR2B. It is also suggested that through the use of bee bread, there is the potential to enhance brain functions as well as inhibit the process of apoptosis (Zulkifli et al., 2023).

12.6 LIMITATIONS AND FUTURE CONSIDERATIONS OF BEE BREAD APPLICATIONS IN NEURODEGENERATIVE DISEASES

The composition of bee bread is different because of a number of factors, which influence whether it will be a dietary supplement and a source of nutraceuticals. The source of chemicals used, season, geography, and soil composition are some of the factors that influence chemical composition. Bees raised in different locations contain more than 300 compounds. These compounds include sugars, minerals, amino acids, organic acids, fatty acids, and polyphenols. The variation in protein, fat, and sugar content of bee bread has also been found in numerous studies. The regional flora also affects bee breed. This is a notable variability that should be taken into account during the investigation of the nutritional and therapeutic properties of bee bread (Dias et al., 2021).

The use of bee bread in the future in functional foods and nutraceuticals can also be feasible and it can be more useful when added with conventional treatments. It is able to improve the fermentation process, or it can be utilized as a food preservative. This is notable due to the rise in consumer interest in health-preserving meals. More research, especially randomized controlled clinical trials, is needed to determine the utility of bee bread and offer evidence-based procedures to use it in diet plans despite its potential in the management of chronic diseases and many health benefits. Its medicinal properties and relevance in the food industry need continuous study (Dias et al., 2021).

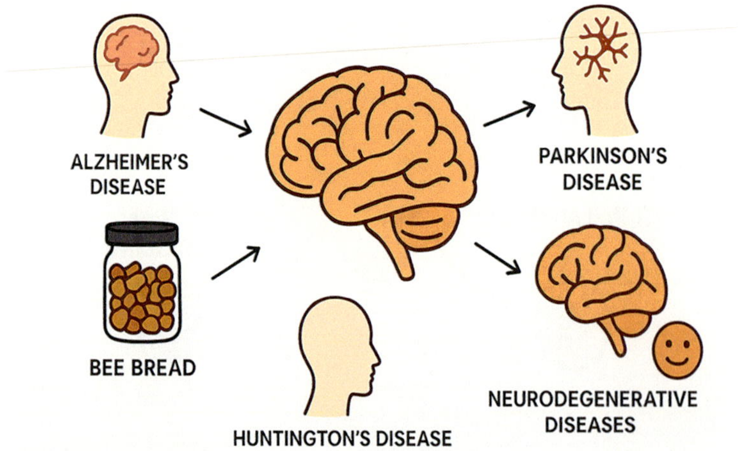

FIGURE 12.3 Bee bread application in neurodegenerative diseases.

12.7 CONCLUSION

Bee bread is a bioactive product with high neuromodulatory potential and a nutrient-rich source. Its carotenoids, phenolic acids, and flavonoids all anti-inflammatory as well as antioxidant properties. New studies indicate that bee bread could be used in the treatment of Alzheimer's, Parkinson's, and other related diseases since it alleviates oxidative stress, inflammation, and neuronal death. Although the preclinical outcomes are promising, clinical validation, product uniformity, and long-term safety assessment still present a problem. Randomized trials and regulatory provisions should be promoted to research to develop bee bread as the supplementary methodology of preventing and managing neurodegenerative diseases.

REFERENCES

Alzheimer's Association. (2023). *2023 Alzheimer's disease facts and figures*. https://www.alz.org

Asuku, A. O., Adeyemo, M. B., Adeyemo, R. O., Agoro, S. A., Odugbesan, O. A., & Ayoola, S. A. (2022). Applications of stem cell therapy in neurodegenerative diseases. *European Journal of Biotechnology and Bioscience*, 10(2), 17–24, 2321–9122.

Asuku, A. O., Ayinla, M. T., Gbonjubola, O. O., Babatunde, S. S., Olajide, T. S., & Oyerinde, T. O. (2024). Neurocognitive aspects of dementia. In N. Moradikor, I. Chatterjee, & W. Mohamed (Eds.), *Nutrition in Brain Aging and Dementia*. Nutritional Neurosciences (pp. 109–130). Springer, Singapore. https://doi.org/10.1007/978-981-97-4117-5_5

Dias, C., Ribeiro, T., Rodrigues, A. C., Ferrante, A., Vasconcelos, M. W., & Pintado, M. (2021). Improving the ripening process after 1-MCP application: Implications and strategies. *Trends in Food Science & Technology*, *113*, 382–396. https://doi.org/10.1016/j.tifs.2021.05.012

Diniz, P., Moecke, E. H. S., & Uchoa, F. D. (2020). Bee products and their potential use in the treatment of neurodegenerative diseases. *Journal of Apicultural Research*, *59*(5), 1–12. https://doi.org/10.1080/00218839.2020.1718190

Ilie, C. I., et al. (2024). Bee bread: A promising source of bioactive compounds—composition and biological activities. *Foods*, *13*, Article. https://www.ncbi.nlm.nih.gov/pmc/articles/PMC10968473/

Khalifa, S. A. M., et al. (2020). Recent insights into chemical and pharmacological studies on bee bread. *Journal of Functional Foods*, 64, 103644.

Kieliszek, M., Piwowarek, K., Kot, A. M., Błażejak, S., Chlebowska-⊠migiel, A., & Wolska, I. (2018). Pollen and bee bread as new health-oriented products: A review. *Trends in Food Science & Technology*, *71*, 170–180. https://doi.org/10.1016/j.tifs.2017.10.021

Li, M., Ye, X., Huang, Z., Ye, L., & Chen C. (2025). Global burden of Parkinson's disease from 1990 to 2021: A population-based study. *BMJ Open*, *15*(4), e095610. https://doi.org/10.1136/bmjopen-2024-095610

Popova, M., Bankova, V., & Trusheva, B. (2021). New trends in apitherapy: Application of bee products in neurodegenerative disorders. *Journal of Functional Foods*, *82*, 104513. https://doi.org/10.1016/j.jff.2021.104513

Silva, R. F. M., Pogačnik, L., & Zupančič, A. (2020). Natural antioxidants in the prevention and treatment of neurodegenerative diseases: Mechanisms and strategies. *Nutrients*, *12*(10), 3104. https://doi.org/10.3390/nu12103104

Urcan, A. C., Marghitas, L. Al, Dezmirean, D. S., Bobis, O., Bonta, V., Muresan, C. I., & Margaoan, R. (2017). Chemical composition and biological activities of beebread – Review. *Bulletin of the University of Agricultural Sciences and Veterinary Medicine Cluj-Napoca. Animal Science and Biotechnologies*, *74*(1), 6. https://doi.org/10.15835/buasvmcn-asb:12646

Uțoiu, E., Matei, F., Toma, A., Diguț⊠, C. F., & Cornea, C. P. (2018). Bee bread: Chemical composition and health benefits—A review. *Annals of the Academy of Romanian Scientists Series on Agriculture, Silviculture and Veterinary Medicine Sciences*, *7*(1), 55–70.

World Health Organization. (2025). *Dementia fact sheet*. https://www.who.int/news-room/fact-sheets/detail/dementia

Zhang, P., Zhang, P., Wu, J., Tao, D., & Wu, R. (2019). Effects of *Leuconostoc mesenteroides* on physicochemical and microbial succession characterization of soybean paste, *Da-jiang*. *LWT*, *115*, 108028. https://doi.org/10.1016/j.lwt.2019.04.029

Zulkifli, N. A., Hassan, Z., Mustafa, M. Z., Azman, W. N. W., Hadie, S. N. H., Ghani, N., & Mat Zin, A. A. (2023). The potential neuroprotective effects of stingless bee honey. *Frontiers in Aging Neuroscience*, *14*. https://doi.org/10.3389/fnagi.2022.1048028

13 Bee Drone Brood Applications in Neurodegenerative Diseases

Abraham Olufemi Asuku, Olawunmi Anthonia Odugbesan, and Swabirah Enimire Sulaiman

13.1 INTRODUCTION

Neurodegenerative disorders are a group of diseases that harm the central nervous system and ultimately result in neurodegeneration. Parkinson's disease (PD), Huntington's disease (HD), Alzheimer's disease (AD), and amyotrophic lateral sclerosis (ALS) are the most prevalent neurodegenerative disorders (Asuku et al., 2022, 2024). It has been demonstrated that the common pathophysiology of neurodegenerative illnesses includes oxidative stress, neuroinflammation, and mitochondrial dysfunction (Ayinla et al., 2020; Akintoye et al., 2023).

A total of 276 million people worldwide suffer from neurodegenerative conditions, and approximately 9 million people die from neurodegenerative disorders annually, according to a comprehensive examination of disorders, injuries, and predisposing factors carried out in 2016 (Feigin et al., 2020). The alarming rate of aging population in the world calls for an urgent solution to the menace of neurodegenerative disease. The potential of natural products as therapy alternatives for neurodegenerative conditions should be explored because of the many adverse effects and high cost of the current options.

Diverse bee products like propolis, honey, royal jelly, beeswax, bee pollen, and bee venom, comprise bioactive molecules that have been used since ancient times by different cultures to treat many disorders (Al Naggar et al., 2021). Investigations on natural products have been more common recently as a means of protection against or treatment of a number of ailments. Bee products and apitherapy have also attracted great interest (Fratellone, 2015; Bogdanov, 2020; Doko et al., 2020). Apitherapy is the application of bees or products derived from them as preventative or therapeutic substances to halt disease progression. Today, in several countries, apitherapy is a component of integrative and complementary medicine. Furthermore, bee products' use as nutraceuticals and dietary supplements has increased because of their nutrient constituents (Pasupuleti et al., 2017; Al Naggar et al., 2021). Research has advanced on the therapeutic properties of products from bees, unveiling several biological characteristics. Moreover, bee products are increasingly utilized in food, cosmetics, and the drug-producing industries (Berretta et al., 2020).

One such product that has been shown to have a variety of bioactive components as well as anti-inflammatory and antioxidant qualities is drone brood, which most beekeepers dispose of in the trash. However, the majority of people would be shocked by the potential health benefits associated with this garbage. This chapter focuses on investigating the potential uses of drone brood for the management of neurodegenerative conditions, emphasizing its antioxidative, anti-inflammatory, and neuroprotective potential, as there is a dearth of information on this.

 DOI: 10.1201/9781003685654-13

13.2 ORIGIN OF DRONE BROOD

Male bees that are formed from unfertilized eggs in the cells of the wax comb through a process called parthenogenesis are referred to as drone brood. Drones take 24 days to grow, compared to 16 and 21 days for workers and queens, respectively (De Oliveira & Engels, 2013). The drones grow in an open cell during the first ten days, and workers gradually feed the larvae. The last 14 days of development are capped. While drone cells are sealed with a convex cap, worker bees seal their own cells, and they also seal the cells of queen bees. Male brood in the comb can be identified thanks to this variation in caps. Pre-pupa and pupa do not feed, undergo metamorphosis, and finish the spermatogenesis process while they are enclosed within the cell (Czekonska & Tofilski, 2010).

Drone brood is sometimes regarded as waste in the beekeeping industry, but it should be utilized because it contains a multitude of important nutrients and bioactive components. However, when drones are frequently harvested, their removal from the colony should increase the production of drones since the quantity of drone brood in the colony controls drone production through a negative feedback process (Jensen et al., 2019).

13.3 FUNCTIONAL COMPOSITIONS OF DRONE BROOD

Royal jelly and fresh drone brood have comparable chemical compositions. Compared to royal jelly, drone brood homogenate has a lower protein and carbohydrate content but a larger water content (Bogdanov, 2016).

The viscosity of drone brood homogenate (apilarnil) is creamy, dense, and milky. It ranges in color from pale gray to white and yellowish (Isidorow, 2013). Apilarnil possesses a special scent similar to royal jelly and a slightly sweet flavor (Kedzia & Hołderna-Kedzia, 2017). It is vital to understand the nutritional components of drone brood because it directly impacts the general vitality and resilience of the colony (Quinlan & Grozinger, 2023).

Drone brood's nutritional composition is also promising for human consumption, for many reasons. It has been reported that drone broods are rich in nutrients. Protein content is the highest (50.4%), followed by fat (20.5%) and carbohydrates (22.1%) (Ghosh et al., 2016). Moreover, they are rich in vitamins and essential minerals; however, the values differ based on the age and type of the drones in comparison to adults.

Antioxidant ability of drone brood is an essential component that is controlled by many factors. Findings have shown the strong antioxidant attributes of honeybee products, like homogenate of drone brood, with sterilized and freeze-dried drone brood demonstrating higher radical-scavenging actions and overall reduction capacity (Sidor et al., 2021). The phytocompounds from the many floral sources that bees graze on could provide health benefits in addition to offering vital nutrients to humans (Eshbah et al., 2021). Drone brood is a great source of essential nutrients because it contains high amounts of phosphorus, calcium, magnesium, and potassium (Ghosh et al., 2016). Drone brood is vital for healthy growth, metabolic activities, and development.

Tongchai et al. (2024) reported that the antioxidant attributes, great mineral contents, and rich phenolic constituents of drone brood make it a viable and sustainable food source that could provide numerous health benefits. Bound and free forms of unsaturated fatty acids, including compounds with sulfhydryl groups, flavonoids, pecial decenic acids, and other phenolic substances, are examples of antioxidants found in drone brood (Mitrofanov et al., 2019). Prokhoda et al. (2019) documented that drone brood contains a high amount of unsaturated fatty acids, with 28% monounsaturated and 10% polyunsaturated fatty acids. Also, it has been reported that drone brood attenuates oxidative stress and protects against heart and vascular disorders (Sawczuk et al., 2018).

Drone brood homogenate, also known as apilarnin, is produced by gathering drone larvae from its cells after hatching from 3 to 11 days (Budnikowa, 2009). Apilarnin, the male form of royal jelly, is extracted from drone brood and then subjected to freeze-drying. More recently, various findings have reported the physical characteristics and uses of drone brood (Isidorov et al., 2016; Jensen et al., 2019; Silici, 2019).

In comparison with other bee products, fresh drone brood is a great source of protein. The protein constituent is 38.5% of the whole drone brood, according to Lazaryan et al. (2003), and 52.3% of fresh drone brood based on Bogdanov's (2011) report. Moreover, drone brood consists of 20 amino acids.

Drone brood contains lipids. According to Stangaciu and Hartenstein (2007), lipids make up 5–8% of drone brood's overall composition. A slightly different value was reported by Barnutiu et al. (2013). In the same vein, Isidorov et al. (2016) reported a value of 3.5% lipid content in drone brood. Free fatty acids, triglycerides, decenoic acids, and fatty acid esters are among these compounds. The fatty acids consist of 50% monounsaturated acids, of which oleic acid makes up 32.3%, and 40% saturated acids, of which stearic and palmitic acids are the most prevalent.

Water-soluble and fat-soluble vitamins are abundant in drone brood. Particularly notable are the high levels of alpha-tocopherol and choline (Czerkasowa & Prochoda, 2006). It has comparatively high levels of calciferol and pantothenic acid (Isidorow, 2013).

13.4 DRONE BROOD AND ITS ROLE IN HUMAN HEALTH

Although drone brood is not well known in Europe, it is a recognized and commonly used medication in some parts of the world, including China and Romania. Royal jelly and fresh drone brood have comparable chemical compositions. The spawn successfully fights cellular aging and numerous diseases because of its rich protein, vitamin, and hormone content. Drone brood injection has been shown to have hepatoprotective effects, to have lower cholesterol and triglyceride levels, and to stimulate the immune system in animal tests (Patel & Ali, 2022). Drone brood DNA has been shown to shield liver tissue against the harmful effects of carbon tetrachloride, buserelin (used to treat endometriosis and prostate cancer), and acetylsalicylic acid (aspirin) (Zhang & Li, 2020). Furthermore, drone brood has demonstrated encouraging outcomes in the treatment of mental health issues and nervous system problems, including promoting neurological health in the elderly (Tanaka & Kuroda, 2023).

One noteworthy observation is that, after 30 days, supplementing dogs with drone brood resulted in a 40% increase in thyroxine (T4) and triiodothyronine (T3) levels and a 37% drop in thyroid-stimulating hormone (TSH) (Singh & Verma, 2021). Additionally, studies using mutant mice with inherited hemolytic anemia revealed that freeze-dried drone brood greatly increased the animals' survival rate, increasing it from two weeks to seven months in half of the individuals. The aforementioned reports show promising therapeutic ability of drone brood for human health (Chen & Zhou, 2022).

Carbohydrates, inositol, biotin, B complex vitamins, proteins, lipids, choline, folic acid, and a large amount of magnesium, potassium, sodium, copper, iron, and selenium are in the list of apilarnin nutritious constituents. It has many free amino acids. Steroid hormones like progesterone, testosterone, prolactin, and estrogen are in high quantity in apilarnil (Isidorov et al., 2016; Sawczuk et al., 2019). The estrogenic activity of apilarnil has been reported in vivo, corroborating its application in traditional medicine and creating avenues for its usage in estrogen insufficiency cases (Seres et al., 2013).

The sexual development of male broilers was accelerated by apilarnil at a young age (Altan et al., 2013). Apilarnil improved the weight of the epididymis and seminal glands, the germ cell density, the volume of ejaculate, and the motility of sperm in pigs (Bolatovna et al., 2015). Apilarnil also attenuates the oxidative process and liver damage in rodents (Doganyigit et al., 2020). Male infertility, ovarian dysfunction, immunological and thyroid problems, and childhood malnutrition are among the ailments and disorders that apilarnil is utilized to manage (Sidor & Dzugan, 2020). Based on findings, apilarnil may be used for andropause issues in males because it has an androgenic effect (Erdem & Ozkok, 2017). Nevertheless, Sidor and Dzugan (2020) reported that further research on bee products is needed due to their many benefits to human health.

13.5 DRONE BROOD APPLICATIONS FOR NEURODEGENERATIVE DISEASES

The rich components of drone brood improve biological activity and have positive effects on human health. Drone brood could halt cellular aging and numerous disorders because of its rich vitamins, protein, and hormonal contents. Due to the reports that aging has been recognized to be the common cause of a plethora of neurodegenerative disorders, drone brood may be used for their management (Asuku et al., 2025a; Asuku et al., 2025b).

Also, drone brood has antioxidant properties. The antioxidant qualities of drone brood were documented by Sidor et al. (2021). The ethanolic and aqueous homogenate extracts of drone brood were investigated for flavonoid and polyphenol content, and antioxidant activity using the FRAP, DPPH, and ABTS methods. The outcomes of the findings show that drone brood contains antioxidant properties comparable to those of royal jelly. These antioxidant attributes are important for scavenging free radicals, which reduces oxidative stress and protects from cellular damage. The Protection against disorders such as neurodegenerative diseases requires these actions. The oxidative process has been connected with the pathophysiology of Parkinson's and Alzheimer's diseases (Ayinla et al., 2020; Asuku et al., 2024), and the antioxidant properties of drone brood could be harnessed for the management of these conditions.

Drone brood contains vitamins, fats, carbohydrates, proteins, minerals, and numerous physiologically active substances including hormones, enzymes, and antioxidants, making it a highly nutritious bee product. It is a promising therapeutic option for neurodegenerative conditions that are associated with inflammation, age-associated decline, and immune system dysfunction, because of its special constituents. Drone brood's immunomodulatory properties are among its most common attributes. It possesses the ability to improve immunity, enhancing the body's capacity to combat infections and heal disorders. Drone brood's anti-inflammatory properties could be explored as a natural cure for neurodegenerative disorders. Chronic inflammation is a common underlying factor for various diseases, especially neurodegenerative disorders (Ayinla & Asuku, 2025). Drone brood could aid in lessening the symptoms and reducing the progression of many disorders by modifying the inflammatory pathways.

Additionally, apilarnil's enzyme inhibitory effects against carbonic anhydrase I and II isoenzymes employing acetylcholinesterase (AChE) and butyrylcholinesterase (BChE) enzymes were reported by Inci et al. (2023). Based on their findings, apilarnil possesses a strong anticholinesterase attributes by inhibiting the two enzymes. The hallmark of dementia, cholinergic deficit, is caused by acetylcholinesterase; but the ability of apilarnil to inhibit the enzymes demonstrates that it could be utilized to treat and manage dementia.

Parkinson's disease (PD) neuropathology is influenced by microglial activation. Therefore, inhibiting M1 while concurrently increasing M2 microglia activity could be a possible PD therapy. In a rat model of Parkinson's disease produced by rotenone, apilarnil has been shown to have neuroprotective benefits and reduce motor impairment by rebalancing M1/M2 microglia polarization and controlling the levels of miR-155 and miR-124 (Salama et al., 2024).

Apilarnil had a neuroprotective effect that prevented apoptosis, decreased the number of deteriorated neurons, and decreased the pro-inflammatory markers high levels brought on by sepsis in male rodents (Hamamci et al., 2020).

Drone brood has also been shown in human clinical trials to improve the mental state of patients with depressive neurosis, fatigue, anorexia, and feelings of helplessness (Iliesiu, 1988); to positively impact neurastemia symptoms (fatigue, weariness, and dizziness) (Meda et al., 2004); to enhance memory and lessen psychomotor issues (Iliesiu, 1988); and to enhance neurovegetative functions in older adults (Bogdanov, 2016).

13.6 CHALLENGES

Despite its potential, drone brood has a difficult time becoming widely accepted in contemporary medicine and in the management of neurodegenerative diseases. Challenges include the need for more scientific proof, healthcare practitioners' lack of understanding, and regulatory obstacles. To guarantee sustainable and responsible use, ethical issues regarding the harvesting of drone brood and the effect on bee populations must also be addressed. Using drone brood for the management of neurodegenerative conditions would necessitate a multidisciplinary approach. To reduce the gap between contemporary medical practices and conventional knowledge, medical practitioners, politicians, researchers, and academics must collaborate. Public education and enlightenment programs would also assist in promoting the awareness and acceptance of drone brood as an alternative therapy for neurodegenerative diseases.

13.7 CONCLUSION

Drone brood is made up of molecules and compounds with anti-acetylcholinesterase, antioxidant, neuroprotective, and anti-inflammatory properties; therefore, it may be used for managing neurodegenerative conditions. This therapeutic potential is beyond personal health gains. Additionally, it highlights the biodiversity value and the bees' role in preservation of ecological balance. Supporting the sustainable beekeeping methods guarantees drone brood availability and promotes larger environmental conservation initiatives. These dual benefits demonstrate the connection between the health of the human brain and our world.

REFERENCES

Akintoye, O.O., Ajibare A.J., Folawiyo M.A., Asuku, A.O., Jimoh Abdulghaffar H.O., Akintoye A.O., Babalola K.T., Idowu O.O., & Oyiza Y.G. (2023). Virgin Coconut Oil-Supplemented Diet Reverses Behavioural Phenotypes of Sodium Benzoate-model of Acetylcholinesterase Dysfunction and Cognitive Impairment: Role of Nrf2/NfKb Signaling. *Niger. J. Physiol. Sci.*, 37, 225–233, https://doi.org/10.54548/njps.v37i2.10

Al Naggar, Y., Giesy, J.P., Abdel-Daim, M.M., Ansari, M.J., Al-Kahtani, S.N., & Yahya, G. (2021). Fighting against the second wave of COVID-19: can honeybee products help protect against the pandemic? Saudi *J. Biol. Sci.* 28, 1519–1527.

Altan, O., Yucel, B., Açikgoz, Z., Seremet, C., Kosoglu, M., Turgan, N., & Ozgonul, A.M. (2013). Apilarnil reduces fear and advances sexual development in male broilers but has no effect on growth. *Br. Poult. Sci.* 54, 355–361.

Asuku A.O., Adeyemo M.B., Adeyemo R.O., Agoro S.A., Odugbesan O.A., & Ayoola S.A. (2022). Applications of stem cell therapy in neurodegenerative diseases. *European J. Biotechnol. Biosci.* 10(2): 17–24, 2321–9122.

Asuku, A.O., Ayinla, M.T., Gbonjubola, O.O., Babatunde, S.S., Olajide, T.S. & Oyerinde, T.O. (2024). Neurocognitive Aspects of Dementia. In: *Nutrition in Brain Aging and Dementia. Nutritional Neurosciences*. Springer, Singapore. https://doi.org/10.1007/978-981-97-4117-5_

Asuku, A.O., Ogungbangbe, G.O., & Olajide, T.S. (2025a). Effect of Happiness and Positive Emotions on Dementia. In *Neuroscience of Feelings. Progress in Brain Research*, Elsevier, https://doi.org/10.1016/bs.pbr.2025.03.001

Asuku, A.O., Ogungbangbe, G.O., & Fajemidagba, G.A. (2025b). Gene therapy for Parkinson's disease: A new frontier in neurodegenerative diseases. *Exploration of Neuroprotective Therapy*, 5, 1004119. https://doi.org/10.37349/ent.2025.100411

Ayinla, M.T., & Asuku, A.O. (2025). The neurotoxic effects of lead acetate and the abrogating actions of 6-gingerol-rich extract of ginger via modulation of antioxidant defence system, pro-inflammatory markers, and apoptotic cascade. *Naunyn-Schmiedeberg's Arch. Pharmacol.*, 398, 9341–9356. https://doi.org/10.1007/s00210-025-03873-x

Ayinla, M.T., Asuku, A.O., Bayo-Olugbami, A., Ayeni, A.O., Abiola, A.A., & Owoyele, B.V. (2020). Melatonin and Vitamin C modulate Cholinergic Neurotransmission and oxidative stress in Scopolamine–induced rat model of memory impairment. *Journal of African Association of Physiological Sciences*. 8(1), 20.

Barnutiu, L.I., Marghitas, L.A., Dezmirean, D., Bobis, O., Mihai, C., & Pavel, C. (2013). Physico-chemical composition of apilarnil (bee drone larvae). *Lucr. ¸Stiin¸ tifice Ser. Zooteh* 59, 199–202

Berretta, A.A., Silveira, M.A.D., Capcha, J.M.C., & De Jong, D. (2020). Propolis and its potential against SARS-CoV-2 infection mechanisms and COVID-19 disease. *Biomed. Pharmacother.* 131, 110622.

Bogdanov, S. (2011). Honey composition. In *The Honey Book*; Bee Product Science: Muethlethurnen, Switzerland, Available online: https://www.researchgate.net/publication/304011775_Honey_Composition (accessed on 20 August 2020).

Bogdanov, S. (2016). Royal jelly and bee brood: Harvest, composition, quality. In *The Royal Jelly Book*; Bee Product Science: Muethlethurnen, Switzerland, Available online: https://www.researchgate.net/publication/304012318_Royal_Jelly_and_Bee_Brood_Harvest_Composition_Quality (accessed on 20 August 2020).

Bogdanov, S. (2020). Antiviral properties of the bee products: A review. *Bee Product Sci.* 2020. www.bee-hexagon.net

Bolatovna, K.S., Rustenov, A., Eleuqalieva, N., Omirzak, T., & Akhanov, U.K. (2015). Improving reproductive qualities of pigs using the drone brood homogenate. *Biol. Med.* 7, 2.

Budnikowa, N.W. (2009). Biologically active compounds in drones. *Beekeeping*, 6, 52–53.

Chen, L., & Zhou, X. (2022). Therapeutic effects of drone brood in mutant mice with hereditary hemolytic anemia. *J Med Res*, 15, 200–210.

Czekonska, K., & Tofilski, A. (2010). Body mass of honey bee drones developing in constant and in changing temperatures. *Apidologie*, 51, 510–518. [CrossRef]

Czerkasowa, A.I., & Prochoda, I.O. (2006). A New Addition of Drone Larvae—An Alternative to Royal Jelly; NFaU: Kharkiv Oblast, Ukraine, pp. 65–69. (In Ukrainian)

De Oliveira, T.S., & Engels, W. (2013). Classification of substages in preimaginal development of honey bee drones (Hymenoptera: Apidae). *Entomol. Gen.* 34, 287–293.

Doganyigit, Z., Okan, A., Kaymak, E., Pandır, D., & Silici, S. (2020). Investigation of protective effects of apilarnil against lipopolysaccharide induced liver injury in rats via TLR 4/HMGB-1/ NF-κB pathway. *Biomed. Pharmacother.* 125, 109967.

Doko, T., Salaric, I., & Bazdarik, K. (2020). Complementary and alternative medicine use among Croatian health studies students - a single center cross-sectional study. *Acta Med. Acad.* 49, 240–248.

Erdem, B., & Ozkok, A. (2017). Can food supplement produced from apilarnil be an alternative to testosterone replacement therapy? *Hacettepe J. Biol. Chem.* 45, 635–638.

Eshbah, H., Mohamed, A., Zedan, O., & Ghanem, A. (2021). Phytochemical components in different types of pollen and their effects on honey bees. *Minia J. Agric. Res. Dev.* 41, 1–13.

Feigin, V.L., Vos, T., Nichols, E., Owolabi, M.O., Carroll, W.M., Dichgans, M., Deuschl, G., Parmar, P., Brainin, M., & Murray, C. (2020). The global burden of neurological disorders: Translating evidence into policy. *Lancet Neurol.* 19(3), 255–265.

Fratellone, P.M. (2015). Apitherapy products for medicinal use. *J. Nutr. Food Sci.* 5, 6.

Ghosh, S., Jung, C., & Meyer-Rochow, V.B. (2016). Nutritional value and chemical composition of larvae, pupae, and adults of worker honey bee, Apis mellifera ligustica as a sustainable food source. *J. Asia-Pac. Entomol.* 19, 487–495.

Hamamci, M., Doganyigit, Z., Silici, S., Okan, A., Kaymak, E., Yilmaz, S., Tokpinar, A., Inan, L.E. (2020). Apilarnil: A novel neuroprotective candidate. *Acta Neurol. Taiwan.* 29, 33–45.

Iliesiu, N. A. (1988). Modern preparation of bees "Apilarnil." *Inf. Reg. Zrzesz. Pszczel. Apipol.* 10, 15–20.

Inci, H., Izol, E., Yilmaz, M.A., Ilkaya, M., Bingöl, Z., & Gülçin I. (2023). Comprehensive Phytochemical Content by LC/MS/MS and Anticholinergic, Antiglaucoma, Antiepilepsy, and Antioxidant Activity of Apilarnil (Drone Larvae). *Chem. Biodivers.*, https://doi.org/10.1002/cbdv.202300654

Isidorov, V.A., Bakier, S., & Stocki, M. (2016). GC-MS investigation of the chemical composition of honeybee drone and queen larva homogenate. *J. Apic. Sci.* 60, 111–120.

Isidorow, W.A. (2013). *The Alchemy of Bees*; S ˛adecki Bartnik: Stró˙ze, Poland, pp. 64–69. (In Polish)

Jensen, A.B., Evans, J., Jonas-Levi, A., Benjamin, O., Martinez, I., Dahle, B., Roos, N., Lecocq, A., & Foley, K. (2019). Standard methods for Apis mellifera brood as human food. *J. Apic. Res.* 58, 1–28.

Kedzia, B., & Hołderna-Kedzia, E. (2017). *Less Known Bee Products*; S ˛adecki Bartnik: Stró˙ze, Poland, 2017. (In Polish)

Lazaryan, D.S., Sotnikova, E.M., & Ertushenko, N.S. (2003). Standardization of bee bread homogenate composition. *Pharm. Chem. J.* 37, 614–616.

Meda, A., Lamien, C.E., & Millogo, J. (2004). Therapeutic uses of honey and honeybee larvae in central Burkina Faso. *J. Ethnopharmacol.* 95, 103–107.

Mitrofanov, D.V., Budnikova, N.V., & Serebryakova, O.V. (2019). *Antioxidant activity of drone brood in the process of stabilization and storage Modern problems of beekeeping and apitherapy eds A Z Brandorf et al.*, Center for Beekeeping, Rybnoe, pp. 264–271.

Pasupuleti, V.R., Sammugam, L., Ramesh, N., & Gan, S.H. (2017). Honey, propolis, and royal jelly: a comprehensive review of their biological actions and health benefits. *Oxida. Med. Cell. Longev.* 2017, 1–21.

Patel, R., & Ali, F. (2022). Cardioprotective effects of drone brood: A natural remedy for Hyperlipidemia. *J. Altern. Med.*, 10, 30–42.

Prokhoda, I.A., Eliseeva, E.V., Katunina, N.P., Laktyushina, O.V., Tachkova, I.A. & Litvin, F.B. (2019). Quality management of the apiproduct from the drone larvae. *IOP Conf. Ser. Earth and Environmental Sci.* 274, 012132. 10.1088/1755-1315/274/1/012132

Quinlan, G.M., & Grozinger, C.M. (2023). Chapter Five—Honey bee nutritional ecology: From physiology to landscapes. In *Advances in Insect Physiology*; Harrison, J.F., Ed.; Academic Press: Cambridge, MA, 64, pp. 289–345.

Salama, R.M., Darwish, S.F., Yehia, R., Eissa, N., Elmongy, N.F., Abd-Elgalil, M.M., Schaalan, M.F., & El Wakeel, S.A. (2024). Apilarnil exerts neuroprotective effects and alleviates motor dysfunction by rebalancing M1/M2 microglia polarization, regulating miR-155 and miR-124 expression in a rotenone-induced Parkinson's disease rat model. *Int. Immunopharmacol.*, 137, 112536.

Sawczuk, R., Karpinska, J., & Miltyk, W. (2018). What do we know and what we would like to know about drone homogenate. *J. Ethnopharmacol.* 10.1016/j.jep.2018.10.042

Sawczuk, R., Karpinska, J., & Miltyk, W. (2019). What do we need to know about drone brood homogenate and what is known. *J. Ethnopharmacol.* 245, 111581.

Seres, A.B., Ducza, E., Bathori, M., Hunyadi, A., Beni, Z., Dekany, M., & Gaspar, R. (2013). Raw drone milk of honeybees elicits uterotrophic effect in rats: evidence for estrogenic activity. *J. Med. Food*, 16, 404–409.

Sidor, E., & Dzugan, M. (2020). Drone brood homogenate as natural remedy for treating health care problem: a scientific and practical approach. *Molecules*, 25, 5699.

Sidor, E., Miłek, M., Tomczyk, M., & Dzugan, M. (2021). Antioxidant activity of frozen and freeze-dried drone brood homogenate regarding the stage of larval development. *Antioxidants*, 10, 639.

Silici, S. (2019). Chemical content and bioactive properties of Drone Larvae (Apilarnil). *Mellifera*, 19, 14–22.

Singh, A., & Verma, S. (2021). Impact of drone brood supplementation on thyroid function in dogs. *Vet. Sci. J.*, 39, 162–168.

Stangaciu, S., & Hartenstein, E. (2007). Medicines from the Bee Pharmacy; KDC: Warsaw, Poland, pp. 62–64. (In Polish).

Tanaka, Y., & Kuroda, S. (2023). Effects of drone brood on neurological and sexual health in elderly populations. *Ageing Res Rev*, 12:56–67.

Tongchai, P., Yadoung, S., Sutan, K., Kawichai, S., Danmek, K., Maitip, J., Ghosh, S., Jung, C., Chuttong, B., & Hongsibsong, S. (2024). Antioxidant capacity, phytochemicals, minerals, and chemical pollutants in Worker Honey Bee (Apis mellifera L.) Broods from Northern Thailand: A safe and sustainable food source. *Foods*, 13, 1998. https://doi.org/10.3390/foods13131998

Zhang, H., & Li, C. (2020). DNA from Drone Brood Protects Against Hepatotoxicity in Rodents. *Toxicol. Rep.*, 7, 125–130.

14 Apitherapy Entrepreneurship
Innovations and Biotechnological Advancement

Abdulkadir Yaradua Imam-Fulani,
Oladimeji Lateefah Olabisi, Buhari Abdulhafeez Oladimeji,
Salaudeen Faidat Sumisola, and Suleiman Muhammed Eze

14.1 INTRODUCTION

Medical conditions, including cancer, autoimmune conditions, chronic pain, and infection by bacteria, fungus, and viruses, are increasing in the world. Meanwhile, the pharmaceutical industry is under serious threat, with the negative effect of artificial drugs on health growing and resistance to the treatment becoming more intense (Kumar et al., 2018). It needs to bring innovative, organic, conventional, and synthetic-free treatments to help deal with various health issues. The former traditional medicine was founded on natural ingredients. Apitherapy is an example of usage of this natural approach (Munstedt & Mannle, 2020). The *Apis mellifera* or honeybees, also known as the Goldin insects, have been utilized by various people to address several health problems. These are fundamental products that are biogenic in nature (Tamura et al., 2015).

Bees have been much cherished due to their contribution in ensuring the ecological balance and the production of the natural substance that enhances human well-being. Derived out of the Greek words *apis* (bee) and *therapeia* (healing), the term *apitherapy* is used to label the use of bee-related products to prevent and treat various health problems. Honey and other products of the bee have been used in traditional medicine over centuries, and these methods proved to be effective due to modern research. State-of-the-art bee products are being purchased by entrepreneurs in different parts of the world and utilized in health foods, cosmetics, and in nutritious and pharmaceutical products. Simultaneously, there are biotechnology innovations, which have increased product extraction, standardization, and sale methods.

In the framework of entrepreneurial/biotechnological environment, apitherapy has been employed to give an insight into the concept of innovation, the introduction of product into the market, and prospects of further bee-derived product use in traditional medicine, to treat wounds, digestive disorders, respiratory disorders, and inflammation (El-Didamony, 2024).

Many believe the Russian physiologist Nikolai M. Artemov to have initiated the scientific use of apitherapy during the 20th century. In the Soviet Union, Artemov started working on using bee venom in medical practice and devised ways of introducing bee stings into approved medical use (Artemov, 1941).

14.2 OVERVIEW OF APITHERAPY

Apitherapy is the procedure that utilizes extracts of bees, including honey, pollen, propolis, royal jelly, bee venom, and wax among others, to treat health problems (Weis et al., 2022; Young & Bundell, 2023). Apitherapy has various hundreds of curing applications and is many years old Nitecka-Buchta et al., 2014). The medicinal properties of products created by bees are listed in some of the ancient works and hieroglyphs of Egypt, China, Korea, and Greece (Wehbe et al., 2019; Noble et al., 2020).

DOI: 10.1201/9781003685654-14

In addition, the practice of apitherapy was initiated by ancient individuals upon realizing the value of products concerning bees, including among the Greeks and Romans (Sanghani et al., 2014). As one of the Ayurvedic therapies in South Asian cultures, especially in India, apitherapy represents a significant portion of Ayurvedic treatment (Shrivastava et al., 2017).

Examples of the products that are based on bee-derived product include honey, venom, royal jelly, propolis, and many more which are rich sources of bioactive compounds that provide a plethora of nutrients. They contain such ingredients as vitamins, minerals, enzymes, and antioxidants and possess antimicrobial and anti-inflammatory effects (Weis et al., 2022). There is a great potential for bee products that have a lot of physiological benefits, including working on the immune system, postoperative wound treatment, and reduction of the symptoms of certain diseases (Gokulakrishnaa & Thirunavukkarasu, 2020).

These advantages, to an extent, are added by the fact that all products derived by bees contain certain elements. Some other ailments that bee venom has been known to treat include arthritis and chronic pain since it has been found to have analgesic and anti-inflammatory properties (Hegazi, 2012). In addition, propolis may be applied against infections because of the proven antibacterial and antiviral effects (Alfaris et al., 2022). Apitherapy uses the bioactive properties that are present in the products of bees, and this improves its curative effects. Among the things are a myriad of physiological activities by the compounds that have been researched and reported (Hegazi, 2012; Sahin et al., 2019).

As an alternative health technique, apitherapy has attracted attention in the recent past with the prospects of having a positive impact on informing the health status of individuals with different health complications (Kavurmaci & Tan, 2019; Popovic et al., 2022). Apitherapy appears to have a bright future. It is expected to be transformed into one of the alternative therapies in the short term, especially in cases of cancer, Parkinson's disease, and Alzheimer's disease (Campos et al., 2021).

14.3 PRODUCTS OF APITHERAPY

14.3.1 Honey

Honey is a well-known product produced by bees as well as a product that is regularly used in apitherapy. Interestingly enough, it has been found to be effective in healing of a surgical wound when other conventional methods of treatment have not worked (Hussein et al., 2011). Honey has already been hailed as possessing various uses, including antifungal, antiviral, antiseptic, anti-inflammatory, and antioxidant uses. Additionally, recent data show that it could be applicable in solving many health problems, including diabetes, cancer, and heart disease (Malone & Tsai, 2016; Hossain et al., 2022). The research is registering many therapeutic benefits of honey, which encompass its effectiveness in the therapy of such diseases as diabetic foot ulcers, oral insensitivity, and respiratory infections (Bucekova, 2019; Hossain et al., 2022).

In the past, honey has been used to treat many medical ailments. Research has revealed that in addition to its antioxidant and anti-inflammatory properties, honey possesses antibacterial, antiparasitic, and anticancer properties, as well as heart-protective properties and various nutritional properties (Helen & Jun, 2008). It is also acidic in nature, with a low pH and a high sugar level, and as such it will not support growth of microorganisms. The final and the most significant factor is hydrogen peroxide (Rodriguez-Wallberg & Oktay, 2012). Mama et al. (2019) also found that the low moisture content of honey and the presence of glucose oxidase of honey components offered potential eradication of the methicillin-resistant *Staphylococcus aureus* (MRSA) bacteria in wounds and prevented bacterial proliferation.

14.3.2 Propolis

Propolis has enormous antibacterial and anti-inflammatory properties, according to a study conducted by Ali and Kunugi (2020). Conventional medicine has incorporated it in the treatment of several health issues, including infections, oral diseases, and skin diseases (Faqihi & Taha, 2022). According to recent studies, propolis acts as an antimicrobial agent against bacteria, viruses, and fungi, but also as an antioxidant, another step that should be taken to justify its potential as a treatment option (Weis et al., 2022).

Propolis can be used as the remedy to respiratory disease, throat infections, and other skin issues due to the different properties it has, as reported by Gokulakrishnaa and Thirunavukkarasu (2020). Honeybees, in trying to preserve their colonies, safely collect propolis or bee glue, which is a collection of buds and new shoots of plants. It is a sticky material that mostly appears dark yellow or brown. Propolis is used by bees to close up the inside parts of the hives—the honeycombs and holes—and to prevent the rot of the intruders in the hive. It serves as a barrier against dangers and helps in the management of many diseases in the hive (Martin, 2012). A number of raw propolis samples were shown to contain approximately 300 varied compounds which comprised of phenolic, flavonoid compounds (including flavones, flavanones, and stilbenoids), chalcones, steroids, and lignans (Martin, 2012; Chlebowski et al., 2002).

Research has therefore established that bee resin has diverse biological activities, which include antioxidant, antimicrobial, and anti-inflammatory activities. Chinese propolis has been observed to have a very high cytotoxicity to human breast cancerous cells (Kroenke et al., 2002). Raw propolis ought to be processed, and then it can be used as a raw material. Ethanol is the most popular way of extracting it since it enables active compounds to dissolve with ease. The traditional ethanolic propolis extract can only be applied due to its alcoholic composition. Trying to improve the bioavailability of the active compounds present in propolis, various methods have been investigated, including encapsulation, the creation of electrospun nanofibers, and the combination of the latter with chemotherapy drugs (Charles et al., 2020).

Incidentally, two researches established that propolis and doxorubicin (DOX) withstood anticancer effects of DOX on MDA-MB-231 cancer cells. The experiment shows that a combination of IC 50 of 10 times lower than the one of DOX alone has an improved cell apoptosis (Rouibah et al., 2021).

Propolis was also transformed into nanofibers, which are utilized as dressings during the wound healing process by using electrospinning technology. To dress the wounds, Alberti (2017) created a dressing that consists of PVA and propolis nano fibers (Yan et al., 2017). The scaffold was fibroblast compatible in a controlled laboratory setting. It also had a higher rate of wound healing compared to the groups which were treated with allantoin solution and those which were not treated (Godfrey et al., 2011).

Propolis is non-alcoholic and can be consumed as capsules. Their effect can be amplified with the help of propaganda beads. Keskin et al. (2019) entrapped ethanolic propolis extract in sodium alginate and dried into alcohol-free beads in order to increase the degree of absorption.

14.3.3 Royal Jelly

Royal jelly is rich in proteins, vitamins, and minerals as well as fatty acids. The special property of this material is that it regulates the immune system and assists in wound repair, owing to its proven anti-inflammatory and antioxidant properties (Ali & Kunugi, 2022). The prospective health value of royal jelly has been considered regarding treatment of skin ailments, immunity system problems, and heart-related illness (Zhu & Wongsiri, 2008; Hegazi, 2012). Notably, royal jelly possesses antibacterial effects on Gram-positive bacteria and secures protection against the harmful effects that nicotine has on cholesterol (Susilowati et al., 2017).

Unlike worker bees, who live about two months on average, the queen can live up to a maximum of four years. The cause of the lifespan difference between the worker bees and the queen is considered to be the aspects of royal jelly. The short-chain hydroxy fatty acids present in the proteins of royal jelly are assumed to be of high impact; still, the mechanism of its impact remains unclear (Nadaf et al., 2018; Lee et al., 2019). The proteins of royal jelly are also known as a secret of life or revitalizer because they are believed to have a great influence, but it is still not fully understood how this influence is being achieved (Nadaf et al., 2018; Lee et al., 2019).

Royal jelly appears in two variations, fresh and freeze-dried. These conditions influence the structure of royal jelly that may be species of bees, feeding behavior, seasonality and mode of production. Royal jelly consists of water (60–70% w/w), proteins (9–18% w/w), carbohydrates (7–18% w/w), and fats (3–8% w/w). Besides, it contains minerals (Ca, Fe, Mn, K, Na, Zn, Mg and Cu), amino acids (Ile, Leu, Val, Thr, Phe, Met, Trp and Lys), and nucleotides, hormones, polyphenols, enzymes, and other small heterocyclic elements, which are also valuable elements. Royal jelly has a pH of 3.6 (Jayaprakasha et al., 2007; Rwigemera et al., 2015).

14.3.4 Bee Venom

Apitoxin, or bee venom, is a product that has a long history of usage in traditional medicine and has been established to be an analgesic and anti-inflammatory agent. Recent reports found that bee venom peptides and enzymes have antibacterial, anti-inflammatory, and immunomodulatory properties (Kokot et al., 2009). Other studies have stated the therapeutic effect of bee venom in the possible treatment of rheumatoid arthritis, multiple sclerosis, and some types of cancer (Wehbe et al., 2019; Bava et al., 2023).

Bee venom (apitoxin) is a fashionable, yellowish, sharp, bitter, and clear acidic mass in the abdomen, which is associated with the stinging procedure of the bee. It is secreted to the maximum in the first two weeks of the honeybee's life cycle. This is a very important protection measure of honeybees. It is only with the assistance of stingers that female bees are able to sting. A honeybee sting has approximately 0.3 milligrams of honeybee venom. When honeybee venom is exposed to air, it dries up and solidifies. During the collection of bee venom, many of its very volatile constituents are easily lost (Donaldson, 2004).

Bee venom contains numerous active ingredients, such as peptides (melittin, adolapin, and apamin), enzymes (hyaluronidase and phospholipase A_2), and a peptide, leading to the degranulation of the mast cell. The phospholipase A_2 (PLA_2) is the second significant component other than melittin that is about 50% of the dried venom (Sengupta et al., 2004).

Diverse therapeutic properties of bee venom are studied both in vitro and in vivo, including anti-arthritic, anti-mutagenic, anti-nociceptive, and radioprotective (Şengü & Vatansev, 2017; Zheng et al., 2019), liver safeguarding, cell safeguarding, free-radical neutralizing, and inflammation-reducing (Şengü & Vatansev, 2017; Zheng et al., 2019).

Apipuncture, using live bees, is administered by a registered practitioner or acupuncturist. Using tweezers, experts pick a live bee and sting it on the area of victimization. Antigenicity of bee venom should be established by an allergy test preceding the process. One or two stings during the initial session should be featured in a progressive form of treatment. Treatment of stings is done both directly and indirectly (Sengupta et al., 2004).

Pigmies are used to hold the bee by the head or the thorax, then carrying it to the place where it is needed. The mark vanishes in a minute, and these are taken in various locations where the need arises. An indirect application, on the other hand, is the application of a pair of forceps in taking the sting out and pressing the same against the body within a very short duration of time. Ten different regions could be applied to use the same stinger (Zheng et al., 2019).

Bee venom therapy (BVT) is an intradermal injection methodology that makes use of the *Apis mellifera* venom. Intradermal injection is more effective in improving the result of injecting with the venom solution, which can be done under the skin to a maximum of 2.0 mL. Such injections need to be provided by and under the supervision of licensed experts in acupuncture. Premratanachai and Chanchao (2014) are first approved by the Korean Food and Drug Authority.

As a localized form of treatment that can be used as cream or ointments, bee venom can now be used in treating various skin disorders. Its use on large parts of the skin might cause serious side effects such as irritations and lesions. In other situations, these side effects can have an acute immune response, which can be anaphylaxis (Wehbe et al., 2014).

14.3.5 Beeswax

Beeswax has a great number of possible therapies. It helps in healing of the wounds and to take care of the skin besides its established antibacterial and anti-inflammatory properties (Weis et al., 2022). Beeswax is a good infection preventive and treatment agent researchers have pointed out that it contains antibacterial properties that combat Gram-positive and Gram-negative bacteria. Beeswax is and tends to be utilized in cosmetic and personal care items due to its moisturizing and skin-enhancing properties.

Beeswax is made in the lower abdomen when still liquid, by eight ceriferous glands known as the wax glands on the young worker bees which are between the ages of 12 and 18 days. On coming into contact with air, it solidifies, and bees cut it with the mandibles into an arrangement of honeycomb, which contains pollen and honey. Raw beeswax is slightly tinting nearly white due to the cocoon; it will turn to different hues of yellow or gold upon interaction with honey and pollen and transform to a deep brown shade after about four years (Eswaraiah et al., 2020).

There is a chance that the fabrication of the wax is severely reliant upon the genetics of the bees and upon the food that the bees ingest; the beeswax structure might be not consistent within the families and species of bees. Nevertheless, beeswax is usually composed of over 300 different compounds with most of these complex wax esters (15–27%), hydrocarbons (12–16%), free fatty acids (12–14%), hydroxyl monoesters (35–45%), fatty ethanol (approximately 1%) and external constituents with traces of propolis, pollen, and a very small fragment of flowers (Lee et al., 2012).

Ancient civilizations were aware of the medicinal value of beeswax. Some of the varied treatments reported by Avicenna in his renowned *Canon of Medicine* include a mixture of beeswax. Furthermore, archeological records have attested to the beeswax salves that date back to the 16th century (Seki et al., 2008). More recently, beeswax has been considered as a natural lipid ingredient to create nanostructured lipid carriers to execute a variety of biological functions (Jia et al., 2013). In addition, beeswax is used in cosmetics; it is a great skincare and beauty wax due to its strength.

Beeswax has desirable longevity that is credited to its solid nature, and this combined with its ability to keep off water makes it a desirable material in the form of coatings, polishes, and waxes that preserve, add luster and improve quality in overall form (Thangapazham et al., 2007). Table 14.1 shows bee products and their bioactivities.

TABLE 14.1
Bee Products and Their Active Biological Components

Bee Product	Major Bioactive Compounds	Therapeutic Functions	References
Honey	Flavonoids, phenolic acids, hydrogen peroxide	Antimicrobial, antioxidant, wound healing	Hegazi (2012); Hossain et al. (2022); Bucekova (2019)
Propolis	Polyphenols, flavonoids, phenolic esters	Anti-inflammatory, antiviral, anticancer	Ali and Kunugi (2020); Rouibah et al. (2021); Weis et al. (2022)
Royal jelly	Proteins, vitamins, lipids, hydroxy lipids	Immunomodulatory, anti-aging, wound repair	Ali and Kunugi (2022); Susilowati et al. (2017); Lee et al. (2019)
Bee venom	Melittin, phospholipase A_2, apamin	Analgesic, anti-arthritic, anticancer	Wehbe et al. (2019); Bava et al. (2023)
Beeswax	Esters, hydrocarbons, fatty acids	Antibacterial, skincare, wound protection	Weis et al. (2022); Eswaraiah et al. (2020); Thangapazham et al. (2007)

14.4 APITHERAPY AND ENTREPRENEURSHIP

In recent years, there has been a rise in the need to promote and support sustainable practices in the beekeeping industry (Vapa-Tankosic et al., 2020) and beekeeping as a form of sustainable rural development (Prodanovic et al., 2024). The new niche of the tourism market has therefore been created—apitourism, which demonstrates the significance of bees and honey tourism, which is essential to human health (Sokhai & Mary, 2024), biodiversity protection (Parreno et al., 2022), and the economic prospects of rural regions (Izquierdo-Gascon & Rubio-Gil, 2023).

We believe in our thesis that, as it is done to other goods like wine or cheese, apitourism encourages the development of genuine experiences within the countryside. The idea of apitourism is the culture and traditions of the rural people, and it could be discussed as one of the most environmentally friendly means of development (Wos, 2014). One of the elements that make apitourism an economic and social practice is the antiqueness of beekeeping and variety of products of bees such as food and health products. According to the experience of the literature on culinary tourism (Hall & Sharples, 2003; Ellis et al., 2018; Dixit, 2021), along with the activities that help develop the knowledge about the work of beekeepers, honey formation, its success, and details, to learn more about other further products of bee association, to see a beehive, and to enjoy the ecological interconnection between people and bees, bee tourism presupposes the visits to bee production and consumption sites (Wos, 2014).

The importance of entrepreneurship in the field of rural tourism is informed by the fact that the study by Madanaguli et al. (2021) has shown that the local environment and conditions of living are the basis of the sustainable practice, even though the body of existing research on the issue at hand is extensive.

The Organization of Economic Co-operation and Development defines entrepreneurs as individuals (business owners) who strive to build and create new ideas in order to improve the economy (Cunha et al., 2020). It is also in the context of their economic achievements in which the future growth of the region depends heavily (Sopo et al., 2017). Entrepreneurship is significant to the food industry due to the need to address the recent challenges that have been observed, such as climate change (McElwee, 2006; Gellynck et al., 2015; Kwilecki et al., 2020). Furthermore, business owners are always willing to diversify the number of products and services (Alsos et al., 2011; Jafari-Sadegh et al., 2022).

This is clear in apitourism. Apitourism is a type of tourism, where individuals get to engage in beekeeping and honey products, which aids rural communities in maintaining their culture (Wos, 2014). This is made possible with apitourism, where guests are able to visit hives via bee-based trails, which are also becoming popular. Apitourism represents one of the keys to sustainable rural development according to Pantoja et al. (2017).

14.5 INNOVATIONS IN APITHERAPY

Apitherapy has experienced a whirlwind of development, due to the advancement of biotechnology, nanotechnology, and the invention of new business strategies. Methods of extraction have been greatly advanced. Such traditional methods as ethanol extraction have been developed or replaced with more advanced methods such as high-performance chromatography and supercritical fluid extraction. These are the contemporary means to obtain more uniform and purified forms of venom, royal jelly, and propolis by means of removing the undesirable impurities (Kurek-Górdecka et al., 2020; Luo et al., 2021).

These are significant steps into improving the feasibility of bioactive products and therefore pharmacological acceptance. An interesting trend in the development of the products is the use of the liposomal delivery system, nanoencapsulation, and nanoemulsions in the process of enhancing absorption, stability, and release control of materials of bee origin. An example is the case of hydrogel and nanofiber dressings, which can help in healing the wound without carrying the risk of

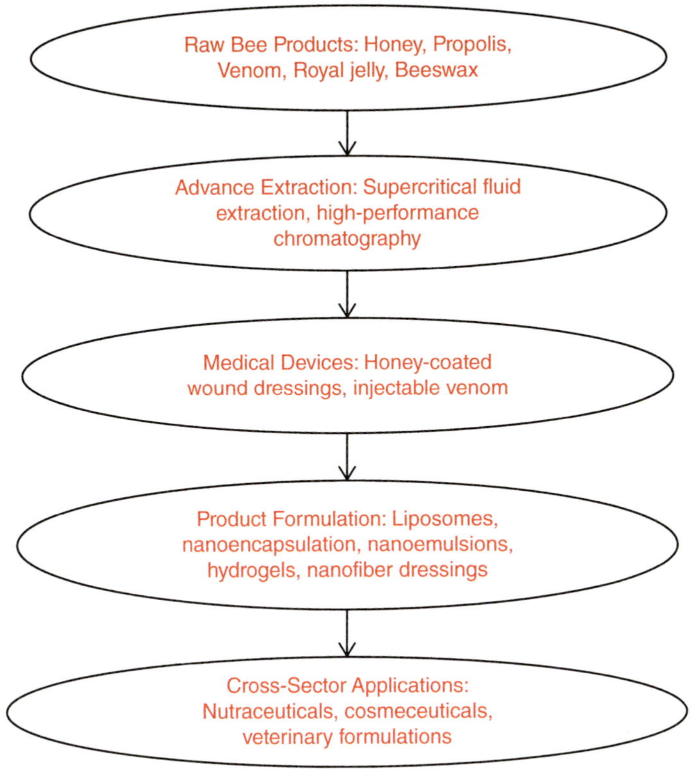

FIGURE 14.1 Innovative pathway in apitherapy.

infections; but in general, more and more areas of the world resort to the application of propolis and honey (Yan et al., 2017; Charles et al., 2020).

These inventions contribute to the efficiency of the bee products-based treatments and expand its usage to health care areas. Another drawback caused by these developments has to do with medical technology. At present, dressings containing honey, which is applied to treat burn, ulcer, and surgical wounds, can be bought. Bee venom developed is a safer form of conventional sting treatment of neurological conditions and arthritis under development where the active ingredient comes in the form of an injectable and controlled delivery system (Wehbe et al., 2019; Putz et al., 2021). These inventions, which combine ancient and new medicine, are making apitherapy treatment options more medically acceptable.

Innovation in other aspects has seen the diversification in apitherapy. The significance of royal jelly and beeswax lies in the skin care products that are used in the treatment of skin in order to enhance its beauty as well as prevent aging; however, propolis and honey are also being utilized as nutraceutical and functional foods. There is active development of products that contain propolis to treat infections of livestock, and this is how versatile the use of bee products can be in different industries, including veterinary medicine (Munstedt & Mannle, 2020). Figure 14.1 shows the innovative pathways in apitherapy.

14.6 APITHERAPY BIOTECHNOLOGICAL DEVELOPMENT

The German scientist Paul Ehrlich came up with the concept of chemotherapy, which he explained as the use of chemicals to fight cancer. The first use of chemotherapy came in the 1940s, when nitrogen mustard was used in the treatment of lymphoma (Altun & Sonkaya, 2018). The study, conducted in the 1950s on the effects of plant alkaloids, established that Vincaro sea alkaloids had been useful in the treatment of individuals with leukemia. The use of the compounds in cancer research has resulted in high survival and low mortality rates due to research on various processes. The last few years have witnessed the development of chemotherapy; some of the drugs are less toxic and react in a more specific manner. A lot has been done to cure cancer, although it is still the number one cause of death by the primary factors being breast and cervical cancer. Although there have been advancements on surgical as well as radiation procedures, chemotherapy has continued to play a very essential role in the management of cancer, especially in cases of early and advanced as well as metastatic cancer.

During chemotherapy of recurrent tumors, traditional cancer treatment drugs are applied. The greatest drawback of chemotherapy, though, is that it typically kills healthy cells along with the cancerous cells, resulting into some of the most excruciating side effects. The chemotherapy agents act via the involvement of the fast-dividing cancer cells, but they may also produce an adverse effect on the healthy tissues in other parts of the body, such as the digestive tract, bone marrow, and hair follicles (Altun & Sonkaya, 2018).

Adverse toxicity of chemotherapy can quite be dramatic and can influence the health and general quality of life of the patient, complicating the cancer treatment. Nevertheless, with the development of cancer treatments, several new techniques have emerged that can tackle the problem of chemotherapy resistance: immunotherapy, thermal therapy, gene therapy, phytotherapy, and apitherapy (Kumar et al., 2012; Munstedt & Mannle, 2020).

The use of drugs in fighting cancer is a vital aspect of the medical practice since it entails chemotherapy. The treatments are designed not to kill as many healthy cells as much as they are intended to primarily destroy cancer cells that are multiplying or growing out of control. A systematic treatment process is also used to manage the disease using chemotherapeutic or alkylating agents; however, it targets the different stages of the cell cycle (Ahmad et al., 2019).

14.6.1 Global Practices and Case Studies

There are several implementations of apitherapy all over the world, with certain applications based on legal provisions and cultural beliefs. It is because of its marketing that Manuka honey has become a multibillion-dollar and worldwide health product (Carter et al., 2016). Apitherapy in China is included in the medical center alongside traditional Chinese medicine, and venom and propolis are included in the pharmacopeia (Wehbe et al., 2019). The European Union, particularly Germany and Romania, passed laws in favor of research and medical use of propolis, products, and medicines of bee venom (Kurek-Górdecka et al., 2020). No matter how, the problem of standardization and quality control remains unsolved, South African, Kenyan, and Nigerian companies are utilizing honey to make cosmetic and nutraceutical products (Bankova et al., 2018).

14.7 CHALLENGES AND OPPORTUNITIES

14.7.1 Challenges

There are also challenges and discussions associated with apitherapy as much as with most other alternative treatments. According to Wehbe et al. (2019), the lack of guidelines and recommendations about the use of bee products in the apitherapy is also one of the obstacles to the problem; this method has numerous therapeutic options in different studies, and it is not easy to compare the results. Medical workers are not necessarily willing to use apitherapy (or even recommend it) because the required empirical evidence is not in place. Patient safety and education should be the priorities of guidelines on apitherapy to cope with this issue. Prior to patients taking any steps toward apitherapy, they are expected to be provided with all the information concerning the potential costs and benefits. Moreover, it is necessary to offer the patients specifications on the usage and storage of bee products. By providing good sources and concise regulations, the medical staff might minimize the risks of apitherapy and encourage the safe and effective usage of bee products (Kavurmaci & Tan, 2019).

14.7.2 Opportunities

Apitherapy projects offer potential in the spheres of biotechnology, business, and medical. To begin with, the increasing popularity of natural and alternative therapy resulted in the necessity to commercialize apitherapy products (Munstedt & Mannle, 2020). In addition, biotechnology gives a chance of personalized medication, and in the case of diseases such as cancer, arthritis, and neuro difficulty, bee-based compounds are also on the anvil (Duarte et al., 2022).

Third, through apitherapy, there is the growth of the global market in cosmetics and nutraceuticals that unlocks the potential of high-quality products, which combine health and beauty (Pasupuleti et al., 2017). In conclusion, e-commerce, online sessions, and integration of telehealth could be employed to promote marketing of apitherapy so as to make it more easeful and efficient globally. Digital entrepreneurship supports this fact (Izquierdo-Gascon & Rubio-Gil, 2023).

14.8 FUTURE PROSPECTS

Apitherapy should be developed through a mix of both old and new research. Development in biotechnology will enhance the state-of-the-art medical methods of identifying therapies on individuals through bee products. The ethics of responsibility and biotech will guarantee responsible production practices. Global alliances, e-commerce, and online marketing are likely to increase the market potential. Apitherapy development has had a positive result, but the study of the ability to cure cancer with the help of bee venom remains in progress. Researchers are investigating bee venom for use as an anticancer treatment due to its ability to delay tumor growth and cause cancerous cells to undergo programmed cell death (Duarte et al., 2022). Additionally, quality assurance of apitherapy could be offered by ensuring beekeeping and product quality. To sum up, apitherapy is an emerging field that can cure various medical diseases. In the case of the appropriate research, education, and regulation, apitherapy can become a helpful supplement to traditional medicine (Kavurmaci & Tan, 2019).

14.9 CONCLUSION

Apitherapy is a business that is a novel combination of biotechnological innovation, new business formulas, and health benefits. With the help of scientific discoveries, the entrepreneurs will be in a position to make bee-based products safe, reliable, and attractive to the global market. The ethical, legal, and ecological issues have to be discussed to promote the sustainable development. Last but not least, apitherapy is a curative treatment but also a source of biodiversity protection and financial developments.

REFERENCES

Ahmad, N., Khan, M.A., & Siddiqui, M.U. (2019). Chemotherapy in cancer treatment: Mechanisms and challenges. *Journal of Cancer Research and Therapeutics*, *15*(3), 457–463.

Alberti, S. (2017). Electrospun nanofiber dressings with propolis for wound healing applications. *Journal of Biomedical Materials Research*, *105*(4), 1236–1243.

Alfaris, N.A., Alshammari, G.M., & Al-Daghri, N.M. (2022). Propolis as an antimicrobial and antiviral natural product: Recent advances. *Frontiers in Pharmacology*, *13*, 842718.

Ali, F., & Kunugi, H. (2020). Propolis, bee honey, and their therapeutic properties: A review. *International Journal of Molecular Sciences*, *21*(12), 4569.

Ali, F., & Kunugi, H. (2022). Royal jelly as an immunomodulatory and anti-inflammatory agent. *Nutrients*, *14* (8), 1624.

Alsos, G.A., Ljunggren, E., & Pettersen, L.T. (2011). Entrepreneurship in the food sector: Innovation and value creation. *British Food Journal*, *113*(7), 1023–1037.

Altun, R., & Sonkaya, A. (2018). Historical development of chemotherapy and modern cancer treatment approaches. *Journal of Oncology*, *2018*, 245–252.

Artemov, N.M. (1941). Therapeutic use of bee venom in medical practice. *Soviet Medical Journal*, *8*(4), 115–121.

Bankova, V., Popova, M., & Trusheva, B. (2018). The phytochemistry of propolis: A review. *Phytochemistry*, *155*, 1–14.

Bava, S.V., Sujith, S., & Anoop, K.R. (2023). Bee venom and its potential role in cancer therapy: A comprehensive review. *Biomedicine & Pharmacotherapy*, *164*, 114842.

Bucekova, M. (2019). Therapeutic potential of honey in clinical medicine. *Frontiers in Pharmacology*, *10*, 1029.

Campos, M.G.R., Fratini, F., & Bonvehí, J.S. (2021). Apitherapy and its potential role in modern medicine. *Journal of ApiProduct and ApiMedical Science*, *13*(2), 45–58.

Carter, D.A., Blair, S.E., Cokcetin, N.N., Bouzo, D., Brooks, P., Schothauer, R., & Harry, E.J. (2016). Therapeutic manuka honey: No longer so alternative. *Frontiers in Microbiology*, *7*, 569.

Charles, S., Nogueira, F., & Silva, M. (2020). Nanoencapsulation of propolis for enhanced bioavailability. *Journal of Pharmaceutical Sciences*, *109*(5), 1458–1467.

Chlebowski, M.M., Pyzik, M., & Wysocka, A. (2002). Chemical composition of propolis and its biological activity. *Journal of Natural Products*, *65*(12), 1681–1685.

Cunha, J., Costa, E., & Ribeiro, C. (2020). Entrepreneurship and regional development: OECD perspectives. *OECD Economic Studies*, *2020*(1), 101–120.

Dixit, S. (2021). *The Routledge handbook of culinary tourism*. Routledge.

Donaldson, S. (2004). Chemical composition and medical uses of bee venom. *Toxicon*, *44*(7), 863–868.

Duarte, A.W., Silva, R.R., & Oliveira, M.L. (2022). Bee products in biotechnology and medicine: Current perspectives and challenges. *Biotechnology Advances*, *55*, 107905.

El-Didamony, H. (2024). Apitherapy: Modern applications of ancient medicine. *Journal of Natural Remedies*, *24*(1), 12–26.

Ellis, A., Park, E., Kim, S., & Yeoman, I. (2018). What is food tourism? *Tourism Management*, *68*, 250–263.

Eswaraiah, S., Rao, G., & Reddy, B. (2020). Composition and industrial applications of beeswax. *Journal of Cosmetic Science*, *71*(5), 403–412.

Faqihi, A.M., & Taha, I.M. (2022). Propolis and its role in oral health care: A systematic review. *BMC Oral Health*, *22*(1), 145.

Gellynck, X., Banterle, A., & Kühne, B. (2015). Food industry entrepreneurship and innovation. *Trends in Food Science & Technology*, *45*(1), 117–128.

Godfrey, K., Williams, D., & Lister, R. (2011). Propolis nanofiber dressings for improved wound healing. *Materials Science and Engineering*, *31*(4), 642–649.

Gokulakrishnaa, M., & Thirunavukkarasu, R. (2020). Bee products in disease prevention and treatment. *Journal of Complementary Medicine Research*, *11*(2), 45–57.

Hall, C. M., & Sharples, L. (2003). The consumption of experiences or the experience of consumption? An introduction to the tourism of taste. In C. M. Hall, L. Sharples, R. Mitchell, N. Macionis, & B. Cambourne (Eds.), *Food tourism around the world: Development, management and markets* (pp. 1–24). Butterworth-Heinemann.

Hegazi, A.G. (2012). Medical importance of bee products. *Asian Pacific Journal of Tropical Biomedicine*, *2*(5), 358–368.

Helen, M., & Jun, Z. (2008). Honey as a natural remedy: A review of its biological properties. *Journal of ApiMedical Science*, *2*(2), 1–12.

Hossain, K.S., Hossain, M.G., & Moni, A. (2022). Therapeutic uses of honey in modern medicine. *Evidence-Based Complementary and Alternative Medicine*, *2022*, 563489.

Hussein, S.Z., Yusoff, K.M., Makpol, S., & Yusof, Y.A.M. (2011). Honey in wound management: Clinical and experimental findings. *International Journal of Medical Sciences*, *8*(3), 213–220.

Izquierdo-Gascon, A., & Rubio-Gil, A. (2023). Apitourism and rural development: New opportunities for sustainability. *Journal of Rural Studies*, *98*, 102–112.

Jafari-Sadegh, M., Ranaei, H., & Li, F. (2022). Entrepreneurial diversification and sustainable food production. *Food Policy*, *107*, 102226.

Jayaprakasha, G.K., Rao, L. J., & Sakariah, K. K. (2007). Antioxidant activities of flavonoids in royal jelly. *Journal of Agricultural and Food Chemistry*, *55*(19), 7429–7434.

Jia, J., Zhu, Y., & Wang, Z. (2013). Beeswax-based nanostructured lipid carriers for biomedical applications. *Colloids and Surfaces B: Biointerfaces*, *109*, 65–71.

Kavurmaci, M., & Tan, A. (2019). Perception and use of apitherapy among health professionals. *Complementary Therapies in Medicine*, *43*, 23–28.

Keskin, D., Ozyurek, M., & Guclu, K. (2019). Encapsulation of propolis for improved bioavailability. *Food Hydrocolloids*, *87*, 192–199.

Kokot, M., Matysiak, J., & Breborowicz, J. (2009). Pharmacological properties of bee venom and its peptides. *Acta Pharmacologica Sinica*, *30*(10), 1440–1452.

Kroenke, J., Udeinya, I.J., & Mu, L. (2002). Cytotoxic activity of Chinese propolis on breast cancer cells. *Cancer Letters*, *185*(2), 173–181.

Kumar, A., Sharma, S., & Tripathi, V.D. (2012). Alternative cancer treatments: Physiotherapy and apitherapy. *Cancer Therapy Reviews*, *38*(7), 891–899.

Kumar, S., Singh, R., & Kumar, N. (2018). Challenges in the pharmaceutical industry and the role of natural products. *Journal of Pharmacy and Pharmacology*, *70*(1), 1–12.

Kurek-Górdecka, A., Rzepecka-Stojko, A., & Stojko, J. (2020). European regulation and research on apitherapy and bee products. *Journal of Ethnopharmacology*, *261*, 113–118.

Kwilecki, D., Nowak, A., & Szopinski, T. (2020). Climate change and entrepreneurship in agriculture. *Environmental Science & Policy*, *106*, 47–56.

Lee, J.H., Park, J.H., & Kim, S. (2012). Chemical characterization and applications of beeswax. *International Journal of Cosmetic Science*, *34*(5), 414–421.

Lee, M., Kim, H., & Lee, D. (2019). Royal jelly proteins and their biological effects. *Food Chemistry*, *276*, 372–381.

Luo, Q., Zhang, J., & Li, B. (2021). Advanced extraction techniques for bee products. *Journal of Food Engineering*, *293*, 110–121.

Madanaguli, T., Kaur, P., Bresciani, S., & Dhir, A. (2021). Entrepreneurship in rural hospitality and tourism: A systematic literature review of past achievements and future promises. *International Journal of Contemporary Hospitality Management*, *33*(8), 2521–2558. https://doi.org/10.1108/IJCHM-09-2020-1121

Malone, L.A., & Tsai, W. S. (2016). Honey and health: Evidence from research. *Food Chemistry*, *197*, 307–314.

Mama, M., Tesfaye, J., & Gemeda, N. (2019). Antimicrobial activity of honey against MRSA. *Journal of Clinical Microbiology*, *57*(6), e01372–18.

Martin, S. J. (2012). The role of *Varroa* and viral pathogens in honey bee colony losses. *Journal of Apicultural Research*, *51*(1), 125–132. https://doi.org/10.3896/IBRA.1.51.1.12

McElwee, G. (2006). Farmers as entrepreneurs: Developing competitive skills. *Journal of Developmental Entrepreneurship*, *11*(3), 187–206.

Munstedt, K., & Mannle, H. (2020). Apitherapy in modern medicine: Potential and limitations. *Evidence-Based Complementary and Alternative Medicine*, *2020*, 435678.

Nadaf, S., Halabi, M.F., & Sheikh, B.Y. (2018). Hydroxy fatty acids in royal jelly: Composition and health benefits. *Journal of Functional Foods*, *42*, 62–70.

Nitecka-Buchta, A., Nowak, J., & Baron, S. (2014). Historical perspectives of apitherapy. *Archives of Medical Science*, *10*(4), 806–810.

Noble, D., Green, H., & Carter, J. (2020). Cultural history of bee product usage in medicine. *Journal of Ethnobiology*, *40*(3), 259–276.

Pantoja, M.T., da Silva, F.L., & Oliveira, R.C. (2017). Sustainable apitourism for rural development. *Tourism Planning & Development*, *14*(4), 437–452.

Parreno, J.R., Gomez, L., & Fernandez, A. (2022). Beekeeping, biodiversity, and sustainable livelihoods. *Sustainability*, *14*(3), 1214.

Pasupuleti, V.R., Sammugam, L., Ramesh, N., & Gan, S.H. (2017). Honey, propolis, and royal jelly: A comprehensive review of their biological actions and health benefits. *Oxidative Medicine and Cellular Longevity*, *2017*, 1259510.

Popovic, M., Jovic, S., & Radulovic, Z. (2022). Apitherapy as an alternative health intervention. *Alternative Medicine Review*, *27*(3), 210–223.

Premratanachai, P., & Chanchao, C. (2014). Bee venom therapy: A review of clinical practices and mechanisms. *Asian Pacific Journal of Tropical Biomedicine*, *4*(2), 93–102.

Prodanovic, R., Boskovic, J., & Pavlovic, D. (2024). Beekeeping as a tool for sustainable rural development. *Journal of Rural Economics*, *12*(1), 88–97.

Putz, N., Gruber, C., & Fischer, T. (2021). Development of bee venom-based injectable therapies. *Biomedical Engineering Letters*, *11*(2), 181–192.

Rodriguez-Wallberg, K.A., & Oktay, K. (2012). Mechanisms of honey's antibacterial activity. *Food Microbiology*, *31*(1), 96–102.

Rouibah, M., Bensouilah, M., & Ozek, T. (2021). Synergistic anticancer effects of propolis and doxorubicin. *Scientific Reports*, *11*, 10892.

Rwigemera, A., Mbah, J., & Ojo, D. (2015). Physicochemical properties of royal jelly and its health effects. *Journal of Food Science and Technology*, *52*(7), 4517–4524.

Sahin, H., Turkez, H., & Aslan, A. (2019). Bioactive components of bee products and their medicinal properties. *Molecules*, *24*(8), 1556.

Sanghani, H., Kothari, V., & Patel, P. (2014). Historical development of apitherapy in ancient civilizations. *Journal of Traditional and Complementary Medicine*, *4*(2), 80–86.

Seki, T., Fujiwara, H., & Matsuura, K. (2008). Ancient and modern uses of beeswax in medicine. *Journal of Archaeological Science*, *35*(2), 234–240.

Şengü, F., & Vatansev, H. (2017). Biological and therapeutic properties of bee venom. *Journal of Applied Biological Sciences*, *11*(3), 75–83.

Sengupta, S., Deb, A., & Das, S. (2004). Pharmacological studies on bee venom and its components. *Indian Journal of Experimental Biology*, *42*(5), 486–491.

Shrivastava, S., Jain, V., & Kumar, D. (2017). Apitherapy in Ayurvedic medicine: A review. *Indian Journal of Traditional Knowledge*, *16*(4), 633–639.

Sokhai, A., & Mary, P. (2024). Apitourism: The rise of bee-based tourism and wellness. *Journal of Tourism and Cultural Change*, *22*(1), 1–18.

Sopo, D., Ricci, F., & Marino, G. (2017). Entrepreneurship and regional economic growth. *Regional Studies*, *51*(6), 845–857.

Susilowati, E., Saputri, F.C., & Mun'im, A. (2017). Antibacterial and lipid-lowering effects of royal jelly. *Food & Function*, *8*(7), 2328–2336.

Tamura, T., Fujii, A., & Kuboyama, N. (2015). Medicinal properties of honeybee products in modern pharmacology. *Journal of ApiResearch*, *54*(1), 15–25.

Thangapazham, R.L., Sharma, A., & Maheshwari, R.K. (2007). Cosmetic applications of beeswax and natural products. *International Journal of Cosmetic Science*, *29*(3), 169–176.

Vapa-Tankosic, J., Ignjatijevic, S., & Prodanovic, R. (2020). Sustainability in beekeeping and apiculture entrepreneurship. *Sustainability*, *12*(5), 1843.

Wehbe, R., Frangieh, J., Rima, M., El-Obeid, D., Sabatier, J.M., & Fajloun, Z. (2019). Bee venom: Overview of composition and therapeutic applications. *Frontiers in Pharmacology*, *10*, 338.

Wehbe, R., Wehbe, M., & Nasser, S.A. (2014). Topical bee venom therapy: Mechanisms and safety considerations. *Dermatologic Therapy*, *27*(3), 165–172.

Weis, M., Kunc, P., & Kvasnicova, M. (2022). Bioactive properties and therapeutic applications of bee products. *Molecules*, *27*(2), 515.

Wos, B. (2014). Apitourism as a sustainable form of rural tourism. *Journal of Environmental Management and Tourism*, *5*(2), 65–73.

Yan, S., Wang, Q., & Li, J. (2017). Electrospun propolis nanofibers for wound dressings. *Materials Letters*, *190*, 249–252.

Young, C., & Bundell, J. (2023). Apitherapy and immune modulation: A review of current evidence. *Journal of Alternative Medicine Research*, *15*(2), 99–112.

Zheng, M., Liu, Y., & Wang, Y. (2019). Anti-inflammatory and anticancer effects of bee venom and its components. *Toxins*, *11*(6), 356.

Zhu, Y., & Wongsiri, S. (2008). Royal jelly in cardiovascular and metabolic health. *Asia Pacific Journal of Clinical Nutrition*, *17*(2), 422–426.

15 Safety and Adverse Effects of Apitherapy

Jamiu Sulaiman Jamiu, Sulaiman Sime Abdullahi, Oluwadamilola Eunice Ajibola, Tomilola Morenikeji Olaseinde, Bisade Esther Adeyeluwa, and Ryu Ki-Jin

15.1 INTRODUCTION

Apitherapy is a type of natural therapy that uses bee products like propolis, honey, bees' pollen, bee venom, royal jelly, and beeswax for healing (Alqarni et al., 2023). The use of bee products for healing goes back to ancient times. Today, honey is seen as a health-promoting food. Research has shown its germ-fighting, inflammation-reducing, and antioxidant effects, studying its role in wound healing (Oryan et al., 2016; Bankova et al., 2019), liver and kidney protection from harmful substances (e.g., CCl_4) (El-Haskoury et al., 2018), and its immune-boosting effects (Ota et al., 2019).

Propolis is a sticky natural material collected by bees, which also contains bee saliva and enzymes. It is utilized for safeguarding against invading insects and microbes, as well as in beehive repair. Findings reveal that propolis shows antibacterial, antifungal, anti-inflammatory, and cell-toxic activities (Rufatto et al., 2017). Bee pollen is formed when flower pollen clumps with nectar, mixed with bee saliva. It is a rich source of active nutrients and energy, and it has many healing and protective benefits: germ-fighting, antifungal, antioxidant, inflammation-reducing, anti-allergy, liver-protecting, and cancer-preventing, along with benefits for the heart, digestion, immunity, and slowing aging (Li et al., 2018). Royal jelly is a substance produced by honeybees that is used to nourish both larvae and adult bee queens. Findings have demonstrated redox regulator, immunomodulatory agent, hepatoprotective, and antitumor properties of royal jelly (Pasupuleti et al., 2017). Notably, royal jelly displays antibacterial effects strength targeting Gram-positive bacteria and provides a protective effect against the adverse influence of nicotine on cholesterol levels (Susilowati et al., 2017).

The limited availability of robust scientific evidence can instill hesitation among healthcare professionals when it comes to recommending or endorsing apitherapy. To address this, emphasis must be placed on patient education and safety within the practice of apitherapy. Through the dissemination of accurate information and clear guidance, healthcare professionals can mitigate risks associated with apitherapy, ensuring that bee products are utilized safely and effectively (Kavurmaci & Tan, 2019).

Significant scientific records to support the use of apitherapy as a prophylactic and treatment for numerous health disorders are available. One key aspect that is important to be monitored is the safeguarding and adverse effects regarding these products. Standard ways should be instituted for all functional foods along with nutritional supplements.

15.2 SAFETY PROFILE OF APITHERAPEUTIC AGENTS

15.2.1 Safety Profile

Apitherapy, a practice of complementary and alternative medicine involving propolis, honey, bee pollen, royal jelly, and beeswax, as well as bee venom for clinical use (Jităreanu et al., 2022). Honeybees (*Apis mellifera*) provide a range of indispensable products that harbor bioactive

DOI: 10.1201/9781003685654-15

constituents to treat various diseases (Cornara et al., 2017). Bee products include propolis, honey, wax, pollen, bee venom, and royal jelly (Waykar & Alqadhi, 2016). Bee pollen comprises close to a hundred vitamins, enzymes, minerals, and amino acids, alongside water, protein, carbohydrates, fatty acids, and antioxidants. Apiary wax is the primary component in topical preparations such as ointments and creams for the treatment for fractures, joint pain, wounds, bruises, and burns (Alshallash et al., 2023).

15.2.2 Apitherapeutic Agents

Bees chemically synthesize royal jelly, bee venom, and beeswax, while bee bread, honey, propolis, and bee pollen are formed through the conversion of plant-derived material by bees (El-Didamony et al., 2024).

15.2.2.1 Honey

Honey is a natural sweet liquid formed by honeybees from the floral nectar or plant secretions, which they collect and transform enzymatically (Stefas et al., 2022). Honey is deposited in comb, dehydrated, and stored until ripening is complete. Its primary constituent includes sugars such as glucose, maltose, fructose, sucrose, threalose, maltulose, melezitose, and maltotriose, as well as H_2O and enzymes like invertase, α- and β-amylase, glucose oxidase, and catalase (El-Didamony et al., 2024). A minor amount of amino acids, proteins, some vitamins (primarily B and C), minerals (sodium, potassium, calcium, magnisium, iron, copper, manganese, and zinc), pollen grains, and other phytochemicals are also present. The presence and activities of these enzymes provide detailed information on the natural state and quality of honey because their activity declines in aged or heat-exposed honey. Honey has antibacterial, antioxidant, anti-inflammatory, and anticancer features influenced by polyphenolic substances (flavonoids and phenolic acids), vitamins C and E, and enzymes (peroxidase, catalase) within it (Kulczyńska et al., 2023).

15.2.2.2 Propolis

Propolis, primarily referred to as bee glue, it is a sticky and thick substance. It appears as a dark-yellow or brownish material that bees harvest from plant buds and shoots to safeguard their hives (Saccardi et al., 2021). Bees utilize propolis to repair the inside walls of the hive, honeycombs, and seal holes and prevent the breakdown caused by intruders inside the hive (Borba et al., 2017). In raw propolis specimens, around 300 distinct compounds have been identified, including polyphenol compounds, flavonoids including flavones, flavanones, and stilbenes, as well as chalcones, steroids, and lignans. Bee glue has been found to exhibit various biological actions, such as antioxidant, anti-inflammatory, and antimicrobial effects against human breast cancer cells (Olegário et al., 2019).

15.2.2.3 Bee Pollen

Bee pollen constitutes a range of bioactive agents, such as fatty acids, proteins, carbohydrates, vitamins, polyphenols, phytosterols, carotenoid pigments, enzymes, and co-enzymes (Thakur & Nanda, 2020). It is also rich in bioactive molecules that have a range of antimicrobial, antioxidant, antiallergic, anticancer, anti-atherosclerotic, chemopreventive, liver-protective, and immunoregulatory effects (Milovanovic & Radovanović, 2021).

15.2.2.4 Bee Venom

Bee venom consists of peptides, such as including melittin, adolapin, and apamin) as well as enzymes phospholipase A_2 and hyaluronidase, and a mast cell degranulating peptide, in addition to non-peptide molecules, parallel to norepinephrine, dopamine, and histamine. Melittin comprises about 50% of the dry venom, acts as the primary makeup, followed by phospholipase A_2, which makes up around 12% (Wehbe et al., 2019). It was discovered that apamine, a BV neuropeptide, acts as a selective inhibitor to the SK family of calcium-activated potassium channels, indicating therapeutic benefits and a novel strategy approch to alleviate attention deficit in individuals with Alzheimer's disease (Park et al., 2020). Bee venom phospholipase 2 demonstrated neuroprotective effects, indicating its potential for treating Parkinson's disease and other neuroinflammation-related disorders (Kim et al., 2019).

15.2.2.5 Royal Jelly

Royal jelly (RJ) appears white. It is a viscous substance with a chemical composition comprising approximately 60% water, 20–40% percent protein, 15–30% carbohydrates, 3–8% lipids, and 1.5–3% minerals and vitamins (Pasupuleti et al., 2017). RJ contains numerous bioactive constituents, comprising 10-hydroxy-trans-2-decenoic acid, which exhibits immunomodulatory action; and it has royalactin, which is a functional component of RJ (Yu et al., 2023). Royal jelly has been proven effective in reducing oxidative stress and promoting neuronal proliferation. Recent research revealed that royal jelly helps counteract neuroinflammation, oxidative stress, and mitochondrial dysfunction, thereby mitigating neuronal injury and enhancing motor function function in Parkinson's disease (Ali & Kunugi, 2020). Additionally, it was found that royal jelly improved insulin resistance, hyperinsulinemia, and hypertension associated with insulin resistance in rats, through indirect dysfunction in vascular mediated by adrenergic and CGRPergic nerves in the hyperinsulinemia state (Weis et al., 2022).

15.3 CLINICAL CASE REPORTS AND EPIDEMIOLOGY

Bee venom therapy has been utilized in treating ailments like arthritis, neurological disorders and pain; however, its therapeutic potential is limited by the high risk of adverse events or allergic reactions that may occur even in individuals without known hypersensitivity (Vahidinia et al., 2024). Evidence indicates that the use of bee venom in bee therapy/apitherapy has an adverse effect, primarily local skin reactions such as redness, swelling, pain, and itching. Again, cases of life-threatening systemic reactions such as anaphylaxis, hypotension, and difficulty breathing have been documented (Yoo & Lee, 2022). Figure 15.1 shows the integration of clinical reports in epidemiological reports. Bee venom contains allergenic proteins like phospholipase A_2 and melittin, which can trigger severe immune responses. Interestingly, anaphylaxis may occur even in individuals without known bee sting allergies. Also, the allergenic components of bee products rely depending on their concentration and frequency of exposure, independent of genetic predisposition. Shafaghat and Zwiener (2018) reported that bee venom causes allergic reactions. Bee bite allergy is linked to anaphylaxis and IgE-mediated allergic responses. Honey allergy is rare, with estimates of the actual incidence of allergy to honey within the general population suggested to be <0.001% (Aguair et al., 2017). However, a clinical study by Di Costanzo et al. (2021) revealed that honey allergy has great impact even in childhood. The clinical symptoms are typically mild—for example, oral allergy syndrome.

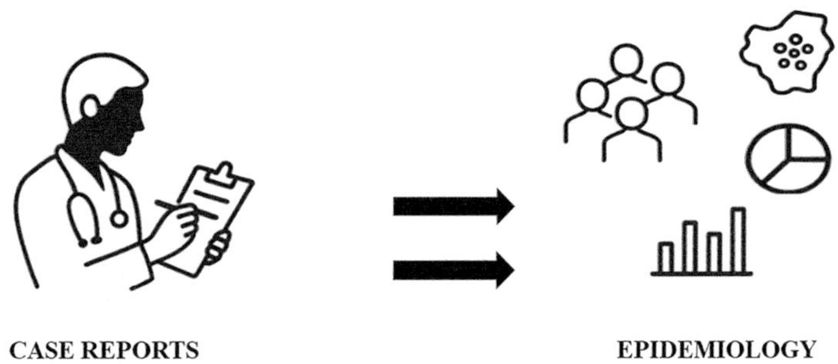

FIGURE 15.1 Integration of clinical case reports in epidemiological research.

TABLE 15.1
Common Adverse Effects, Severe Reactions, Pathophysiology of Anaphylaxis Induced by Apitherapeutic Products, and Risk Mitigation Strategies in Apitherapy Administration

Bee Product	Common Adverse Effects	Severe Reactions	Risk Mitigation Strategies
Honey	Gastrointestinal upset, allergy	Botulism in infants	Avoid in infants; use pasteurized honey
Bee venom	Local pain, swelling, erythema	Anaphylaxis	Allergy testing, emergency kit readiness
Propolis	Contact dermatitis, stomatitis	Systemic hypersensitivity	Patch testing, dose control
Propolis	Skin rash, asthma	Anaphylaxis	Pre-screening for asthma/allergy
Propolis	Photosensitivity, urticaria	Acute renal failure	Allergy testing, moderate dosing

The pathophysiology in anaphylaxis induced by different bee venom is shown in Table 15.1. Also, it has been revealed that increased consumption of honey in health food may also elevate the risk of allergenic reaction. Furthermore, it has been shown that propolis allergic reaction can appear as contact cheilitis, stomatitis, labial oedema, oral pain, eczema, and dyspnea (Oršolić, 2022).

15.4 DOCUMENTED MECHANISMS OF THE ADVERSE EFFECTS OF APITHERAPY

While apitherapy is advertised for many therapeutic uses, its safety concerns stem from a range of adverse effects. The common adverse effects of apitherapeutic products are shown in Table 15.1, and Figure 15.2 pictures the different adverse effects of apitherapy. Reported outcomes vary from mild local reactions to serious systemic toxicity, most of which can be linked to specific active components of honeybee venom (HBV) and their molecular targets. (Park et al., 2015; Stela et al., 2024).

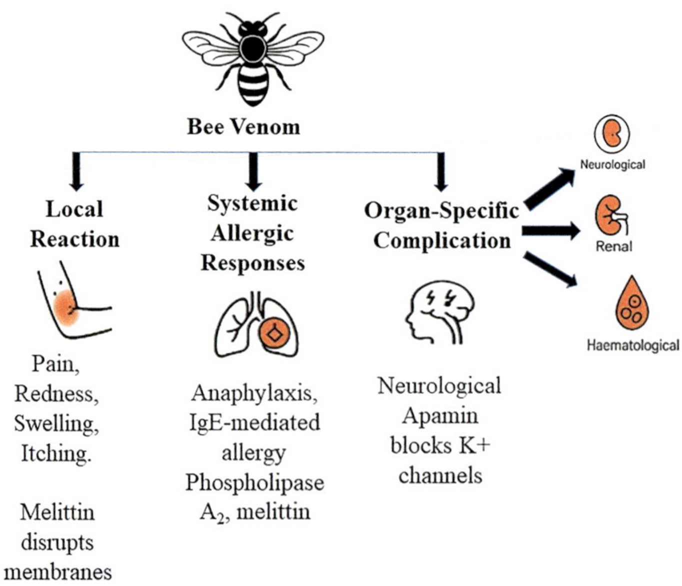

FIGURE 15.2 Adverse effects of apitherapy.

15.4.1 Local Reactions

Localized pain, redness, swelling, hives, and itching are the most common side effects (Stela et al., 2024; Bistué-Rovira et al., 2025). These reactions are caused by vasoactive amines like histamine, dopamine, and norepinephrine. These substances increase blood vessel permeability and activate pain-sensing nerve cells. Additionally, peptides such as melittin (Api m 4) damage cell membranes by forming pores, which adds to cell death and inflammation at the injection site (Yu et al., 2023).

15.4.2 Systemic Allergic Responses

Severe allergic reactions throughout the body, including shortness of breath, low blood pressure, and anaphylaxis, have been commonly reported (Cui et al., 2021). At the molecular level, these reactions are driven by IgE responses to specific allergens in bee venom, primarily phospholipase A_2, hyaluronidase, and acid phosphatase (Burzyńska & Piasecka-Kwiatkowska, 2021).

The allergenic profile of HBV is very complex. At least 12 compounds capable of inducing allergic reactions have been identified in *Apis mellifera* by the WHO/IUIS Allergen Nomenclature Sub-Committee. Out of these, 11 are proteins derived from venom, and two are from royal jelly products. HBV includes over 113 identified proteins and peptides, with functional differences further increased by glycosylation variants (Burzyńska & Piasecka-Kwiatkowska, 2021).

The main allergens linked to hypersensitivity are:

- Phospholipase A_2 (Api m 1): a strong cytotoxic enzyme and the main IgE-binding allergen.
- Hyaluronidase (Api m 2): increases tissue permeability and helps spread allergens.
- Acid phosphatase (Api m 3): adds to pro-inflammatory responses.
- Melittin (Api m 4): a peptide that can cause hemolysis and has potential cardiotoxic effects.

Other venom components, such as vasoactive amines (histamine, dopamine, catecholamines), contribute to blood vessel leakage, low blood pressure, and local cell damage, especially after multiple exposures (Burzyńska & Piasecka-Kwiatkowska, 2021).

15.4.3 Organ-Specific Complications

Neurological Complications: Neurological side effects such as peripheral neuropathy have been reported following bee venom acupuncture (Mathis et al., 2022). These effects are partly mediated by apamin, a neurotoxic peptide in HBV, which blocks the small-conductance Calcium-activated potassium channels in neurons, causing sustained excitability as well as potential neurotoxicity (Kuzmenkov et al., 2022). Experimental studies further show that repeated subcutaneous administration of melittin induces depression-like behavior in mice in a dose-dependent manner. Mechanistically, melittin impairs mitochondrial oxidative phosphorylation by inhibiting complex I activity, leading to oxidative stress injury in hippocampal neurons. This mitochondrial dysfunction is compounded by impaired synaptic plasticity via disruption of the BDNF/TrkB/CREB signaling pathway, highlighting a direct link between melittin exposure, hippocampal damage, and neurobehavioral deficits (Sun et al., 2023). Together, these findings suggest that bee venom exerts its neurotoxic effects through both ion channel dysregulation and mitochondrial/synaptic injury, ultimately contributing to serious neurological adverse outcomes.

Renal Toxicity: Acute kidney injury (AKI) linked to bee venom exposure is due to rhabdomyolysis and hemolysis (Rauf et al., 2021). This matc$_2$hes the lytic properties of melittin and phospholipase A_2, which destabilize red blood cell and skeletal muscle membranes. This process releases myoglobin and hemoglobin (Silva et al., 2017). Both substances are toxic to the kidneys and contribute to tubular blockage and oxidative stress.

Hematological Effects: Bee venom components can also disrupt hematological balance. Clinically, this manifests as hemolysis, thrombocytopenia, and bleeding disorders (Abdulsalam et al., 2016).

Mechanistically, melittin intercalates into red blood cell membranes, forming pores that lead to osmotic lysis, while phospholipase A_2 hydrolyzes membrane phospholipids, amplifying erythrocyte destruction and releasing pro-inflammatory lipids (Stela et al., 2024). In platelets, phospholipase A_2 alters phospholipid composition and impairs normal clotting cascades, contributing to coagulopathy. In addition, venom proteases such as dipeptidyl peptidase IV and Api m 5 can cleave coagulation factors, directly weakening hemostasis and predisposing to bleeding complications (Zolfagharian et al., 2015).

15.4.4 Bee Venom Components Involved in Adverse Effects

Bee venom is a compound secretion made mostly of enzymes, peptides, and biogenic amines. Several of these are directly linked to toxic and allergic effects (Burzyńska & Piasecka-Kwiatkowska, 2021). While some of these molecules have therapeutic potential under controlled conditions, high concentrations can lead to adverse effects. In this section, we review the major components, focusing on their molecular properties, known adverse effects, and the pathways involved.

15.4.4.1 Melittin

Melittin constitutes the largest part of bee venom, making up about 40–60% of its dry weight. This peptide remains one of the most biologically active peptides present. This 26-amino acid peptide can easily insert into phospholipid bilayers and clump together to create pores that cross cell membranes. This process leads to an imbalance of ions, hemolysis, and eventually cell lysis (Rady et al., 2017). Besides its direct ability to destroy cells, melittin works well with phospholipase A_2. The creation of pores increases the availability of the substrate, allowing for the breakdown of phospholipids by enzymes. In addition to damaging membranes, melittin activates signaling pathways inside cells, including nuclear factor kappa-light-chain-enhancer of activated B cells (NF-κB) and mitogen-activated protein kinase (MAPK) (Cui et al., 2024). This process encourages the activation of substances inducing inflammation like tumor necrosis factor-alpha (TNF-α), interleukin-1 beta (IL-1β), cyclooxygenase-2 (COX-2), and inducible nitric oxide synthase (iNOS). Furthermore, melittin can trigger cell death through the activation of caspases and changes in Bcl-2 family proteins. However, uncontrolled exposure may instead cause necrotic cell death (Choi et al., 2016). Reports of DNA breaks and chromosomal changes at high concentrations highlight its potential to damage genetic material (Chauhan et al., 2016).

15.4.4.2 Phospholipase A_2

Phospholipase A_2 (PLA_2) makes up about 10–12% of bee venom and is another key factor in its harmful effects (Burzyńska & Piasecka-Kwiatkowska, 2021). PLA_2 is a 16 kDa enzyme that is held together by several disulfide bridges. Its catalytic activity focuses on the 2-acyl ester bonds found in membrane phospholipids. When these bonds are broken, there is a release of arachidonic acid and lysophospholipids, which are substrates for strong inflammatory substances such as prostaglandins and leukotrienes (Muntean & Florea, 2025). This enzymatic activity not only adds to inflammation but also causes direct cell damage, including hemolysis and the breakdown of platelets (Darwish et al., 2021).

Moreover, PLA_2 is the most important allergen contained in bee venom (Burzyńska & Piasecka-Kwiatkowska, 2021). Immunoglobulin E (IgE) antibodies exclusive to PLA_2 are present in almost all individuals allergic to venom. This makes it the leading cause of systemic hypersensitivity. When IgE binds to basophils and mast cells, it results in a rapid release of histamine, increased vascular permeability, and can potentially trigger life-threatening anaphylaxis (Darwish et al., 2021).

15.4.4.3 Apamin

Apamin is a small but potent component of bee venom. This 18-amino acid peptide blocks small-conductance Ca^{2+}-activated K^+ (SK) channels (Kuzmenkov et al., 2022). These channels normally help regulate how neurons fire by controlling after-hyperpolarization. When these channels are inhibited, it disrupts neuronal firing patterns. This leads to increased excitability and hyper-responsiveness (Chen et al., 2019). On a molecular level, apamin changes dendritic structure and hinders normal synaptic integration. This explains its use in research as a neuropharmacological tool and its potential to cause excitotoxicity. Although some research explores its potential to improve cognitive functions, its narrow therapeutic range and strong neurotoxic effects limit its practical use (Kuzmenov et al., 2022). Peptide, mast cell degranulating (MCD), hyaluronidase, and biogenic amines are also bee venom components involved in adverse effects (Carpena et al., 2020).

15.5 RISK FACTORS AND VULNERABLE POPULATIONS

Although apitherapy has gained interest as a complementary treatment for various conditions, its use in clinical settings raises safety concerns. The effects of bee venom and other bee products differ among individuals. Susceptibility to adverse reactions depends on a person's immune history, other health issues, and overall health status. In this context, certain groups are especially at risk.

15.5.1 Sensitized or Allergic Individuals

People who have been sensitized to bee venom or have a history of atopy are the most at risk. For these individuals, exposure to bee venom triggers an IgE-mediated response (Carli et al., 2025). Antibodies attached to mast cells and basophils link together, causing degranulation and releasing histamine, leukotrienes, and other inflammatory substances. This means that even small doses can lead to serious reactions, such as hives, trouble breathing, or, in severe cases, heart failure (Dispenza et al., 2023). Patients with atopic conditions like asthma, allergic rhinitis, or eczema usually have a Th2-skewed immune profile, making them more likely to experience intense hypersensitivity reactions (Carli et al., 2025).

15.5.2 Children

Children are a high-risk group because their immune responses are not fully developed and their overall physiological reserves are limited. Evidence from pediatric dentistry suggests that children may be more vulnerable to hypersensitivity reactions from bee products, although clinical data is still scarce (Mori et al., 2024). Data on pediatric cases is limited; however, reports of lethal effects to bee bites in children show that therapeutic exposure to venom carries a significant risk in this age group. Cancer patients, the elderly population, and autoimmune and immunocompromised populations are also affected by bee venom (El Boussaadni & Oulmaati, 2022).

15.6 REGULATORY AND ETHICAL CONSIDERATIONS

15.6.1 Regulatory Considerations

Regulatory compliance in apitheraphy sets rules, policies, and legal requirements enforced by government or health agencies to protect patients, ensure their safety, and promote proper apitheraphy practices. Regulations guiding apitherapy are often ill-defined, making it difficult to be accepted and used in normal healthcare systems. Across different regions, apitherapy practices and products are not classified under existing regulatory rules designed for pharmaceutical drugs or medical devices. Hence, apitherapy products like honey typically don't get the same rigorous testing for safety, effectiveness, and quality that other modern medicines do. The American Apitherapy Society, for example, says that apitherapy is mostly unregulated and is typically subject to the same rules as dietary and nutritional supplements (AAS, 2024).

Bee products (honey, propolis, bee pollen, royal jelly) are classified as nutritional supplements in the United States. Hence, they do not require pre-market approval from the Food and Drug

Administration (U.S. Food and Drug Administration, 2024). In the European Union (EU), apitherapy products are treated as food supplements. This means they have to meet safety standards and carry proper labels, and they cannot make health claims unless there is solid proof to back them up (Dwyer et al., 2015). In Nigeria, however, the National Agency for Food and Drug Administration and Control (NAFDAC) is responsible for regulating herbal and natural products. Yet there are still no specific guidelines for apitherapy. Because of this, many apitherapy products fall outside the scope of regulatory oversight, raising concerns about safety, consistency, and proper labeling (Eruaga et al., 2024).

The International Federation of Apitherapy (IFA, 2022) has been working to establish legal policies and certification systems for trained practitioners, but these are not yet widely accepted or legally enforced. In some regions, regulations are slowly emerging. For instance, in Türkiye, apitherapy is formally recognized within the framework of traditional and complementary medicine (TCM). However, awareness of these regulations among healthcare professionals remains low, highlighting the need for greater dissemination and stronger enforcement (Gökkaya et al., 2025). This includes developing clear rules for product manufacturing, labeling, and advertising, as well as defining the scope of practice for apitherapists and introducing proper training and certification programs.

15.6.2 Ethical Consideration

Apart from the challenges linked to regulations, apitherapy also raises important ethical concerns. These mainly relate to bee welfare, its environmental sustainability, and, importantly, patient safety. Since apitherapy depends wholly on bees and their products, this means we have to think carefully about how our actions affect these incredible insects and their homes. A major ethical issue centers on the well-being of the bees. For example, bee venom is typically harvested by stimulating bees to sting a specially designed surface, from which the venom is then collected. This makes us wonder if we're causing them stress or harm. While some believe the bees aren't seriously hurt and can go on with their lives, others worry about the distress we might be causing them just for our own benefit. According to Omar and Elsaiegh (2023), harvesting bee venom can influence colony biological characteristics. Recent experimental research shows that electric-stimulus bee venom harvesting significantly alters colony behavior, enhancing hygienic actions by 22% in one study and increasing it markedly in another, while reducing hoarding behavior in others (El-Saeady et al., 2016). These behavioral disruptions suggest real welfare and ecological concerns, underscoring the need for ethical consideration when conducting venom extraction. Hence, sustainable and thoughtful beekeeping is essential. Ethical apitheraphy should support beekeeping methods that are good for the environment, protect the diversity of bees, and ensure their population stays healthy. This implies harvesting bee products responsibly, not taking too much, and supporting beekeepers that treat their bees well (MDPI, 2022). The idea of "spiritual beekeeping" encourages a deeper, more respectful connection between humans and bees, valuing them for more than just what they can produce (Eshelman, 2025).

15.7 RISK MITIGATION AND BEST PRACTICES

15.7.1 Risk Mitigation

The most significant and immediate risk in apitheraphy using bee venom is the potential for a severe allergic reaction. The immune system in its effort to protect the body can sometimes overreact to foreign substances, and bee venom is a potent allergen. Before the therapeutic application of bee venom, a systematic assessment (allergy testing) is not just a recommendation; it is an ethical and clinical procedure.

Initial Consultation: The procedure begins with a detailed patient history. A practitioner must thoroughly inquire about any previous reactions to bee or other insect stings, a personal or family history of allergy (such as hay fever or asthma), and any existing autoimmune conditions. This step is crucial for identifying potential risks before any physical testing is performed (Rosman et al., 2019).

1. *The skin prick test*: This step is crucial for allergy screening and should be administered by a qualified allergist or a practitioner with specific training in allergy testing. It involves a controlled amount of bee venom (applying for a minute) to the skin and observing for a reaction, providing a clear indication of hypersensitivity (Stoevesandt & Trautmann, 2022).
2. *Graded dosing*: For individuals with no prior history but who may be at risk, a practitioner might employ a graded challenge or test dose. This involves administering an extremely low dose of venom at the start of the first session and monitoring the patient closely for at least 30–60 minutes for any reaction before proceeding. This is the "start low, go slow" approach to confirm tolerance in a controlled setting (Park et al., 2015).

The risk mitigation strategies in apitherapy administration are shown in Table 15.1. Research has also emphasized how important it is for any studies on apitherapy to be approved by ethics committees and have official permission.

15.7.2 Best Practices

The therapeutic benefits of bee products depend greatly on both their purity and the way they are administered. Ensuring integrity "from hive to bottle" is the first and most essential step. Because bee products are natural, they are vulnerable to environmental contamination, which makes sourcing from trusted apiaries absolutely critical for reducing the risk of harmful residues such as pesticides or heavy metals (Kieliszek et al., 2018). To safeguard quality, responsible suppliers should also provide a Certificate of Analysis (COA), offering third-party confirmation that the product has been tested for both purity and potency. More recent reviews emphasize the advancement of tailored apitherapy delivery systems, such as nanofibers and scaffolds that improve therapeutic control and minimise risks (El-Didamony et al., 2024). These innovations effectively support the apitherapy principle of personalized, cautious dosing, akin to "start low and go slow," in the absence of a one-size-fits-all protocol. By combining careful attention to product quality with individualized dosing, apitherapy moves beyond simply using bee products. It becomes a practice that respects both the hive and the human body, transforming treatment into a responsible, safe, and truly healing art.

Expertise in practitioner-led apitherapy is essential for safe and trusted practice. Ethnographic research highlights how experienced beekeepers in their roles as knowledge guardians ensure safety, cultural authenticity, and patient trust in apitherapy (Kalle et al., 2025). Hence, when seeking a practitioner, one should look for evidence of formal training, certification from a reputable apitherapy organization, and a strong background in the healthcare field. Moreover, it is the practitioner's obligation to inform the patient with comprehensive and unbiased information about the procedure.

15.8 CONCLUSION

Traditional knowledge emphasizes the therapeutic use of bee products, and modern studies confirm their diverse pharmacological activities, including anti-inflammatory, antimicrobial, antioxidant, and neuroprotective effects. However, differences in flora, geography, and climate influence their chemical composition, creating variability in therapeutic outcomes and emphasizing the urgent need for standardization in harvesting, preparation, and formulation. Experimental models (laboratory studies and animal studies) have uncovered significant biological mechanisms, but clinical trials remain the definitive way to establish efficacy and safety.

Evidence indicates that bee products such as bee venom, bee pollen, propolis, royal jelly, and honey can enhance health outcomes, with propolis, honey, and venom being most extensively examined. Notably, combining bee products with conventional drugs has shown synergistic effects, improving efficacy and reducing side effects (Oliveira et al., 2016, 2019; Bernardino et al., 2018).

Therapeutic applications are expanding: propolis and royal jelly derivatives show promise for reducing neuroinflammation, oxidative stress, and mitochondrial dysfunction in neurodegenerative conditions such as Parkinson's disease; bee venom acupuncture (BVA) and sting acupuncture (BSA) have potential in musculoskeletal disorders, arthritis, cancers, bacterial infections, and kidney disease (Zhang & Chen, 2017; Jin et al., 2018; Moga et al., 2018).

Novel formulations such as allergen-reduced "essential bee venom," melittin delivery, and propolis inhalation further suggest safer, innovative approaches. Nevertheless, adverse effects from mild to severe remain a concern, particularly with venom therapy, necessitating strict quality control, practitioner training, adverse event monitoring, and adherence to international reporting standards. Advancing apitherapy will require global standardization, rigorous clinical trials, and multidisciplinary collaboration to bridge traditional practices with evidence-based medicine.

REFERENCES

Abdulsalam, M. A., Ebrahim, B. E., & Abdulsalam, A. J. (2016). Immune thrombocytopenia after bee venom therapy: a case report. *BMC Complementary and Alternative Medicine*, *16*, 107.

Aguair, R., Duarte, F. C., Mendes, A., Bartolomé, B., & Barbosa, M. P. (2017). Anaphylaxis caused by honey: a case report. *Asia Pacific allergy*, *7*(1), 48–50.

Ali, A. M., & Kunugi, H. (2020). Apitherapy for Parkinson's disease: A focus on the effects of propolis and royal jelly. *Oxidative Medicine and Cellular Longevity*, *2020*(1), 1727142.

Alqarni, S. A., Owayss, A. A., Mahmoud, A. A., & Hannan, M. A. (2023). Therapeutic applications of honeybee products for apitherapy. *Saudi Journal of Biological Sciences*, *30*(8), 103818.

Alshallash, K. S., Abolaban, G., Elhamamsy, S. M., Zaghlool, A., Nasr, A., Nagib, A., & Taha, I. M. (2023). Bee pollen as a functional product–chemical constituents and nutritional properties. *Journal of Ecological Engineering*, *24*(2), 173–183.

American Apitherapy Society (AAS). (2024). *Position statement on apitherapy's role in healthcare systems*. American Apitherapy Society. https://apitherapy.org/en/position-statement/

Bankova, V., Popova, M., Trusheva, B., & Kunimasa, K. (2019). Recent trends and important developments in propolis research. *Evidence-Based Complementary and Alternative Medicine*, *2019*, 1–29. https://doi.org/10.1155/2019/4725935

Bernardino, P. N., Bersano, P. R. O., Neto, J. F. L., & Sforcin, J. M. (2018). Positive effects of antitumor drugs in combination with propolis on canine osteosarcoma cells (spOS-2) and mesenchymal stem cells. *Biomedicine & Pharmacotherapy*, *104*, 268–274.

Bistué-Rovira, À., Solé, M., Anguera-Tejedor, M., Pérez, B., García-Tercero, L., Díaz-Pérez, A., Martínez-Benitez, Z., Delgado-Hernández, R., & Jiménez-Altayó, F. (2025). Bee venom disrupts vascular homeostasis: Apitoxin and melittin trigger vascular cell toxicity and aortic dysfunction in mice. *Toxicological Sciences*, *207*(1), 196–207.

Borba, R. S., Wilson, M. B., & Spivak, M. (2017). Hidden benefits of honeybee propolis in hives. In *Beekeeping–from science to practice* (pp. 17–38). Springer International Publishing.

Burzyńska, M., & Piasecka-Kwiatkowska, D. (2021). A review of honeybee venom allergens and allergenicity. *International Journal of Molecular Sciences*, *22*(16), 8371.

Carli, T., Locatelli, I., Košnik, M., Bevk, D., & Kukec, A. (2025). Epidemiology and risk factor analysis of systemic allergic reaction to bee venom in the Slovenian population of beekeepers. *Zdravstveno varstvo*, *64*(1), 40–48.

Carpena, M., Nuñez-Estevez, B., Soria-Lopez, A., & Simal-Gandara, J. (2020). Bee venom: An updating review of its bioactive molecules and its health applications. *Nutrients*, *12*(11), 3360.

Chauhan, V. V. S., Kumar, A., Raj, P., & Chaudhary, R. K. (2016). Melittin induced cytogenetic damage, oxidative stress and changes in gene expression in human peripheral blood lymphocytes. *Toxicon*, *110*, 56–67.

Chen, T., Zhu, J., Hang, C. H., & Wang, Y. H. (2019). The Potassium SK Channel Activator NS309 protects against experimental traumatic brain injury through anti-inflammatory and immunomodulatory mechanisms. *Frontiers in Pharmacology*, *10*, 1432.

Choi, H. J., Jeon, Y. J., Park, M. K., & Hwang, D. H. (2016). Anti-inflammatory applications of melittin, a major component of bee venom: Detailed mechanism of action and adverse effects. *Molecules*, *21*(5), 616.

Cornara, L., Biagi, M., Xiao, J., & Burlando, B. (2017). Therapeutic properties of bioactive compounds from different honeybee products. *Frontiers in Pharmacology*, *8*, 412.

Cui, L., Xu, Y.-Y., Li, L.-S., Wang, Z. X., Bian, S.-N., Han, Q.-J., Li, W.-Y., Li, Y., Lanier, B. Q., Chang, C., & Guan, K. (2021). Bee venom allergy in Chinese beekeepers and patients under live nee apitherapy. *Allergy, Asthma & Clinical Immunology*, *17*, 14. https://doi.org/10.1186/s13223-021-00515-5

Cui, Z., Zhou, Z., Sun, Z., Duan, J., Liu, R., Qi, C., & Yan, C. (2024). Melittin and phospholipase A2: Promising anti-cancer candidates from bee venom. *Biomedicine & Pharmacotherapy*, *179*, 117385. https://doi.org/10.1016/j.biopha.2024.117385

Darwish, D. A., Masoud, H. M. M., Abdel-Monsef, M. M., Helmy, M. S., Zidan, H. A., & Ibrahim, M. A. (2021). Phospholipase A2 enzyme from the venom of Egyptian honey bee Apis mellifera lamarckii with anti-platelet aggregation and anti-coagulation activities. *Journal, Genetic Engineering & Biotechnology*, *19*(1), 10.

Di Costanzo, M., De Paulis, N., Peveri, S., Comberiati, P., & Peroni, D. G. (2021). Anaphylaxis caused by artisanal honey in a child: A case report. *Journal of Medical Case Reports*, *15*(1), 235.

Dispenza, M. C., Metcalfe, D. D., & Olivera, A. (2023). Research advances in mast cell biology and their translation into novel therapies for anaphylaxis. *Journal of Allergy and Clinical Immunology: In Practice*, *11*(7), 2032–2042.

Dwyer, J. T., Coates, P. M., & Smith, M. J. (2015). Dietary supplements: Regulatory challenges and research resources. *Nutrients*, *7*(1), 109–127.

El Boussaadni, Y., & Oulmaati, A. (2022). Bee stings in children: Not always benign. *IOSR Journal of Nursing and Health Science*, *11*(3, Ser. VI), 37–39.

El-Didamony, S. E., Gouda, H. I., Zidan, M. M., & Amer, R. I. (2024). Bee products: An overview of sources, biological activities and advanced approaches used in apitherapy application. *Biotechnology Reports*, *44*, e00862.

El-Haskoury, R., Al-Waili, N., Kamoun, Z., Makni, M., Al-Waili, H., & Lyoussi, B. (2018). Antioxidant activity and protective effect of Carob Honey in CCl4-induced Kidney and Liver injury. *Archives of Medical Research*, *49*(5), 306–313.

El-Saeady, A., Diab, A., Shehata, I., Nafea, E., & Metwaly, A. (2016). Effect of bee venom collecting on the behavior of honeybee colonies. *Journal of Plant Protection and Pathology*, *7*(6), 347–351. https://doi.org/10.21608/jppp.2016.50576

Eruaga, M. A., Itua, E. O., & Bature, J. T. (2024). The role of regulatory authorities in the regulation and control of herbal medicines: A case study of NAFDAC. *International Journal of Science and Research Archive*, *11*(2), 207–211.

Eshelman, J. (2025). *Spiritual beekeeping: An ethnographic exploration of sacred apiculture in present-day United States* (Master's thesis, University of Wales Trinity Saint David). University of Wales Trinity Saint David.

Gökkaya, İ., Renda, G., Ayhan, F., & Yazici, N. (2025). Awareness of apitherapy: A study of healthcare professionals in Türkiye. *Journal of Pharmacopuncture*, *28*(1), 43–51.

International Federation of Apitherapy (IFA). (2022). *Statutes of the International Federation of Apitherapy*. International Federation of Apitherapy. https://api-terra.org/statutes/

Jin, Z., Yao, J., & Xie, N. (2018). Melittin constrains the expression of identified key genes associated with bladder cancer. *Journal of Immunology Research*, *2018*, 5038172.

Jităreanu, A., Caba, I. C., & Agoroaei, L. (2022). Toxic risks associated with apitherapy products. *Carpathian Journal of Food Science and Technology*, *14*(1), 149–158.

Kalle, R. et al. (2025). Beekeepers as guardians of apitherapeutic knowledge in Estonia, SW Ukraine, and NE Italy. *Journal of Ethnobiology and Ethnomedicine*, *21*, Article 15. https://doi.org/10.1186/s13002-025-00764-6

Kavurmaci, M., and Tan, M. 2019. Determination of knowledge and attitudes of nurses about apitherapy. *Complementary Therapies in Clinical Practice*, *36*, 39–42.

Kieliszek, M., Piwowarek, K., Kot, A. M., Błażejak, S., Chlebowska-Śmigiel, A., & Wolska, I. (2018). Pollen and bee bread as new health-oriented products: A review. *Trends in Food Science & Technology*, *71*, 170–180.

Kim, K. H., Lee, S. Y., Cho, H. J., Hong, J. T., & Yoon, H. S. (2019). Bee venom phospholipase A2 ameliorates Parkinson's disease-related pathology by modulating neuroinflammation in a mouse model. *Journal of Neuroinflammation*, *16*(1), 1–13.

Kulczyńska, A., Stuper-Szablewska, K., & Kocira, S. (2023). Honey-The source of active compounds with biological and therapeutic activity. *Molecules*, *28*(9), 3831.

Kuzmenkov, A. I., Peigneur, S., Nasburg, J. A., Mineev, K. S., Nikolaev, M. V., Pinheiro-Junior, E. L., Arseniev, A. S., Wulff, H., Tytgat, J., & Vassilevski, A. A. (2022). Apamin structure and pharmacology revisited. *Frontiers in Pharmacology*, *13*, 977440.

Li, Q. Q., Wang, K., Marcucci, M.C., Frankland Sawaya, A. C. H., Hu, L., Xue, X.-F., Wu, L. M., & Hu, F. L. (2018). Nutrient-rich bee pollen: A treasure trove of active natural metabolites. *Journal of Functional Foods*, *49*, 472–484.

Mathis, S., Carla, L., Duval, F., Nadal, L., Solé, G., & Le Masson, G. (2022). Acute peripheral neuropathy following animal envenomation: A case report and systematic review. *Toxicon*, *214*, 1–8.

MDPI. (2022). Definition and identification of honey bee welfare practicees within the Five Domains framework for sustainable beekeeping. *Applied Sciences*, *14*(24), Article 11902.

Milovanovic, M., & Radovanović, B. (2021). Functional foods and bioactive compounds in health promotion and disease prevention. *Journal of Hygienic Engineering and Design*, *34*, 69–81.

Moga, M. A., Dimienescu, O. G., Arvatescu, C. A., Ifteni, P., & Ples, L. (2018). Anticancer activity of toxins from bee and snake venom-an overview on ovarian cancer. *Molecules*, *23*(3), 692.

Mori, F., Saretta, F., Riscassi, S., Caimmi, S., Bottau, P., Liotti, L., Franceschini, F., Bianchi, A., Valluzzi, R. L., Crisafulli, G., & Caffarelli, C. (2024). Risk factors for drug hypersensitivity reactions in children. *Italian Journal of Pediatrics*, *50*(1), 127.

Muntean, M., & Florea, A. (2025). Phospholipase A2—A significant bio-active molecule in honeybee (*Apis mellifera L.*). Venom. *Molecules* (Basel, Switzerland), *30*(12), 2623.

Olegário, L. S., Andrade, J. K. S., Andrade, G. R. S., Denadai, M., Cavalcanti, R. L., da Silva, M. A. A. P., & Narain, N. (2019). Chemical characterization of four Brazilian brown propolis: An insight in tracking of its geographical location of production and quality control. *Food Research International*, *123*, 481–502.

Oliveira, L. P. G., Conte, F. L., Cardoso, E. O., Conti, B. J., Santiago, K. B., Golim, M. A., Cruz, M. T., & Sforcin, J. M. (2016). Immunomodulatory/inflammatory effects of geopropolis produced by Melipona fasciculata Smith in combination with doxorubicin on THP-1 cells. *Journal of Pharmacy and Pharmacology*, *68*, 1551–1558.

Oliveira, L. P. G., Conte, F. L., Cardoso, E. O., Conti, B. J., Santiago, K. B., Golim, M. A., Feltran, G. S., Zambuzzi, W. F., & Sforcin, J. M. (2019). A new chemotherapeutic approach using doxorubicin simultaneously with geopropolis favoring monocyte functions. *Life Sciences*, *217*, 81–90.

Omar, A. S., & Elsaiegh, M. S. (2023). Effect of bee population and bee venom milking during spring and early summer in some honey bees (*Apis mellifera L.*) colonies biological characters. *NTU Journal of Agriculture and Veterinary Science*, *3*(1). https://doi.org/10.56286/ntujavs.v3i1.320

Oršolić, N. (2022). Allergic inflammation: Effect of propolis and its flavonoids. *Molecules*, *27*(19), 6694. https://doi.org/10.3390/molecules27196694

Oryan, A., Alemzadeh, E., & Moshiri, A. (2016). Biological properties and therapeutic activities of honey in wound healing: A narrative review and meta-analysis. *Journal of Tissue Viability*, *25*(2), 98–118.

Ota, M., Ishiuchi, K., Xu, X., Minami, M., Nagachi, Y., Yagi-Utsumi, M., Tabuchi, Y., Cai, S-Q., Makino, T. (2019). The immunostimulatory effects and chemical characteristics of heated honey. *Journal of Ethnopharmacology*, *228*, 11–17.

Park, J. H., Yim, B. K., & Lee, J.-H. (2015). Risk associated with bee venom therapy: A systematic review and meta-analysis. *PLoS ONE*, *10*(5), e0126971.

Park, J., Jang, K. M., & Park, K. K. (2020). Apamin suppresses LPS-induced neuroinflammatory responses by regulating SK channels and TLR4-mediated signaling pathways. *International Journal of Molecular Sciences*, *21*(12), 4319.

Pasupuleti, V. R., Sammugam, L., Ramesh, N., & Gan, S. H. (2017). Honey, propolis, and royal jelly: A comprehensive review of their biological actions and health benefits. *Oxidative Medicine and Cellular Longevity*, *2017*(1), 1259510.

Rady, I., Siddiqui, I. A., Rady, M., & Mukhtar, H. (2017). Melittin, a major peptide component of bee *venom*, and its conjugates in cancer therapy. *Cancer Letters*, *402*, 16–31.

Rauf, A. K., Vijayan, A., Hashitha, V. M., John, S. T., & Peringat, J. (2021). Rhabdomyolysis and acute kidney injury following multiple bee stings in a child: A case report. *Journal of Pediatric Critical Care*, *8*(3), 252–254.

Rosman, Y., Nashef, F., Cohen-Engler, A., Meir-Shafrir, K., Lachover-Roth, I., & Confino-Cohen, R. (2019). Exclusive bee venom allergy: Risk factors and outcome of immunotherapy. *International Archives of Allergy and Immunology*, *180*(2), 128–134.

Rufatto, L. C., dos Santos, D. A., Marinho, F., Henriques, J. A. P., Ely, M. R., & Moura, S. (2017). Red propolis: Chemical composition and pharmacological activity. *Asian Pacific Journal of Tropical Biomedicine*, *7*(7), 591–598.

Saccardi, L., Schiebl, J., Weber, K., Schwarz, O., Gorb, S., & Kovalev, A. (2021). Adhesive behavior of propolis on different substrates. *Frontiers in Mechanical Engineering*, *7*, 660517.

Shafaghat, F., & Zwiener, R. D. (2018). Immunology of bee venom. *Clinical Reviews in Allergy & Immunology*, *54*(3), 386–396.

Silva, G. B. D., Vasconcelos, A. G., Rocha, A. M. T., Vasconcelos, V. R., Barros, J., Neto, Fujishima, J. S., Ferreira, N. B., Barros, E. J. G., & Daher, E. F. (2017). Acute kidney injury complicating bee stings - a review. *Revista do Instituto de Medicina Tropical de Sao Paulo*, *59*, e25.

Stefas, D., Gyftokostas, N., Kourelias, P., Nanou, E., Tananaki, C., Kanelis, D., . . . & Couris, S. (2022). Honey discrimination based on the bee feeding by Laser Induced Breakdown Spectroscopy. *Food Control*, *134*, 108770.

Stela, M., Cichon, N., Spławska, A., Szyposzynska, M., & Bijak, M. (2024). Therapeutic potential and mechanisms of bee venom therapy: a comprehensive review of apitoxin applications and safety enhancement strategies. *Pharmaceuticals*, *17*(9), 1211.

Stoevesandt, J., & Trautmann, A. (2022). Risk factors in bee and Vespula venom allergy: A state-of-the-art review. *Allergo Journal International*, *31*, 1–10.

Sun, D., Li, S., Huang, H., & Xu, L. (2023). Neurotoxicity of melittin: Role of mitochondrial oxidative phosphorylation system in synaptic plasticity dysfunction. *Toxicology*, 497–498, Article 153628.

Susilowati, H., Murakami, K., Yumoto, H., Amoh, T., Hirao, K., Hirota, K., Matsuo, T., & Miyake, Y. (2017). Royal jelly inhibits *Pseudomonas aeruginosa* adherence and reduces excessive inflammatory responses in human epithelial cells. *BioMed Research International*, *2017*, 3191752.

Thakur, M., & Nanda, V. (2020). Composition and functionality of bee pollen: A review. *Trends in Food Science & Technology*, *98*, 82–106.

U.S. Food and Drug Administration. (2024, March 6). *Questions and answers on dietary supplements*. https://www.fda.gov/food/information-consumers-using-dietary-supplements/questions-and-answers-dietary-supplements

Vahidinia, Z., Barati, S., Tameh, A. A., Bagheri-Mohammadi, S., & Garshasebi, A. (2024). Bee venom as a promising therapeutic strategy in central nervous system diseases. *Neuropeptides*, *107*, 102451.

Waykar, B., & Alqadhi, Y. A. (2016). Beekeeping and bee products; boon for human health and wealth. *Indian Journal of Pharmaceutical and Biological Research*, *4*(3), 20.

Wehbe, R., Frangieh, J., Rima, M., El Obeid, D., Sabatier, J. M., & Fajloun, Z. (2019). Bee venom: Overview of main compounds and bioactivities for therapeutic interests. *Molecules* (Basel, Switzerland), *24*(16), 2997.

Weis, W. A., Ripari, N., Conte, F. L., Honorio, M. da S., Sartori, A. A., Matucci, R. H., & Sforcin, J. M. (2022). An overview about apitherapy and its clinical applications. *Phytomedicine Plus*, *2*(2), 100239.

Yoo, J., & Lee, G. (2022). Adverse events associated with the clinical use of bee venom: A review. *Toxins*, *14*(8), 562.

Yu, X., Jia, S., Yu, S., Chen, Y., Zhang, C., Chen, H., & Dai, Y. (2023). Recent advances in melittin-based nanoparticles for antitumor treatment: from mechanisms to targeted delivery strategies. *Journal of Nanobiotechnology*, *21*(1), 454.

Zhang, S. F., & Chen, Z. (2017). Melittin exerts an antitumor effect on nonsmall cell lung cancer cells. *Molecular Medicine Reports*, *16*, 3581–3586.

Zolfagharian, H., Mohajeri, M., & Babaie, M. (2015). Honey bee venom (*Apis mellifera*) contains anticoagulation factors and increases the blood-clotting time. *Journal of Pharmacopuncture*, *18*(4), 7–11.

Index

Pages in *italics* refer to figures and pages in **bold** refer to tables.